Liquid Chromatography/Mass Spectrometry, Spectrometry, MS/MS and Time-of-Flight MS

ACS SYMPOSIUM SERIES **850**

Liquid Chromatography/Mass Spectrometry, MS/MS and Time of Flight MS

Analysis of Emerging Contaminants

Imma Ferrer, Editor
U.S. Geological Survey

E. M. Thurman, Editor
U.S. Geological Survey

Sponsored by the ACS Division of Environmental Chemistry, Inc.

American Chemical Society, Washington, DC

Chemistry Library

SEP/AE
CHEM

Library of Congress Cataloging-in-Publication Data

Liquid chromatography/mass spectrometry , MS/MS and time-of-flight MS : analysis of emerging contaminates / Imma Ferrer, editor, E. M. Thurman, editor.

   p. cm.—(ACS symposium series ; 850)

   Includes bibliographical references and index.

   ISBN 0–8412–3825–1

   1. Liquid Chromatography—Congresses. 2. Mass spectrometry—Congresses. 3. Time-of-flight mass spectrometry—Congresses. 4. Environmental chemistry—Congresses.

   I. Ferrer, Imma, 1970- II. Thurman, E. M. (Earl Michael), 1946- III. American Chemical Society. Division of Environmental Chemistry, Inc. IV. American Chemical Society. Meeting (223rd : 2002 : Orlando, Fla.) V. Series.

QD79.C454L55365   2003
543′.0894—dc21   2003041829

The paper used in this publication meets the minimum requirements of American National Standard for Information Sciences—Permanence of Paper for Printed Library Materials, ANSI Z39.48–1984.

ac

PRINTED IN THE UNITED STATES OF AMERICA

Foreword

The ACS Symposium Series was first published in 1974 to provide a mechanism for publishing symposia quickly in book form. The purpose of the series is to publish timely, comprehensive books developed from ACS sponsored symposia based on current scientific research. Occasionally, books are developed from symposia sponsored by other organizations when the topic is of keen interest to the chemistry audience.

Before agreeing to publish a book, the proposed table of contents is reviewed for appropriate and comprehensive coverage and for interest to the audience. Some papers may be excluded to better focus the book; others may be added to provide comprehensiveness. When appropriate, overview or introductory chapters are added. Drafts of chapters are peer-reviewed prior to final acceptance or rejection, and manuscripts are prepared in camera-ready format.

As a rule, only original research papers and original review papers are included in the volumes. Verbatim reproductions of previously published papers are not accepted.

ACS Books Department

Contents

Emerging Contaminants: Pharmaceuticals

Emerging Contaminants: Pesticides

Emerging Contaminants:
Surfactants and Natural Products

Indexes

Preface

This is a book of methods and approaches to discover unknowns and not a book of theory and explanation. Several beautiful books have come before to explain the operation of liquid chromatography/mass spectrometry/mass spectrometry (LC/MS/MS) instruments and the breakthrough of modern electrospray LC/MS/MS. Examples include Cole (*Electrospray Ionization Mass Spectrometry;* Cole, R. B., Ed.; John Wiley & Sons, Inc.: New York, 1997; 577 pp), Willoughby et al. (Willougby, R.; Sheehan, E.; Mitrovich, S. *A Global View of LC/MS, Second Edition;* Global View Publishing: Pittsburgh, PA, 2002; 518 pp), and the recent historical accounts of mass spectrometry and its applications (*Measuring Mass;* Grayson, M. A., Ed.; Chemical Heritage Press: Philadelphia, PA, 2002; 149 pp), which was published at the 50th anniversary meeting of the American Society of Mass Spectrometry.

Our book focuses on emerging contaminants and environmental issues using today's modern state-of-the-art techniques, which are LC/MS/MS. and time-of-flight mass spectrometry (TOF-MS). This book follows the ground breaking environmental LC/MS book of Barceló (*Applications of LC-MS in Environmental Chemistry;* Barceló, D., Ed.; Elsevier: Amsterdam, The Netherlands, 1996, 543 pp). With our volume, the bar is raised to a new level of environmental puzzle solving with the power of three LC/MS/MS instruments: quadrupole ion trap, triple quadrupole, and quadrupole/time-of-flight. Although these instruments have been known for some short time among mass spectrometrists, the instruments are just now being recognized in the environmental sciences.

For the most part, the pioneers of these methods have not written chapters in this book. Rather a younger group of scientists, who perhaps will lead the way in environmental organic analysis, are showcasing their work. The symposium from which the book follows has been a joy to organize. The 225th American Chemical Society (ACS) meeting in Orlando, Florida was the setting for this symposium, *Analysis of Emerging Contaminants Using LC/MS/MS*. We thank the authors for all their efforts in presenting stimulating talks and interesting chapters to our book. The sense of "nerves" that seemed to haunt all of us at the meeting must have been the high expectations that each of us felt for the level of our presentations. Thanks also to several authors who contributed chapters, who were not at the meeting, but who helped to give a more well-rounded account of the emerging-contaminant issue.

The topics in this volume are wide ranging from pesticides, pharmaceuticals, surfactants, aromatic hydrocarbons, humic substances, plasticizers, steroids, and hormones, to disinfection by-products and their degradates or metabolites. Thus, the book should appeal to a wide audience of researchers in environmental chemistry, graduate students, and those in related fields who may gain new insights and ideas for methods' research.

The book focuses on LC/MS/MS and TOF-MS. Sample preparation and solid-phase extraction are left for other books already published; thus, we made an effort not to cover the same "old ground" again. Instead this volume brings together, under one cover, the many applications of LC/MS/MS and TOF-MS for puzzle solving in the field of environmental analysis. Our goal was discovery of environmental contaminants in water, sediments, and soil using LC/MS/MS. We hope to have made a reasonable contribution to this fascinating new topic (written on McClure Pass, 8755 feet, August 24, 2002).

Imma Ferrer[1]
E.M. Thurman[2]
U.S. Geological Survey
P.O. Box 25046, MS 407
Denver, CO 80225

[1]Current address: immaferrer@menta.net
[2]Current address: U.S. Geological Survey, 4821 Quail Crest Place, Lawrence, KS 66049 (email: ethurman@usgs.gov).

Liquid Chromatography/Mass Spectrometry, MS/MS and Time-of-Flight MS

Overview

Chapter 1

Analysis of Emerging Contaminants

Imma Ferrer[1,2] and E. M. Thurman[1,3]

[1]U.S. Geological Survey, P.O. Box 25046, MS407, Denver, CO 80225
[2]Current address: immaferrer@menta.net
[3]Current address: U.S. Geological Survey, 4821 Quail Crest Place, Lawrence, KS 66049

In this overview chapter we will focus on the importance of emerging contaminants and their presence in the environment, as well as giving a description of this book. The environmental issue of emerging contaminants is tied to the analysis of wastewater samples using the new analytical methods of the last decade, especially LC/MS and LC/MS/MS. Thus, this book focuses on applications of LC/MS/MS to analyze unknowns in the aquatic environment. Emerging contaminants are defined as compounds that are not currently covered by existing regulations of water quality, that have not been previously studied, and that are thought to be a possible threat to environmental health and safety. In particular the compounds that are addressed include pharmaceuticals, personal care products, surfactants, and pesticide degradates. The book has three main sections that address these issues with special attention to the analytical methods of mass spectrometry (such as triple quadrupoles, ion-traps, and quadrupole time-of-flight analyzers). All the different important aspects on today's emerging contaminants have been covered throughout the chapters.

Introduction

The exponential growth in human populations over the past century has put a demand on our freshwater supplies. Equally important is the discharge of wastewater from cities and farm areas to this fragile freshwater. These wastewaters have been impacted over the past 50 years with an ever diverse and increasing number of compounds that offer improvements in industry, agriculture, health, medical treatment, and household conveniences. These compounds may inadvertently have "hidden" environmental issues, such as endocrine disruption. The recent book, *Our Stolen Future* by Colburn et al. (*1*) has caused a paradigm shift from "do compounds in our water cause cancer" to "do compounds in our water supplies cause endocrine disruption." In particular the focus has shifted from chlorinated organic compounds, such as pesticides, to hormones and to compounds that imitate hormone action (i.e. some surfactants and plasticizers). Since the publication of *Our Stolen Future* in 1996 there have been many papers reporting possible endocrine disruption in fish as expressed by female characteristics in both male and female fish (*2-3*). The fear that this process may also be occurring in humans was acerbated by the publication of a controversial Danish study (*4*) that sperm counts in males have been dropping over the past twenty years. These events have accelerated the study of emerging contaminants and water quality during the last five years.

During this same time frame of the past decade, the analytical methods of liquid chromatography and mass spectrometry have come together because of the improvements in atmospheric pressure ionization, especially electrospray (*5*). This culminated in the Nobel Prize in Chemistry for 2002 being given to John Fenn for the discovery of electrospray of large molecules in the late 1980s (*6*). During the decade of the 1990s instrument manufacturers have developed sensitive and reliable methods for liquid chromatography/mass spectrometry/mass spectrometry (LC/MS/MS). A variety of instruments and design types have been marketed with great success. The majority of this work has focused on the new field of proteomics (*7*), which is the study of proteins using enzyme hydrolysis linked to liquid chromatography/mass spectrometry and computer modeling. The field of environmental mass spectrometry has been following the activities of LC/MS/MS in biochemistry with great interest and there is now a growing group of researchers actively working in environmental mass spectrometry, especially LC/MS/MS. The recent review of environmental mass spectrometry by Richardson (*8*) points out the importance of mass spectrometry and its ability to focus on emerging contaminants and current issues.

The blending then of emerging contaminants and LC/MS/MS is the perfect couple for issues of environmental water quality. The sensitivity of these mass spectrometry techniques are such that it is possible to look at the concentrations in water of many emerging contaminants especially in wastewater. A recent symposium volume by the American Chemical Society edited by Daughton and Jones-Lepp (9) has focused the attention on pharmaceuticals and personal care products in the environment. This book was published in 2001 and has outlined many environmental water quality issues. Two symposia recently organized by the National Water Well Association in Minneapolis dealing with pharmaceuticals in water supplies were the outgrowth of research in Europe in the early 1990s, which showed that pharmaceuticals were entering groundwater from wastewater application (10). Following these symposia was a reconnaissance study of the U.S. Geological Survey for pharmaceuticals, hormones, antibiotics, and personal care products in water resources of the United States (11), which found that 80% of the streams samples contained measurable concentrations of a variety of pharmaceuticals and other wastewater related compounds. Thus, the time seemed appropriate for the organization of a meeting to deal with the analytical issues and methods development techniques for emerging contaminants.

Our symposium, which was organized in late 2001 and presented in April of 2002, deals with the coupling of LC/MS/MS with issues of emerging contaminants. This book contains 22 chapters that are divided into three sections. Section one deals with overview issues and includes the Introduction, a comparison of MS/MS instrumentation as they apply to emerging contaminants, and an overview chapter on identification criteria used by the European Union to identify emerging contaminants in the aquatic environment. The chapter on identification criteria by Stolker et al. shows how the European Union has come to terms with the concept of identification of unknowns using both GC/MS and LC/MS and MS/MS. The chapter is instructional in its structured approach to identification. The second section deals with the identification of unknown compounds using LC/MS/MS and TOF/MS. Section three addresses the applications of LC/MS/MS to emerging contaminants and is subdivided into pharmaceuticals, pesticides, surfactants and natural products.

A Decade of LC/MS/MS History

There have been several excellent books that discuss LC/MS/MS directly, its history and development. They include *A Global View of LC/MS* by Willoughby et al. (12), *Measuring Mass* (7), which is a history of mass spectrometry just published as part of the 50th anniversary meeting of the American Society of Mass Spectrometry and contains many excellent figures and discussion on LC/MS, *Electrospray Ionization Mass Spectrometry* (5),

Liquid Chromatography-Mass Spectrometry (*13*), and *Applications of LC-MS in Environmental Chemistry* (*14*). A handy book on terminology and definitions in mass spectrometry is *Mass Spectrometry Desk Reference* (*15*). A considerably older volume, which deals with GC/MS/MS, but may be useful, is *Mass Spectrometry/Mass Spectrometry* (*16*) published in 1988. Two other volumes with the same title that are of interest include *Liquid Chromatography/Mass Spectrometry* (*17*) and *Liquid Chromatography/Mass Spectrometry* (*18*) both published in 1990.

The book edited by Cole (*5*) contains several excellent chapters that deal with topics related to this book, they are Chapters 5-9 (*19-23*) dealing with double focusing magnetic sector MS/MS instruments, time-of-flight MS, ion-trap MS/MS, Fourier transform ion cyclotron resonance MS, and the chapter on combining liquid chromatography and mass spectrometry. One of the interesting readings in this book is the Foreword by John Fenn on the history of electrospray, which has made LC/MS/MS so popular and powerful these last 15 years.

Thus, this short synopsis of important books dealing with LC/MS/MS sets the stage for this book on development and application of LC/MS/MS and TOF/MS to environmental issues, such as emerging contaminants.

LC/MS/MS and TOF/MS: Identification of Unknowns

The analysis of emerging contaminants, many of which were unknown until recently, is a poignant environmental issue. Identification of pharmaceutical compounds and pesticide degradates in the environment (*11, 24*) are two excellent examples that have followed the advent of LC/MS instrument technology. In these examples, unequivocal identification by GC/MS has not been possible, mainly because the compounds are not volatile in the inlet of the GC.

LC/MS, on the other hand, will ionize many compounds in either electrospray or atmospheric pressure chemical ionization. In this case identification typically involves two steps. Firstly, there is the production of a mass spectrum from an authentic standard to create a user library. Secondly, there is identification by matching chromatographic retention time and mass spectrum for molecular ion and fragment ions. This procedure was considered adequate for final identification until recently, when this level of identification has been challenged for complex samples containing many compounds (*25*). Because LC/MS relies on fragmentation using collision induced dissociation (CID) in the source of the mass spectrometer, co-eluting peaks in the mass spectrum will interfere with each other resulting in multiple CID spectra that are superimposed on one another. LC/MS/MS removes this problem and results in a much higher degree of certainty in identification of unknowns.

However, with LC/MS/MS, there still exists the problem of identification of an unknown peak from only the mass spectrum. Because there are no libraries of unknowns available at this time, except of course for user created libraries, it is difficult to identify a compound from only a mass spectrum. The first section of this book contains 5 chapters where the authors have addressed this problem with a series of techniques that allow the identification of compounds that are true unknowns. The use of accurate mass, both with TOF/MS and sector instruments, are used in conjunction with LC/MS/MS to solve these problems of unknown identification.

For example, in Chapter 4 by Gilbert et al., they identify complex natural compounds that may be developed into commercial products using various LC/MS/MS techniques. Their method involves the use of both ion trap and Q-TOF MS/MS methods to determine the fragments of various ions and piece together the natural products. They have developed not only a library of spectra from their identifications, but perhaps more importantly, a library of fragmentation and pathways of fragmentation (ion trap) that may be used to identify unknowns. Accurate mass from the Q-TOF is used to give elemental composition from both the molecular ion and the fragment ions. The power of ion trap and Q-TOF is exemplified in this paper.

Chapter 5 deals with a different type of unknown identification, where a pesticide is treated with iron and UV light (Fenton reaction) and the newly generated compounds are identified by a combination again of ion trap and time-of-flight mass spectrometry. Fernadez-Alba et al. apply first ion trap from the known fragmentation of diuron and work out the structural fragments. Then these structures are used in the identification of the degradates. In several cases, where the ion trap was not adequate for identification, TOF was used for the elemental composition, which then gave the elemental composition of the unknown compounds, which when combined with ion-trap fragmentation gave the molecular structure.

Chapter 6 by Vanderford et al. shows the third use of combining MS/MS and accurate mass measurements. In this chapter, the authors determine the structure of an unknown contaminant generated in the source of the mass spectrometer. The approach here involves the use of elemental composition, the nitrogen rule (i.e. even electron ions have even mass), isotope patterns, and logical reasoning to determine the structure of a sulfonamide plasticizer. They use accurate mass from a sector instrument rather than TOF/MS to generate the elemental composition needed for unknown identification.

Chapter 7 by Benotti et al. compares time-of-flight mass spectrometry and triple quadrupole MS/MS to identify many pharmaceuticals in wastewater including compounds such as, sulfamethoxazole (an antibiotic), diltiazem (an antihypertensive), and ranitidine (an antiulcerative). This paper shows that TOF analysis is quickly becoming an important analysis tool both separately and in unison with LC/MS/MS for the analysis of pharmaceuticals in water.

Chapter 8 by Thurman et al. addresses the discovery of ethane-sulfonic-acids degradates of the chloroacetanilide herbicides using a combination of TOF/MS and ion trap MS/MS, and a "procedure of discovery". The procedure involves four steps including (1) hypothesis, (2) diagnostic and molecular-ion identification, (3) synthesis of standards, and (4) determination of herbicide degradates in groundwater samples. The compounds are shown to be herbicide degradates found frequently in groundwater of the Midwestern United States.

LC/MS/MS of Pharmaceuticals

Pharmaceuticals have taken the spotlight in emerging contaminants chiefly because of the direct impact that they may have on man. Several salient issues include: hormone releases to the environment that affect both man and beast (i.e. fish especially have been affected by hormones and their mimics in the aquatic environment), antibiotic releases to the environment and the possibility of antibiotic resistance development in native bacteria and the potential of this resistance creating environmental disasters, and lastly the release of many medicinal pharmaceuticals to the natural aquatic environment and subsequent uptake by man. These medicines are the common over-the-counter medications, as well as prescriptions, and other medicinal compounds used in the treatment of disease. Concern arises here that the pharmaceuticals, although low in concentration (less than 1 µg/L), may target binding sites in man and cause unwanted pharmaceutical effects.

Thus, Chapters 9-13 deal with analysis of pharmaceuticals as emerging contaminants. Chapter 9 by Cardoza et al. use a combination of LC/NMR and LC/MS/MS (both Q-TOF and triple quadrupole) to determine the structure of degradates of the antibiotic, cyprofloxacin, which were formed by degradation of the parent compound in pond water. Cyprofloxacin, recently in the news because of its use for treating anthrax, has been found in environmental water samples (*11*) and therefore was the object of study in this chapter. The power of LC/NMR when combined with LC/MS/MS and accurate mass determination was pivotal in its ability to determine the structure of the new degradates.

Chapter 10 by Snow et al. discusses the occurrence of antibiotics in the aquatic environment from their use in animal husbandry, especially as they apply to swine. The tetracycline family of antibiotics are found in hog wastewater at concentrations reaching a mg/L. At these concentrations the danger of resistance to microorganisms is possible. Snow et al. use ion trap

MS/MS to analyze these compounds after first isolating them by solid-phase extraction.

Chapter 11 by Furlong et al. addresses pharmaceuticals in sediments rather than water. In this chapter, ion trap MS/MS is used to identify pharmaceuticals that are found on sediments. Trimethoprim (an antibacterial) and thiabendazole (an antihelmenitic) were found in sediments taken in Denver, Colorado. This chapter also has examples of unknown identification for surfactants in surface-water samples containing agricultural runoff.

Chapter 12 by Bratton et al. uses a combination of Q-TOF and triple quadrupole MS/MS to identify trace levels of antihypertensives in river water, wastewater, and tap water. They found approximately 1 ng/L of several pharmaceuticals in tap water of Lawrence, Kansas, where the work was completed at the University of Kansas. The method here used the power of multiple reaction monitoring (MRM) to detect extremely low concentrations of pharmaceuticals in tap water. MRM uses the fact that both quadrupoles are operated in selected ion monitoring to obtain low background counts and high sensitivity.

Chapter 13 by Zavitsanos uses ion trap MS/MS to identify hormones and ovulation inhibitors (a class of compounds that is important to the study of emerging contaminants because of their effect on fish and other susceptible species) in water. The sensitivity of the ion trap was used to isolate the molecular ion of each of the hormones and ovulation inhibitors and then to fragment them resulting in a characteristic spectrum that can be used for the unequivocal identification of these compounds in the environment.

LC/MS/MS of Pesticides

The analysis of pesticides by LC/MS has a rather long history now (14) beginning in the early 1990s and continuing to the present time. Many classes of pesticides are easily analyzed by LC/MS and a more challenging task is to identify the degradates of pesticides (26). The importance and formation of pesticide degradates was outlined as an environmental issue in the early 1990's by a series of publications identifying the degradates of atrazine and metolachlor, two of the most commonly used herbicides in the United States (27). This work included the discovery that pesticide degradates are commonly detected in shallow groundwater whereas the parent compound is not found (28). Since this time environmental research in water quality has focused on the degradates of pesticides and this section of the book deals with two major classes of herbicides. They are Chapters 14-16 that deal with the acetanilide herbicides and Chapter 17, which deals with the herbicide trifluralin.

For example, Chapters 14 and 16 by Vargo and Vargo et al. focus on the ionic degradates of the acetanilide herbicides, both the ethane sulfonic acid and oxanilic acids of alachlor, metolachlor, acetochlor, and dimethanamid. Here is shown the reliability of using MRM techniques for monitoring of water quality in the Midwestern United States. Chapter 14 describes a triple quadrupole method for the detection of this family of compounds using MRM techniques. Chapter 16 shows a comparison study of LC/MS and LC/MS/MS methods for this same family of compounds. These chapters and Chapter 15 by Fuhrman and Allan compare various triple quadrupole methods. Fuhrman and Allan show that MRM methods may be extremely sensitive, such that direct analysis of water samples may be accomplished with no sample preparation, and the elimination of methods like solid-phase extraction. This paper shows the power of triple quadrupole LC/MS/MS for routine monitoring of emerging contaminants and the sensitivity that is available by MRM.

Chapter 17 by Lerch et al. uses ion trap MS/MS to identify 8 degradates of the common soybean and cotton herbicide, trifluralin. One of the analytes was a degradate that was hypothesized from the literature but had never been detected in soil, a hydroxymetabolite of trifluralin. Using the ion trap MS/MS, the molecular ion for the hydroxymetabolite of trifluralin was tentatively identified along with a fragmentation characteristic of the trifluralin family. From here Lerch et al. synthesized the standard and confirmed the identification of the degradate. This chapter emphasized again the power of the ion trap to identify unknowns within a family of compounds.

LC/MS/MS of Natural Products and Surfactants

This last section of the book discusses the identification of natural products and surfactants. Natural products, as such, are not emerging contaminants; however, the study of natural compounds and the role they play in the interaction of contaminants has a long and interesting history (29). Humic substances, the natural organic acids from soil and plant material, have long been associated with contaminants both in metal complexation (29) and in their role to co-solubilize contaminant organic compounds (30). Thus, the first chapter of this section deals with the study of these natural organic acids.

Man-made surfactants, such as the polyethoxylates, are important as emerging contaminants because they have been discovered to mimic hormones and are thought to contribute to endocrine disruption in fish (31). These compounds are common components of wastewater and their degradates as well.

Chapter 18 by Leenheer deals with the MS/MS analysis of fulvic and humic acids in an attempt to understand structural components by using ion trap MS^3 or MS^4. The paper shows that the sequential losses of water are common in these mixture of compounds and points the way for a better understanding of their underlying structure.

Chapter 19 by Young deals with the MS/MS analysis of hydrocarbons and their oxidation products. Upon oxidation these degradates have a surfactant like structure that affects their chromatography and mass spectral fragmentation. This chapter shows the approach to method development to identify unknowns associated with the groundwater contamination by this family of compounds.

Chapter 20 by Petrovic and Barceló deals with identification of chlorinated and brominated surfactants in the environment. These compounds are generated by the chlorination of wastewater that contains trace levels of bromide. Chlorine reacts with bromide to generate bromine, which is more reactive than chlorine toward organic compounds. Compounds such as nonylphenol and ethoxylates formed with this starting material have the potential to react with bromine and form new degradates. They show the LC/MS/MS spectra of this family of compounds and outline their fragmentation pathways.

Chapter 21 by Zwiener et al. deals with water treatment and chlorination and the generation of lower molecular weight unknowns from the larger natural or humic substances in water. This chapter demonstrates that various carbonyls may be identified using LC/MS/MS after derivatization of the functional groups. The power of LC/MS/MS is then used to quantify the importance of these functional groups in forming new compounds from the chlorination of water.

Chapter 22 by Ferrer et al. uses ion trap for the identification of unknown homologue surfactants in wastewater. Here the approach is to use the ion trap at MS^2 and MS^3 to see various fragment ions, which are diagnostic of a certain chemical structure. Polyethylene glycols are first fragmented to give diagnostic ions of 89, 133, and 177 m/z, which are the result of fragmentation and formation of characteristic ions of polyethylene glycol. These ions become the double diagnostic ions of two other classes of surfactants, the nonylphenolethoxylates and the linear alkylethoxylates, when they appear in the MS^3 spectrum. The unique approach of Ferrer et al. has many applications beyond surfactants and shows the power of understanding the structure of ion formation in LC/MS/MS.

Future Trends in LC/MS/MS of Emerging Contaminants

It is difficult to predict the exciting future that lies ahead for LC/MS/MS and TOF/MS of emerging contaminants. Surely the methods of LC/MS/MS will continue to be used for unknown identification. Nonetheless, several important trends in LC/MS/MS appear obvious in this field.

LC/MS/MS

1. Development of a searchable library for unknown identification using both LC/MS and LC/MS/MS.
2. Continued and improved use of TOF and Q-TOF for elemental composition of unknown molecular ions and fragment ions.
3. Routine use of triple quadrupole MRM for routing monitoring of emerging contaminants in water and wastewater.
4. Replacement of sample preparation techniques for many applications of LC/MS/MS by direct analysis of water using triple quadrupole MRM analysis.
5. Library of characteristic fragmentations and diagnostic ions, including smart software, to help in the ion trap MS/MS analysis of unknowns.
6. Marriage of ion trap and TOF into a Trap-TOF for unknown identification by LC/MS/MS.

Emerging Contaminants

1. Analysis of hormones and ovulation inhibitors will continue to be a hot topic for methods development and monitoring.
2. Discovery of a new list of possible endocrine disruptors based on the analysis of wastewater samples by LC/MS/MS.
3. A continued and lengthy list of pesticide degradates will be found from the top 40 pesticides applied annually to soils in the United States and Europe. The top 40 are pesticides with a greater than 1 million kilograms annual use.

References

1. Colborn, T.; Dumanoski, D.; Myers, J.P. *Our Stolen Future*; Penguin Books, USA Inc., 1996; 306p.
2. Jobling, S.; Nolan, M.; Tyler, C.R.; Brighty, G.; Sumpter, J.P. *Environmental Science and Toxicology* **1998**, *32*, 2498-2506.
3. Kime, D.E., 1998, *Endocrine disruption in fish*: Kluwer Academic Publishers, Boston, MA, 396 p.

4. Carlsen, E.; Giwercman, A.; Keiding, N.; Skakkebaek, N. *British Medical Journal* **1992**, v. 305, 609-613.
5. *Electrospray Ionization Mass Spectrometry;* Cole, R.B., Ed.; John Wiley & Sons, Inc., 1997; p 577.
6. Fenn, J.B.; Mann, M.; Meng, C.K.; Wong, S.F.; Whitehouse, C.M. *Science* **1989**, *246*, 64.
7. *Measuring Mass*; Grayson, M.A., Ed.; Chemical heritage Press; Philadelphia, PA, 2002; p 149.
8. Richardson, S.D., *Anal. Chem.* **2002**, *74*, 2719-2742.
9. *Pharmaceuticals and Personal Care Products in the Environment*; Daughton C.G. and Jones-Lepp, T.L., Eds.; American Chemical Society Symposium Series 791, 2001, 395p.
10. Ternes, T.A. *Water Res.* **1998**, *32*, 3245-3260.
11. Kolpin, D.W.; Furlong, E.T.; Meyer, M.T., Thurman, E.M.; Zaugg, S.D.; Barber, L.B.; Buxton, H.T. *Environ. Sci. Technol.* **2002**, 1202-1211.
12. Willoughby, R.; Sheehan, E.; Mitrovich, S. *A Global View of LC/MS, Second Edition;* Global View Publishing, Pittsburg, PA, 2002; 518p.
13. Niessen, W.M.A. *Liquid Chromatography-Mass Spectrometry;* Marcel Dekker, Inc.: New York, NY, 1999.
14. *Applications of LC-MS in Environmental Chemistry;* Barcelo, D., Ed.; Elsevier: Amsterdam, 1996, p 543.
15. Sparkman, O.D. *Mass Spectrometry Desk Reference;* Global View Publishing: Pittsburg, PA, 2000.
16. *Mass Spectrometry/Mass Spectrometry*; Busch, K.L.; Glish, G.L., McLuckey, S.A., Eds.; VCH Publishers Inc.: New York, NY, 1988; p 333.
17. Liquid Chromatography/Mass Spectrometry: Applications in agricultural, pharmaceutical, and environmental chemistry, ACS Symposium Series 420; Brown, M.A. Ed.; American Chemical Society: Washington, D.C., 1990; p 298.
18. *Liquid Chromatography/Mass Spectrometry: Techniques and Applications;* Yergey, A.L.; Edmonds, C.G.; Lewis, I.A.S.; Vestal, M.L., Eds.; Plenum Press: New York, NY, 1990; p 306.
19. McEwen, C.N.; Larsen, B.S. Electrospray ionization on quadrupole and magnetic-sector mass spectrometers. *In: Electrospray Ionization Mass Spectrometry;* Cole, R.B., Ed.; John Wiley & Sons, Inc., 1997; Chapter 5, pp 177-202.
20. Chernushevich, I.V.; Ens, W.; Standing, K.G. Electrospray ionization time-of-flight mass spectrometry. *In: Electrospray Ionization Mass Spectrometry;* Cole, R.B., Ed.; John Wiley & Sons, Inc., 1997; Chapter 6, pp 203-234.

21. Bier, M.E.; Schwartz, J.C. Electrospray ionization quadrupole ion-trap mass spectrometry. *In: Electrospray Ionization Mass Spectrometry;* Cole, R.B., Ed.; John Wiley & Sons, Inc., 1997; Chapter 7, pp 235-290.

22. Laude, D.A.; Stevenson, E.; Robinson, J.M. Electrospray/Fourier transform ion cyclotron resonance mass spectrometry. *In: Electrospray Ionization Mass Spectrometry;* Cole, R.B., Ed.; John Wiley & Sons, Inc., 1997; Chapter 8, pp 291-320.

23. Voyksner, R.D. Combining liquid chromatography with electrospray mass spectrometry. *In: Electrospray Ionization Mass Spectrometry;* Cole, R.B., Ed.; John Wiley & Sons, Inc., 1997; Chapter 9, pp 323-342.

24. *Herbicide Metabolites in Surface Water and Groundwater,* Meyer, M.T. and Thurman, E.M., Eds., American Chemical Society Symposium Series 630, 1996, 318p.

25. Vargo, J.D. *Anal. Chem.* **1998**, *70*, 2699-2703.

26. Ferrer, I.; Barcelo, D.; Thurman, E.M. *Environ. Sci. Technol.* **2000**, *34*, 714-718.

27. Thurman, E.M.; Goolsby, D.A.; Meyer, M.T.; Kolpin, D.W. *Environ. Sci. Technol.* **1991**, *25*, 1794-1796.

28. Kolpin, D.W., Thurman, E.M., Linhart, S.M., **1998**, *Archives of Environmental Contamination and Toxicology*, *35*, 385-390.

29. Thurman, E.M., 1985, *Organic Geochemistry of Natural Waters*: Martinus Nihjoff/Dr. W. Junk Dordrechdt, 497p.

30. Chiou, C.T., 2002, *Partition and adsorption of organic contaminants in environmental systems,* John Wiley and Sons, Inc., New York, 257p.

31. Jobling, S.; Sheahan, D.; Osborne, J.A.; Matthiessen, P.; Sumpter, J.P.; *Environmental Toxicology and Chemistry* **1996,** *15*, 194-202.

Chapter 2

Comparison of Quadrupole Time-of-Flight, Triple Quadrupole, and Ion-Trap Mass Spectrometry/Mass Spectrometry for the Analysis of Emerging Contaminants

E. M. Thurman[1] and Imma Ferrer [2,3]

[1]U.S. Geological Survey, 4821 Quail Crest Place, Lawrence, KS 66049
[2]U.S. Geological Survey, Denver Federal Center, Box 25046, Denver, CO 80225
[3]Current address: immaferrer@menta.net

Unique types of structural information for pharmaceutical and pesticide degradates are derived from quadrupole-time-of-flight (Q-TOF), triple quadrupole, and quadrupole ion-trap mass spectrometry/mass spectrometry (MS/MS) instruments for the analysis of emerging contaminants in water. This chapter explains the unique features of the three instruments and gives examples of their complimentary nature. For example, the Q-TOF MS/MS is unique in its ability to give accurate mass measurements (1 to 2 millimass units) of the fragment ions that are ejected from the collision chamber, which give a high assurance of correct identification of unknowns, as well as an empirical formula of fragment ions. The triple quadrupole MS/MS has the unique feature of neutral loss, which allows both quadrupoles 1 and 3 to scan in tandem and is used for identifying unknowns in the chromatogram that are structurally related to one another by fragmentation losses within the molecule. Finally, the unique feature of the quadrupole ion-trap MS/MS is its ability to do MS to the n, which typically is MS^3 or MS^4 for most unknowns. This feature is used for structural elucidation by tracing the pathway of fragmentation within ion fragments. This chapter gives several examples of emerging-contaminant analysis that exemplify the unique features of these three instruments for the identification of unknown compounds.

Introduction

The analysis of emerging contaminants is a major environmental issue over the past few years because of the discovery of pharmaceutical compounds and pesticide degradates in the environment (*1-4*). Emerging contaminant issues have been highlighted by several recent scientific meetings on this topic and a series of papers that deal with these compounds in the environment (two symposia by the National Water Well Association in Minneapolis and several symposia by the American Chemical Society including this book). Most recently is the finding of widespread pharmaceuticals in surface water of the United States (*3*). The results of this reconnaissance by the U.S. Geological Survey shows that 80% of all surface water had detectable concentrations of pharmaceutical compounds. Approximately 82 compounds were detected including steroids, antibiotics, analgesics, heart medications, and other compounds. Typically the concentrations were in the sub-microgram-per-liter range. The majority of the pharmaceuticals identified were detected using liquid chromatography/mass spectrometry (*3*).

In the case of both pharmaceuticals and pesticide degradates, unequivocal identification of these unknowns typically involves two steps. Firstly, there is the tentative identification of the compound using either gas chromatography/mass spectrometry (GC/MS) or liquid chromatography/mass spectrometry (LC/MS) with spectral and library searching and matching. Secondly, there is identification using an authentic standard and matching chromatographic retention time and mass-spectral matching for molecular ion and fragment ions. This procedure is considered adequate for final identification. Recently this level of identification has been challenged for complex samples containing many compounds, and a higher level of certainty has been advocated for identification of unknowns, such as pesticide degradates (*5*). In many cases, the more specific level of identification requires liquid chromatography/mass spectrometry/mass spectrometry (LC/MS/MS).

This chapter will compare three types of LC/MS/MS instrumentation for environmental analysis of emerging contaminants: quadrupole-time-of-flight (Q-TOF), triple quadrupole, and ion-trap MS/MS. The chapter will focus the comparison on the identification of unknowns and give a thumbnail sketch of the unique features of each of the three instruments with examples using emerging contaminants from the pharmaceutical and pesticide families. The chapter is meant to show the complimentary nature of the types of LC/MS/MS instrumentation currently available for analysis of emerging contaminants and the advantages and disadvantages of each of the instruments.

Experimental Methods

The Q-TOF MS/MS described herein was a Micromass model Q-Tof™ 2 (Micromass Limited, Manchester, United Kingdom) fitted with the Z-spray source and operated with a resolution of 10,000 at full peak width at one half maximum (FWHM). The instrument was operated in electrospray positive mode with a cone voltage of 40 V. Calibration was carried out with cesium iodide at the same pusher frequency as when the data were acquired. Exact mass corrections were made with dioctylphthalate (background ion in mobile phase with nominal mass of 391), which was the lock mass at an intensity of 300 to 500 counts/second. The mobile phase was 90% methanol and 10% water with 0.5% formic acid. All standards were injected in flow injection mode. Q-Tof™ 2 was controlled by Micromass' MassLynx™, SampleCentric™, Mass-Informatics™ running under Microsoft Windows NT®. The mass accuracy specification of the Q-TOF MS/MS was operated at 1 millimass unit based on initial calibration with dioctylphthalate.

The triple quadrupole LC/MS/MS system described herein was a Micromass Quattro Ultima (Micromass Limited, Manchester, United Kingdom) fitted with the Z-spray source and operating in electrospray positive mode with a cone voltage of 40 V. The collision gas was argon at a pressure of 1.7×10^{-3} bar. The mobile phase was 80% methanol and 20% water with 5 mM ammonium formate. All standards were injected in flow injection mode. Work on the Q-TOF and triple quadrupole MS/MS were carried out in the Mass Spectrometry Laboratory at the University of Kansas, Lawrence, Kansas.

The quadrupole ion-trap mass spectrometer (LC/MS/MS) described herein was an Esquire LC/MS/MS (Bruker Daltonics, Bellerica, MA) system equipped with an orthogonal electrospray-ionization (ESI) source running in positive-ion mode. Operating conditions of the MS system were optimized in full-scan mode (m/z scan range: 50 to 400) by flow-injection analysis of selected compounds at 10-μg/mL concentration or by using liquid chromatography. The maximum accumulation time value was set at 200 ms. in positive-ion mode of operation. The analytes were separated by using a series 1100 Hewlett Packard liquid chromatograph (Palo Alto, CA) equipped with a reverse-phase C-18 analytical column (Phenomenex RP18, Torrance, CA) of 250 x 3 mm and 5-μm particle diameter. Column temperature was maintained at 60° C. The mobile phase used for eluting the analytes from the HPLC column consisted of acetonitrile and 10-mM ammonium formate buffer at a flow rate of 0.3 mL/min.

Results and Discussion

Three MS/MS instruments are compared in this study. These three instruments represent three different design types, and their comparison shows the unique capabilities of each of these instruments for the analysis of emerging contaminants. Table I shows the first simple comparison of the three instruments. The discussion will follow the order of Table I including a discussion of each instrument type, its unique features, advantages, and disadvantages on the basis of analysis of unknowns and emerging contaminants, such as pharmaceuticals and pesticide degradates.

Table I. Comparison of MS/MS Capabilities of Q-TOF, Triple Quadrupole, and Ion-Trap Mass Spectrometers

Instrument	Unique Features	Advantages	Disadvantages
Q-TOF MS/MS	Accurate mass of fragment ions	Most sensitive in MS/MS for precursor-ion scans and accurate mass for fragment identification, always gives full-scan data of the precursor ion.	Most expensive, quasi-selected reaction monitoring, neutral loss not possible.
Triple Quadrupole MS/MS	Neutral loss	Sensitive for quantitation with multiple-reaction monitoring (MRM).	Moderately costly and least sensitive in scan.
Quadrupole Ion-Trap MS/MS	MS^n	Sensitive in scan, least expensive cost, pathway of fragmentation deduced easily	Quantitation less reliable than MRM, neutral loss not possible.

Q-TOF MS/MS

The Q-TOF MS/MS is a quadrupole/orthogonal-acceleration time-of-flight tandem mass spectrometer that was developed in the last 5 years (Figure 1). The Q-TOF mass spectrometer combines the simplicity of a quadrupole MS with the ultra high efficiency of a time-of-flight (TOF) mass analyzer. The model used in this study is the second generation Q-Tof™ 2, which is approximately 3 years old and delivers a resolution of ~10,000. The resolving power (defined below) enables mass measurement accuracy for small molecules, charge state identification of multiply charged ions, and greater differentiation of isobaric species (two different compounds with the same integer mass but different elemental compositions and, therefore, different exact masses).

Figure 1 shows the operation of the Q-Tof 2. The sample is introduced through the Z-spray interface, and ions are focused using the hexapole ion bridge into the quadrupole MS, which is MS 1. Here the precursor ion is selected for later fragmentation and analysis with a mass window of approximately 3 mass units, which is a typical window to preserve the isotope envelopes in the product ion spectra. The ions are ejected into the hexapole collision cell where argon is used for fragmentation. From this point, the ions are collected into the TOF region of the MS/MS. The introduction of ions is such that the flight path of the ions changes 90 degrees, which is called an orthogonal TOF. The purpose of the change in direction is to optically focus the kinetic energy of the ions so that their kinetic energies are as similar as possible. The ions then are accelerated by the pusher and travel down the flight tube approximately 1 meter to the reflectron. The purpose of the reflectron is to slow down ions of equal mass but higher kinetic energy and then to focus this beam of ions at the detector such that ions of the same exact mass but slightly different energies arrive at the detector at exactly the same moment. This process results in the mass accuracy of the Q-TOF MS/MS.

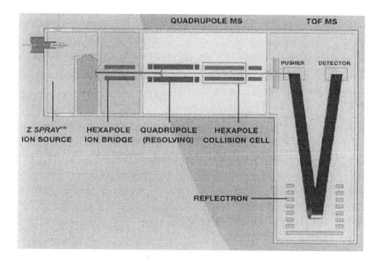

Figure 1. Diagrammatic view of a Q-TOF MS/MS. Published with permission of Micromass Limited, Manchester, United Kingdom.

Thus, the TOF side of the Q-TOF MS achieves simultaneous detection of ions across the full mass range at all operation times. This continuous full-scan mass spectrum is in contrast to the tandem quadrupoles that must scan over one mass at a time, and for this reason the Q-TOF MS/MS is more sensitive when scanning the TOF side of the instrument (estimates are 10 to100 times in product literature) in scan mode than the third quadrupole of the triple quadrupole MS/MS (see Table I). However, it is important to remember that the TOF side of the Q-TOF MS/MS has the same sensitivity in scan mode and in selected-ion mode, which is not true for the triple quadrupole MS/MS, which has increased sensitivity in multiple reaction monitoring (MRM) compared to scan mode. MRM is a kind of selected-ion mode, see next section on triple quadrupole MS/MS.

The Q-TOF MS/MS system is considered a high resolution instrument capable of 10,000 resolution expressed at full peak width at one-half maximum (FWHM), which is shown in Figure 2. Resolution is defined as shown in equation (1), where M is the mass of the ion and ΔM is the width of the peak at the half height of the peak (Figure 2).

M/ΔM = Resolution at full peak width at one-half maximum (FWHM)   *(1)*

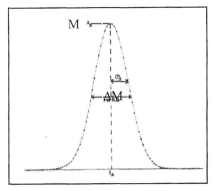

Figure 2. Definition of FWHM (full peak width at one-half maximum in standard deviation units, which is calculated from the Gaussian distribution function, $f(x) = Ce^{(-x^2/2\sigma^2)}$ and resolution, which is defined above in equation 1.

Because the TOF mass detector collects data in a gaussian-shaped peak, it is possible to express the resolution in terms of standard deviation units. Thus, the peak width at one-half height is 2.355 σ (Figure 2), defined below. The resolution using Q-TOF MS/MS is expressed as the mass of the analyte, typically reserpine (609 nominal mass), divided by the ΔM value, which is equal to 2.355 σ in mass units. This value for resolution is different than the values

20

used for a magnetic-sector high resolution GC/MS. The definition for resolution in these systems is shown in Figure 3, where the resolution is defined as in equation (2)

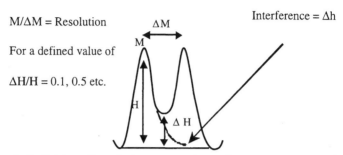

M/ΔM = Resolution

For a defined value of

ΔH/H = 0.1, 0.5 etc.

Interference = Δh

Figure 3. Definition of resolution commonly used in magnetic-sector high resolution GC/MS (6).

$M/\Delta M$ = *Resolution for a defined value of $\Delta H/H$ = 0.1 or 0.5.*   (2)

Note here that equations 1 and 2 are identical, but the definition of ΔM is different. In Figure 3, ΔM is the difference between two closely related mass spectral peaks with a valley between them that is defined by ΔH. Typically the ΔH/H value for magnetic-sector high resolution GC/MS is 0.1. This method is called the doublet method (6) versus the singlet method of TOF (6). The resolution equation may be calculated at any height using the gaussian-distribution function and expressing the width in standard deviation units (see equation 3).

$$f(x) = Ce^{(-x^2/2\sigma^2)}$$ (3)

Because the doublet method requires that two peaks overlap at a value of 0.1, the value of H (i.e., the height at the point of intersection of the two mass-spectral curves) is 0.05 for each of the two peaks. Solution of equation 3 for a f(x) of 0.05 gives a peak width from the mean to the point of intersection of 2.45σ. Because of the definition of ΔM in Figure 3 (i.e., the difference between the two mass-spectral peaks), ΔM is equal to twice the 2.45σ value or 4.9σ. Thus if reserpine were used for a resolution calculation, then the mass of reserpine would be divided by the mass width of 4.9σ rather than by the 2.355σ value used by the singlet method:

$M_{reserpine}/4.9\sigma,$ (4)

which is approximately one-half the calculated resolution obtained by the TOF method. This calculation means that a singlet resolution of 10,000 ≠ 10,000 resolution by the doublet method. In fact, the difference is approximately a factor of 2 less resolution with the singlet method versus the doublet method. Table II shows the effect of calculating resolution by the two methods.

Table II. Calculation of Resolution in Mass Spectrometry.

Type of Instrument	ΔH	ΔM (amu)	Difference Factor	Defined Resolution
Q-TOF MS/MS	--	2.3σ	1.0	20,000
Sector GC/MS	0.1	4.9σ	0.5	10,000

Note: See equations 1 and 2 for definition of resolution.

Thus, a resolution of 10,000 for a sector GC/MS method is equal to 20,000 for Q-TOF MS/MS. Nonetheless, resolution of 10,000 by Q-TOF MS/MS is suitable for accurate measurements on complex samples as shown in the following examples.

For example, the unique capability of the Q-TOF MS/MS compared to the triple quadrupole and ion-trap MS/MS instruments lies then in its ability to determine accurate mass on the ion fragments generated in the collision cell. Because the quadrupole allows ions of nominal mass to pass through the quadrupole, there may be masses that interfere with the determination of the molecular ion. Interfering ions are much less likely for the fragment ions, which help in the determination of accurate mass by lowering mass interferences and increasing accuracy with the same resolution.

This unique ability of the Q-TOF MS/MS is demonstrated in Figures 4-6, which show the mass accuracy for the determination of two antibiotics, oxytetracycline and sulfathiazole, in a hog-waste pond. Figure 4 shows the diode array detector (DAD) signal for the waste pond with milliabsorbance units going to 400 mAU, which is a large interfering signal. The quadrupole first is set at 461 to capture the molecular ion of oxytetracycline, an antibiotic added to swine feed and suspected to be present in hog-pond wastewater. All 461 m/z ions, ± 0.5 m/z, are accepted into the collision chamber for fragmentation. Figure 5 shows the fragmentation of the 461 m/z ion of oxytetracycline followed by accurate mass TOF analysis of the ammonia loss to the 444.1294 m/z and the loss of both ammonia and water to the 426.1189 m/z. The measurement of the two fragment ions were within –1.3 and –2.0 millimass units (μ). This analysis was done in flow injection mode without separation by liquid chromatography. A secondary method of analysis (7) confirmed these results for the identification of oxytetracycline.

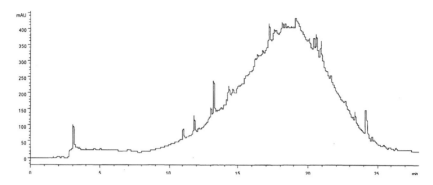

Figure 4. Diode array signal for hog wastewater suspected of containing antibiotics.

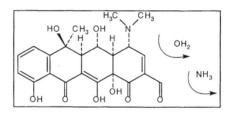

	Measured	True Mass	Error
-NH₃	444.1281	444.1294	-1.3 μ
-NH₃-H₂O	426.1169	426.1189	-2.0 μ

Figure 5. Q-TOF MS/MS analysis of hog wastewater without chromatography and identification of oxytetracycline by accurate mass analysis of fragment ions.

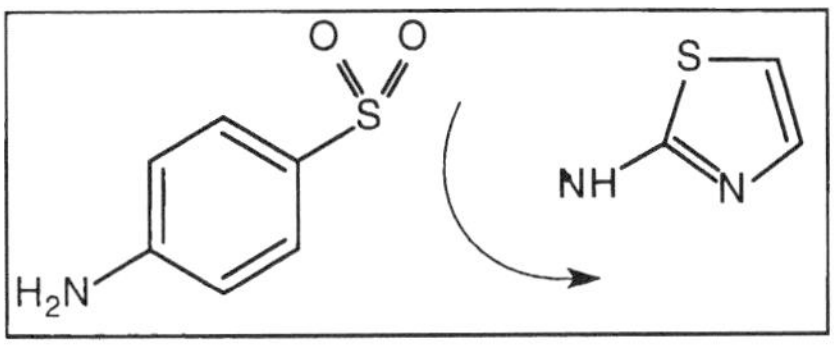

	Measured	True Mass	Error
-thiazole	156.0121	156.0119	+0.2 μ

Figure 6. Q-TOF MS/MS analysis of hog wastewater without chromatography showing the identification of sulfathiazole by accurate mass analysis of a major fragment ion.

In similar fashion, sulfathiazole was identified in the same hog-waste pond. This time the protonated molecule of 256 m/z was filtered through the quadrupole and fragmentation was carried out to give the characteristic fragment ion of 156.0119 m/z with an accuracy of +0.2 m/z (Figure 6). The accuracy was less than 1.0 μ, and a second confirmation was carried out using another LC/MS method (7). Thus, the unique power of the Q-TOF MS/MS lies in its ability to do accurate mass identification of fragment ions.

Several recent papers that further report the use of the Q-TOF MS/MS for emerging contaminant issues include work on identification of arsenosugars that originate from algal extracts (8) and these same compounds may be present in clams (9). The authors elegantly demonstrate that the arsenosugars may be identified without the use of standards, which shows the power and usefulness of the Q-TOF MS/MS.

Triple Quadrupole MS/MS

The triple quadrupole MS/MS consists of two single quadrupoles with a collision cell in between (Figure 7). The ions are directed from the Z-spray source into the first quadrupole, where the precursor ion is selected. A hexapole collision cell is next followed by the final quadrupole and the photomultiplier detector. This configuration is nearly identical for most triple quadrupole MS/MS instruments within the limits of patented configurations.

Standard triple quadrupole MS/MS instruments have four major modes of operation (shown in Table III). They are product-ion scan, multiple reaction monitoring, constant neutral loss, and the precursor-ion scan. These four modes of operation correspond to the options of running quadrupole 1 and 3 in either scan mode or in selected-ion monitoring.

In the product-ion scan, the first quadrupole (MS1) selects the molecular ion of an unknown or suspected emerging contaminant and sends only this ion to the collision cell where it undergoes fragmentation to generate the product-ion spectrum. This is the fundamental example of LC/MS/MS and is commonly the first step in the identification of an unknown or the identification of a spectrum for a standard compound, which will be compared with the unknown spectrum.

Table III. Modes of Triple Quadrupole MS/MS operation

Experiment Type	MS 1	MS 2
Product-ion scan	Static precursor ion	Scanning
Multiple reaction monitoring	Static precursor ion	Static product ion
Constant neutral loss	Scanning	Scanning
Precursor-ion scan	Scanning	Static product ion

The second mode of operation is multiple reaction monitoring (MRM), which is sometimes called selected reaction monitoring (SRM). In this mode, MS 1 selects the molecular ion of interest and sends it to MS 2 for collision and subsequent fragmentation. A specific product ion is selected and monitored in MS 3.

In Figure 8 there is an example of atrazine and its precursor ion of 216 m/z $(M + H)^+$ which undergoes loss of propylene (-42) to give the product ion (174). The MRM is written as 216 → 174 m/z. For isotope-dilution measurements, the deuterated atrazine also can be monitored, which gives the d5-atrazine transition of 221 → 179 m/z, and gives excellent quantitation because it removes any possibilities of matrix suppression. The MRM is a useful technique that is most effective in the triple quadrupole. The MRM experiment when coupled with deuterated standards gives both excellent sensitivity and quantitative results because of the fact that any matrix interaction that may occur in MS 1 occurs to both the analyte and its deuterated standard. Furthermore, the low background noise in this mode of operation gives rise to low detection limits for emerging contaminants in water. For example in the chapter by John Fuhrman on the analysis of ethane sulfonic acid (ESA) degradates of acetochlor and alachlor, it

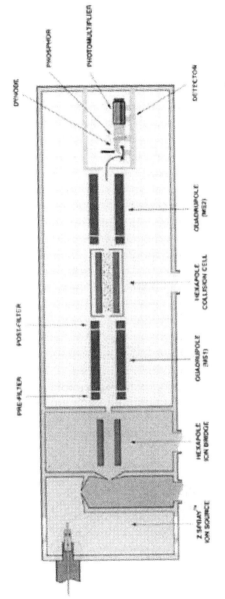

Figure 7. Diagram of triple quadrupole MS/MS. Published with permission of Micromass Limited, Manchester, United Kingdom.

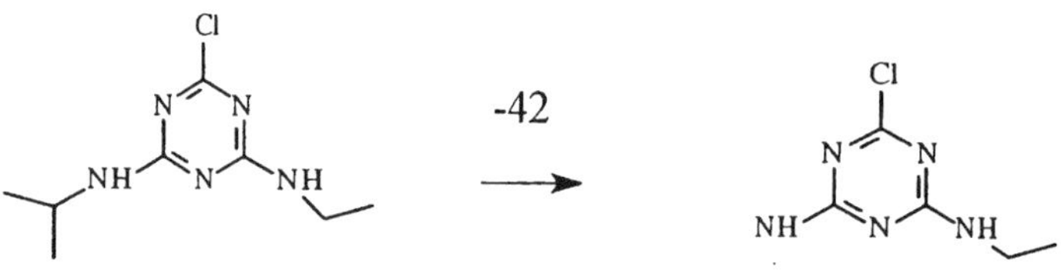

Atrazine [M + H $^{+}$] = 216

Ion 174

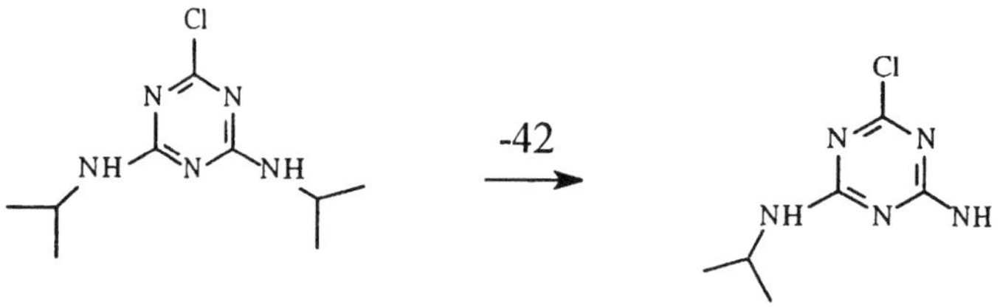

Propazine [M + H $^{+}$] = 230

Ion 188

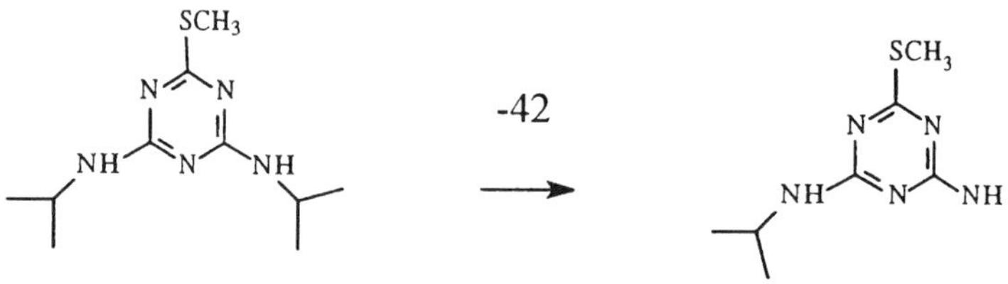

Prometryn [M + H +] = 242

Ion 200

Figure 8. MRM transitions for three triazine herbicides showing the loss of the propylene group with a mass loss of 42 mass units. These examples also show the neutral loss of 42 mass units.

is possible to detect the ESA degradates of these three herbicides in the water sample directly without solid-phase extraction clean up and preconcentration of the sample. The sensitivity of the various MS/MS systems are now such that trace levels, as little as 0.1 pg, may be detected with sensitivity of signal to noise ratio of 10:1.

The third method of operation is the neutral loss experiment in triple quadrupole MS/MS (Table III). In this mode of operation, MS 1 and MS 3 scan in synchronization so that neutral losses from molecular ions detected in MS 1 can be linked to their appropriate spectra. The instrument then calculates the mass loss between MS 1 and MS 3 and finds all peaks of the specified mass loss. An example is shown in Figure 8 where the loss of 42 m/z (propylene) is scanned for several triazine herbicides in water samples. The triazines, atrazine, propazine, and prometryn are detected by their common loss of 42 m/z (propylene). This technique may be useful for the detection of unknown compounds from the same family by the neutral-loss experiment.

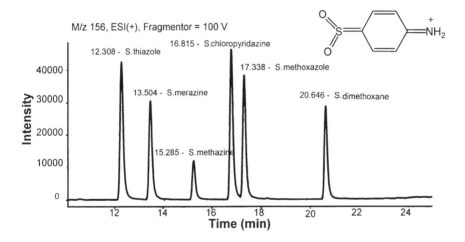

Figure 9. Example of the reconstructed ion chromatogram for the 156 m/z ion that is characteristic of the sulfonamide antibiotics (20).

The last mode of operation of the triple quadrupole is the precursor-ion scan. In this mode of operation, MS 1 is scanned and MS 3 looks at a single ion (selected ion). This operation may be diagnostic when the fragment ion produced in the collision cell is specific for a family of compounds. An example shown in Figure 9 is the sulfonamide antibiotics that undergo loss of their substitutent to yield the same 156 m/z ion. In the example in Figure 9, all of the sulfonamide compounds give the 156 m/z ion because of their common structural feature. The reconstructed-ion chromatogram then shows the peaks originating from the 156 m/z ion and draws the chromatogram. Each of these peaks then can be re-examined at full spectrum with the possibility of identifying unknowns. The full spectrum of each of the peaks in the reconstructed 156 m/z ion chromatogram is the precursor-ion spectrum. More information on the operation of the triple quadrupole MS/MS operation may be found in the classic reference by Yost and Johnson (10), although it deals exclusively with GC/MS/MS operation and the book by Busch et al. (11).

Quadrupole Ion-Trap MS/MS

In quadrupole ion-trap MS/MS, the instrument operates by trapping the full spectrum of ions present in the sample (Figure 10). The isolation in the trap is carried out by first focusing the ions into the trap and holding them by various combinations of RF voltages, in a kind of "electronic beaker" for ions. The usefulness of the quadrupole ion-trap MS/MS lies in several applications. Because it does MS/MS in time rather than in space, scan mode is sensitive, although not as sensitive as the Q-TOF MS/MS, but more sensitive in scan mode than the triple quadrupole MS/MS. The quadrupole ion-trap MS/MS is not able to do the MRM experiment, which limits its abilities in quantitation as compared to the triple quadrupole MS/MS. However, the quadrupole ion-trap has the feature of MS^n or until sensitivity disappears by first trapping a specific ion, then fragmenting it, then trapping a new product ion and fragmentating it, etc. This tool in MS/MS is unique to the quadrupole ion-trap MS/MS and is useful for the identification of unknowns, such as emerging contaminants (12).

An example of the use of MS^n is shown in Figure 11. Here the compound is a chlorination degradation-product of diazinon. The structure of the new

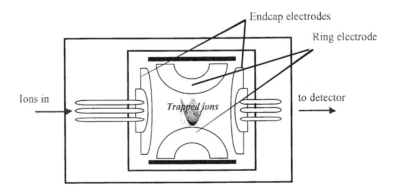

Figure 10. Operation of the quadrupole ion-trap MS/MS.

compound was thought to be diazoxon, whose structure also is shown in Figure 11. By carrying out a MS^4 experiment it was possible to assign the structural fragments by the losses of ions shown in Figure 11. The two ions at 289 and 311 m/z are assigned as the $(M+H)^+$ and the $(M+Na)^+$ for diazoxon. When the 289 m/z ion is isolated and fragmented in the trap it gives rise to the 261 m/z ion, which is the loss of 28 mass units or the ethylene group. Isolation of the 261 m/z and MS^3 gives the 233 m/z ion, which is a second loss of 28 mass units (ethylene group). MS^4 of the 233 m/z ion gives the 153 m/z ion, which is the pyridinol of diazoxon. The loss of 80 mass units when going from 233 to 153 m/z shows the loss of the PO_3H group. This examples shows the usefulness of the quadrupole ion-trap at MS^4.

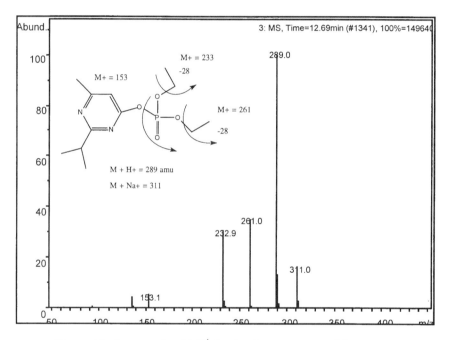

Figure 11. An example of MS^4 for the determination of diazoxon.

The diazoxon was synthesized in the laboratory by treatment of diazinon with a 10-ppm solution of hypochlorite. The diazoxon has been implicated as a toxic degradation product of diazinon that may occur during the chlorination of drinking water. This example shows how the quadrupole ion-trap MS/MS was used to identify the degradation product of diazinon for future pathway and toxicity studies.

Further references on the use of quadrupole ion-trap MS/MS include the review from Chapter 7 (*13*) in the book on electrospray MS edited by Cole (*14*).

Comparison of MS/MS Instruments

Table I shows the comparison of the three instrument types with the unique features of each. The Q-TOF MS/MS has the unique feature of accurate mass of the fragment ions. The advantages of the instrument are that it is most sensitive in MS scan mode for the product ions of the precursor ion and it gives accurate mass so that each fragment may be given an empirical formula. These empirical formulae are useful for the identification of the ion fragments. The disadvantages of the Q-TOF MS/MS include its cost, which may be prohibitive for routine analysis in many laboratories, and the fact that this instrument is not capable of MRM and the neutral-loss experiment.

The triple quadrupole MS/MS has the unique capability of the neutral-loss experiment. This feature may be used to identify unknown compounds that are related to one another, a feature that is useful to find related compounds or metabolites of parent compounds. The triple quadrupole MS/MS is also excellent for quantitation of compounds, especially in difficult matrices, using the MRM feature (Table III). It is the most popular instrument for analysis of emerging contaminants, but it has the weakness of not obtaining full-scan spectra with the same sensitivity of the ion trap and Q-TOF MS/MS instruments. Nonetheless, the triple quadrupole MS/MS is the instrument most commonly chosen for LC/MS/MS analysis of unknowns and emerging contaminants.

Finally, the quadrupole ion-trap MS/MS system is unique in its ability to obtain MS^3 and higher. It does give sensitive full-scan spectra and is a rugged and low-cost instrument. The disadvantages of the quadrupole ion-trap MS/MS is that the MRM experiment is not possible, which limits quantitation as compared to the triple quadrupole MS/MS, especially for difficult matrices and the neutral-loss experiment is not possible.

References

1. Halling-Sorensen, B.; Nielson, S.N.; Lanzky, P.F.; Ingerslev, F.; Holten Lutzhoft, J.; Jorgensen, S.E. *Chemosphere.* **1998,** *35,* 357-393.
2. Daughton, C.G.; Ternes, T.A. *Environ. Health Perspect.* **1999,** *107(Supplement 6),* 907-938.
3. Kolpin, D.W.; Furlong, E.T.; Meyer, M.T., Thurman, E.M.; Zaugg, S.D.; Barber, L.B.; Buxton, H.T. *Environ. Sci. Technol.* **2002,** 1202-1211.
4. *Herbicide Metabolites in Surface and Groundwater;* Meyer, M.T.; Thurman, E.M., Eds.; ACS Symposium Series 630; American Chemical Society: Washington, D.C., 1996, p 318.
5. Vargo, J.D. *Anal. Chem.* **1998,** *70,* 2699-2703.
6. Willoughby, R.; Sheehan, E.; Mitrovich, S. *A Global View of LC/MS, Second Edition;* Global View Publishing, Pittsburg, PA, 2002; p 518.

7. Lindsey, M.E.; Meyer, M.T.; Thurman, E.M. *Anal. Chem.* **2001,** *74,* 4640-4646.
8. Pergantis, S.A.; Wangkarn, S; Francesconi, K.A.; Thomas-Oates, J.E.. *Anal. Chem.* **2000,** *72,* 357-366.
9. McSheehy, S.; Szpunar, J.; Lobinski, R.; Haldys, V.; Tortajada, J.; Edmonds, J.S. *Anal. Chem.* **2002,** *72,* 2370-2378.
10. Yost, R.A.; Johnson, J. *Anal. Chem.* **1985,** *57,* 758A-767A.
11. *Mass Spectrometry/Mass Spectrometry;* Busch, K.L.; Glish, G.L.; McLuckey, S.A., Eds.; VCH Publishers Inc.: New York, NY, 1988; p 333.
12. Ferrer, I.; Furlong, E.T. *Anal. Chem.* **2002,** *74,* 1275-1280.
13. Bier, M.E.; Schwartz, J.C. In: *Electrospray Ionization Mass Spectrometry;* Cole, R.B., Ed.; John Wiley & Sons, Inc., New York, NY, 1997; Chapter 7, pp 235-290.
14. *Electrospray Ionization Mass Spectrometry;* Cole, R.B., Ed.; John Wiley & Sons, Inc., New York, NY, 1997; p 577.

Acknowledgments: Todd Williams, Mass Spectrometry Center, University of Kansas, Lawrence, KS, and Ed Furlong, U.S. Geological Survey, National Water Quality Laboratory, Lakewood, CO. The use of brand names is for identification purposes only and does not constitute endorsement by the U.S. Geological Survey.

Chapter 3

Identification of Residues by LC/MS/MS According to the New European Union Guidelines

Application to the Trace Analysis of Veterinary Drugs and Contaminants in Biological and Environmental Matrices

Alida A. M. Stolker[*], Ellen Dijkman, Willem Niesing, and Elbert A. Hogendoorn

Laboratory for Organic-Analytical Chemistry, National Institute for Public Health and the Environment, P.O. Box 1, Ant. van Leeuwenhoeklaan 9, 3720 BA Bilthoven, The Netherlands
·[*]Corresponding author: Fax: 31–30-2744424, email: linda.stolker@rivm.nl

New EU guidelines for identification and quantification of organic residues and contaminants are established to guarantee efficient and reliable control of food. The new EU criteria are based on the use of identification points (IPs), a new approach to set up quality criteria for the spectrometric identification and confirmation of organic residues and contaminants. For a legal compound it is necessary to collect 3 IPs. For the identification and confirmation of a prohibited compound it is necessary to collect 4 IPs. The number of IPs "earned" by the detection of a product ion depends on the technique used. For example, the detection of an ion in low resolution mass spectrometry 1 IP is earned, by high resolution mass spectrometry 2 IPs. The main advantage of using IP is the fact that the verification of the identity can be done in a well described and internationally accepted way.

Introduction

Although many LC-MS papers dealing with confirmation or analyte identification have been published over the years, the mass spectrometric identification criteria used are not always clear. Often, a criterion with three diagnostic ions and the retention time are used. However in other cases 5 diagnostic ions and three ion ratios are considered necessary. The conditions that have to be fulfilled to meet criteria for identification are not unambiguous and are hardly discussed in the literature. Geerdink *et al.* (*1*) apply specific criteria, developed for the legislation of environmental analyses based on GC-EI-MS, to the identification of some triazines analysed by flow injection analysis MS-MS (FIA-MS-MS). Because of the absence of retention time information (contrary to GC-MS analysis) one additional diagnostic ion was monitored, i.e., for FIA-MS-MS four diagnostic ions should be present, whereas for GC-MS three diagnostic ions and a retention time have to fulfil specific criteria. Although the approach was successfully applied to the identification of different triazines in samples of water, the application of this approach is limited to the specific FIA-MS-MS technique. To monitor four diagnostic ions in a regular LC-MS method using mild (chemical) ionisation is very difficult, because in most cases only two or three diagnostic ions are available.

In the European Union the European Commission regulates the inspection of animals and fresh meat for the presence of residues of veterinary drugs and specific contaminants. The identification and confirmation criteria for the analysis of such residues are described in a series of Commission Decisions (*2,3*). These legislative decisions are revised periodically to take into account current scientific knowledge and the latest technical improvements. Due to the complex nature of the revision process, in May 1998 the Commission designated a working group to draft new or revised criteria. A very important limitation of the existing EU criteria is that the application is limited to GC-MS. Nowadays mass spectrometric detection can be carried out by recording full mass spectra, for example by MS^n techniques (e.g. ion traps), or by selected ion monitoring (SIM) and selected reaction monitoring (SRM) (e.g. in quadrupoles). Other MS or MS^n techniques in combination with separation techniques (column liquid chromatography (LC) or GC) and a variety ionisation modes (e.g. EI, CI, atmospheric pressure chemical ionisation, electrospray ionisation (ESI)). The criteria for mass spectrometry in 93/256/EEC are not really adequate given the technical advances that have occurred recently.

During the discussions in the EU working group for revising the EU criteria document, it became clear that the diversity and dynamics of the compounds to be tested for, being either prohibited or legal, ask for a flexible system of developing and validating analytical methods. The solution was found in setting up sets of minimum performance characteristics, which have to be fulfilled by methods to be used for residue control of (veterinary) drugs, natural occurring toxic compounds and environmental contaminants. Mass spectrometry, either in combination with gas chromatography (GC-MS) or liquid chromatography (LC-MS) plays a key role in residue analysis. For confirmatory (reference) purposes, the revised criteria discriminate between prohibited and legal compounds. Confirmatory methods are methods that provide full or complementary information enabling the analyte to be identified unequivocally at the level of interest. Although the EU criteria as described in (4,5) are primarily used in the field of identification of veterinary drugs and specific contaminants in animals and fresh meat, the approach is universal applicable to the identification of organic residues and contaminants.

Next to common criteria for confirmatory analysis (e.g. criteria for specificity of the method, criteria for LC retention time windows, criteria for the quantitative aspects of the analysis) the EU criteria document describes specific criteria for the verification of the identity of the analyte. This paper discusses the criteria for verification of the identity of small molecules and demonstrates their general applicability. Therefore all the contaminants studied in this paper are treated as 'illegal compounds' so 4 IPs have to be earned. The EU identification criteria are applied to the LC-MS/MS trace analysis of a) corticosteroids in samples of bovine urine, b) nicotine in samples of rat plasma and c) drugs in surface water.

Method

Mass spectrometric methods are suitable for consideration as confirmatory methods only following either an on-line or an off-line chromatographic separation (4,5). In the EU guideline for data evaluation the concept of IPs for confirmatory purposes was introduced. For the identification and confirmation of a legal (e.g. registered drug) compound or contaminant it is necessary to collect 3 IPs. For the identification and confirmation of an illegal (prohibited) compound it is necessary to collect 4 IPs. This increase in number of IPs reflects the seriousness of the finding and the extent of the consequences. The number of IPs "earned" depends on the technique used. If full scan spectra are recorded in single MS, a minimum of four ions must be present with a relative intensity of $>10\%$.

If fragment ions are measured using other than full-scan techniques, a system of IPs must be used to interpret the data. Table I shows the number of IPs that each of the basic MS techniques can earn.

Table I. MS techniques and identification points to be earned

MS technique	Identification Points earned per ion
Low resolution :	
MS	1.0
MS^n precursor ion	1.0
MS^n each transition ion	1.5
High resolution :	
MS	2.0
MS^n precursor	2.0
MS^n each transition ion	2.5

Remarks: However, in order to qualify for the IPs required for confirmation:
- A minimum of at least one ion ratio must be measured, and
- All measured ion ratios must meet the criteria and,
- A maximum of three separate techniques can be combined to achieve the minimum number of IPs
- Transition ion = (grand)daughter ion

The allowable tolerances for the ion ratios are summarised in Table II.

Table II. Maximum permitted tolerances for relative ion intensities using a range of mass spectrometric techniques

Relative intensity (% of base peak)	EI-GC-MS (relative)	CI-GC-MS, GC-MS-MS^n, LC-MS, LC-MS-MS^n
> 50%	± 10%	± 20%
> 20% - 50%	± 15%	± 25%
> 10% - 20%	± 20%	± 30%
≤ 10%	± 50%	± 50%

The main advantage of using IPs is the fact that the verification of the identity can be done in a well-described and internationally accepted way. The arithmetic approach of adding IPs also is helpful in evaluating the total information gathered in those cases where more than one measurement is necessary. Several LC-MS methods use the detection of a (de)protonated molecular ion during the first analyses resulting in 1 IP (no ratio). This value of 1 can be added to the numerical result of further measurements using other techniques like LC-MS/MS and LC-MS^n.

Experimental

The applicability of the criteria for LC-MS/MS and LC-MS^n methods was evaluated for several analytical methods routinely used in our laboratory.

Confirmation of residues of corticosteroids in samples of bovine urine

After enzymatic deconjugation with a beta-glucuronidase/sulfatase mixture (suc d'Helix Pomatia, Brunschwig Chemie, Amsterdam, Netherlands) in order to deconjugate the analytes, the sample is extracted on a SPE (Oasis-HLB) column (Waters, Etten-Leur, Netherlands) (6). The column is conditioned with 3 ml of methanol and 3 ml of water and washed twice with 3 ml of methanol-0.02 mol/l NaOH (40:60, v/v) and finally, with 3 ml of water. The analytes are eluted with 1.5 ml of methanol. After evaporation of the methanol, the final extract is dissolved in 250 µl of LC-solvent of which 100 µl is injected into the LC-MS system (ion-trap LC-MS^n (Finnigan LCQ-system, ThermoQuest, Breda, Netherlands). HPLC-conditions: mobile phase acetonitrile - water gradient; column: LichroCART® 125-4 mm, Superspher® 100 RP-18, 4 µm particles endcapped (Waters). Acquisition parameters: vaporiser 500°C; capillary 150°C; nitrogen (high purity), 70 p.s.i.; ions are detected in the APCI (+)-MS^2 and MS^3 mode. One MS^2 and one MS^3 ion are monitored and the relative intensity between the two-recorded ions is calculated and compared with the ratio as obtained for the corresponding standards. The ions monitored for respectively dexamethasone (dex), flumethasone (flu) and triamcinolone acetonide (tria) are m/z [MS^2-373; MS^3-355] from parent ion m/z 393, [MS^2-391; MS^3-371] from parent ion m/z 411 and [MS^2-415; MS^3-357] from parent ion m/z 435.

Confirmation of nicotine residues in samples of rat plasma

Before extraction of the residues of nicotine and cotinine from the sam̲
of rat plasma, trichloroacetic acid is added for protein precipitation (7). Afte̲
washing the sample with dichloromethane the pH is increased by the addition of
sodium hydroxide and the analytes are extracted by dichloromethane. The
dichloromethane is evaporated by a stream of nitrogen and the final extract is
dissolved in 100 µl of water. 50 µl of the extract is injected in the LC-MS/MS
system (Micromass Quatro Ultima; Almere, Netherlands). HPLC-conditions:
mobile phase methanol - 50 mM ammonium acetate (5:95;v/v); column: XTerra
RP18 3.5µm; 2.1x100 mm. Acquisition parameters: desolvation temperature
500°C; Cone gas flow 80 l/hr; desolvation gas flow 350 l/hr; source temperature
120°C; ions are detected in the APCI (+) mode. Corona 0.50 µA and Cone
voltage 30V. Two MS/MS-ions are monitored and the relative intensity
between the two ions is calculated and compared with the ratio as obtained for
the corresponding reference standards. Parent ion m/z 163; two MS/MS ions
monitored m/z 163→132 and m/z 163→106.

The confirmation of residues of drugs in surface water

After adjusting the pH to 3.0 by 2 M HCl solution a 100 ml of sample is
extracted on a SPE column (Oasis MCX 6 ml; Waters). The column is
conditioned with 5 ml of acetone and 5 ml of water at pH=3.0 and washed with
6 ml of water at pH=3.0. The analytes are eluted with 8 ml methanol-16 M
ammonia (95:5, v/v). After evaporation of the eluate, the final extract is
dissolved in 50 µl of methanol and 475 µl of water of which 25 µl is injected in
the LC-MS system (triple quad system; Micromass Quatro Ultima; Almere,
Netherlands) (8,9). HPLC-conditions: mobile phase methanol water gradient
buffered with 2 mM ammonium acetate; column: Waters XTerra® 100-2,1 mm,
RP-18, 3,5 µm particles with a 1 cm pre-column. Acquisition parameters:
vaporiser 375°C; Cone gas flow 80 l/hr; desolvatation gas flow 350 l/hr; source
temperature 120°C; ions are detected in the ESI (+) mode. Capillary voltage: 3.2
kV and Cone voltage 30 V. Two MS/MS-ions are monitored and the relative
intensity between the two-recorded ions is calculated and compared with the
ratio as obtained for the corresponding standards. Table III shows the analytes in
combination with the parent ion and the corresponding transition ions
monitored.

Results and Discussion

…spect to the verification of the identity of small …clusive the extension to multiple MS, as described in …ecently. The concept of multiple ion monitoring for …however, is already used for more than a decade in …pplication of the draft revised criteria was verified in …ur laboratory for a number of different methods, using a different type of LC-MS method. Examples are from the field of the detection of the presence of residues of illegal growth promoting compounds as well as examples from the field of the detection of the presence of residues of contaminants and drugs in environmental and biological samples.

It has to be emphasized that the revised criteria are based on consensus obtained by the members of the 'EU working group of experts'. There is no fundamental chemometrical basis for the criteria. The need for 3 IPs for confirming the presence of a legal drug and 4 IPs for an illegal drug reflects the implications of such a finding in terms of subsequent legal actions. Also the earning of 1 IP using LRMS and 2 IPs using HRMS is only based on the qualitative knowledge that HRMS is more powerful in terms of identification. For reasons of practical simplicity there is no differentiation made between ionisation techniques in respect to the amount of IPs earned in other words, no difference between EI and CI.

Table III. Ions monitored in the LC-MS/MS analysis of drugs

Analyte	Parent ion m/z	First MS/MS ion m/z	Second MS/MS ion m/z
Sulfamethoxazole	254	156	92
Paracetamol	152	110	93
Metoprolol	268	116	191
Carbamazepine	237	194	192
Bezafibrate	362	316	276
Dehydroerythromycine	716	158	558
Diclofenac	296	215	250
Clofibrate	243	87	169
Fenofibrate	361	233	139
Acetylsalicylic acid*	137	93	-
Clofibric acid*	213	127	85
Ibuprofen*	205	161	-

*Ionisation mode ESI(-)

The confirmation of residues of corticosteroids in samples of bovine urine

Within the European Union the use of synthetic corticosteroids as growth promoters in livestock breeding is prohibited. Corticosteroids are often added illegally for growth promoting purposes. Bovine urine samples are monitored for residues of these compounds. Confirmation of the identity of the analytes is very important because of its legal consequences. Figure 1a shows the chromatograms obtained for LC-MS^2/MS^3 analysis of a sample of bovine urine fortified at a level of 1μg/L with dexamethasone, flumethasone and triamcinolone acetonide. Figure 1b shows the chromatograms obtained for a suspected sample of bovine urine. Comparison of the MS^2 and MS^3 data for the standards and samples clearly show the presence of dexamethasone, and the absence of the other two corticosteroids. The signals observed in the traces of flumethasone and triamcinolone acetonide do not have the proper retention times and are expanded noise signals. No signals for the specific ions of Figure 1 were monitored when a blank urine sample was analysed. The evaluation of the ratios is summarized in Table IV.

According to Table I the detection of two MS^n transition products yield 3 IPs (1.5 points each). However, since the precursor ion also must have been present, an additional IP can be added. Therefore, the data shown are adequate for the confirmation of the identity of dexamethasone at the level concerned.

The confirmation of nicotine and cotinine in samples of rat plasma

For a toxicological study to the influence of nitric oxide on the intake of nicotine, a set of approximately 200 samples of rat plasma had to be analysed to measure the concentration of nicotine. For the confirmation of the identity of the measured analyte the EU criteria were applied. Two transition ions were measured and the ratio of the abundance's of the ions the samples of rat plasma were compared with the ratio obtained for the nicotine reference sample. Figure 2 shows the chromatogram of a) reference standard of nicotine and b) rat plasma sample. The evaluation of the ratios is summarised in Table IV.

According to Table I the detection of two transition products yields 3 IPs (1.5 IP each) and an additional IP for the parent ion resulted in 4 points. Therefore, the data shown are adequate for the confirmation of the identity of nicotine at the level concerned.

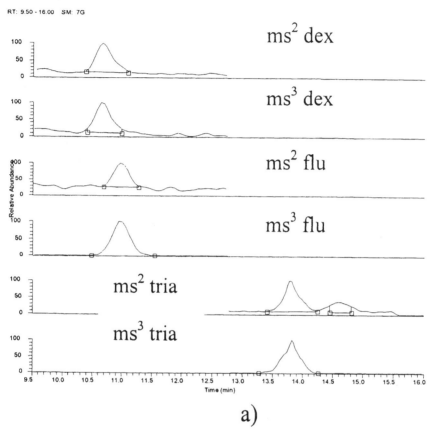

Figure 1[1]. LC-MS[2]/MS[3] Chromatograms of (a) blank urine fortified with 1 μg/l of each corticosteroid, track 1 MS[2] ion of DEX m/z 373, track 2 MS[3] ion of DEX m/z 353, track 3 MS[2] ion of FLU m/z 391, track 4 MS[3] ion of FLU m/z 371, track 5 MS[2] ion of TRIA m/z 415, track 6 MS[3] ion of TRIA m/z 357 (b) sample of urine positive for DEX (1.3 μg/l) analysed after hydrolysis. LC conditions see Experimental.

(Reproduced with permission from reference 6. Copyright 2000 Elsevier Science.)

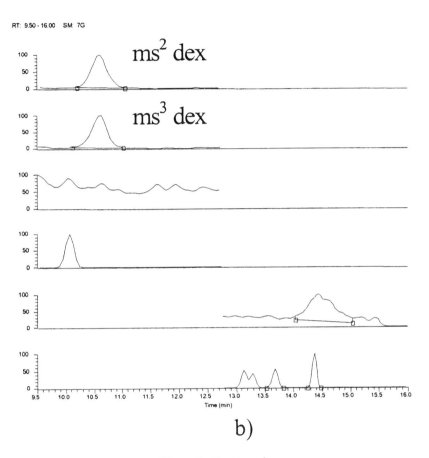

b)

Figure 1. *Continued.*

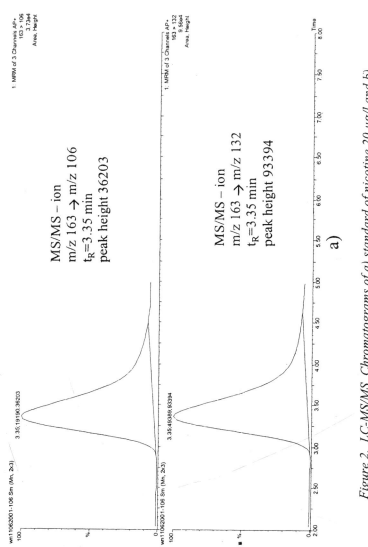

Figure 2. LC-MS/MS Chromatograms of a) standard of nicotine 20 µg/l and b) a sample of rat plasma. For the LC-MS/MS conditions see Experimental.

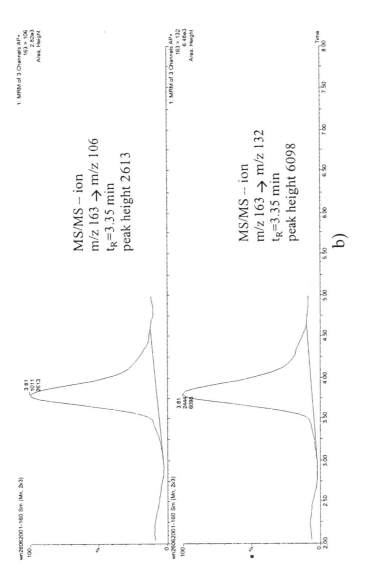

Figure 1. *Continued.*

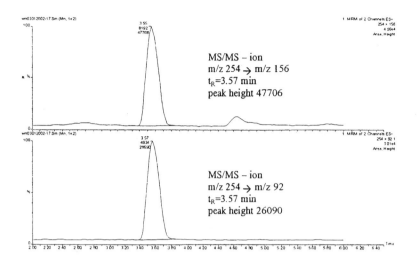

MS/MS – ion
m/z 254 → m/z 156
t_R=3.57 min
peak height 47706

MS/MS – ion
m/z 254 → m/z 92
t_R=3.57 min
peak height 26090

a)

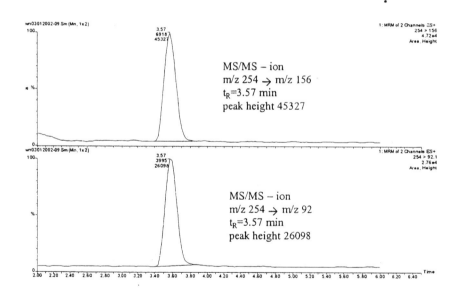

MS/MS – ion
m/z 254 → m/z 156
t_R=3.57 min
peak height 45327

MS/MS – ion
m/z 254 → m/z 92
t_R=3.57 min
peak height 26098

b)

Figure 3. LC-MS/MS chromatogram of a) blank sample of water fortified with sulfamethoxazole 100 ng/l b) and c) suspected samples; for LC-MS/MS conditions see Experimental.

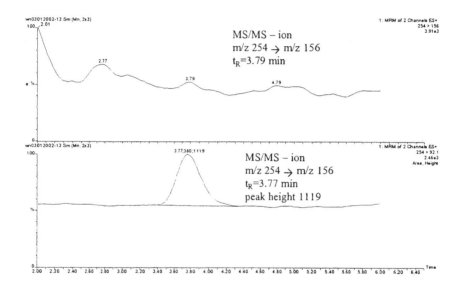

c)

Figure 3. *Continued.*

Table IV. Intensity ratios of reference standard and unknown

	Ratio reference	Tolerance conform Table II	Ratio unknown	Result
MS^3/MS^2 ratio of Dexamethasone Sample:1.3 µg/l (Figure 1b)	0.245	0.245±25% (0.184-0.306)	0.250	Confirmed
MS/MS ion ratio of Nicotine Sample: 5 µg/kg (Figure 2b)	0.388	0.388±25% (0.291-0.485)	0.44	Confirmed
MS/MS ion ratio of Sulfamethoxazole Sample: 96 ng/l (Figure 3b)	0.547	0.547±20% (0.438-0.656)	0.585	Confirmed
MS/MS ion ratio of Sulfamethoxazole Sample: - (Figure 3c)	0.547	0.547±20% (0.438-0.656)	One ion no ratio	Not Confirmed
MS/MS ion ratio of Diclofenac Sample: 20 ng/l	0.658	0.658±20% (0.526-0.790)	0.676	Confirmed

The confirmation of presence of residues of drugs in surface water

To monitor the concentration of drug residues in surface water and ground water samples a LC-MS/MS method was developed. The used analytical strategy is a combination of a screening method and a confirmatory method. During screening one single transition ion of each analyte of interest (see Table III) is monitored. When a sample of water contains one or more of the screening ions and if the monitored signal S/N>3, then the suspected sample is reanalysed with LC/MS/MS for the identification and confirmation of the analyte. During the confirmatory analysis, two transitions ions (two MS/MS ions) are monitored and the ratio between the ions of the suspected sample is compared with the ratio of the ions in the reference sample. Thirty samples of water were analysed and in the screenings analysis 11 samples contained residues of one or more of the analytes of interest. In 11 samples of water the identity of the analyte was confirmed conform the EU criteria. The following drug residues were identified at concentration levels ranging from approximately 10-100 ng/l: sulfamethoxazole, metoprolol, carbamazepine, diclofenac, bezafibrate, erythromycin, fenofibrate and clofibric acid. In Table IV some examples are given of the data evaluation for the confirmation of analytes using two MS/MS ions.

Figure 3 shows chromatograms of the monitored transition ions (LC-MS/MS) for confirmation of sulfamethoxazole in an extract of surface water. The corresponding chromatogram of the blank sample did not show any response. Again two transition products resulting in 3 IPs and 1 IP for the precursor ion make 4 IPs. For one sample (Figure 3b) two transition ions were monitored and the ratio is calculated. The evaluation of the ratio is summarized in Table IV. For the second sample (Figure 3c) only one ion is monitored and the identity of the analyte was not confirmed.

From Table III it can be concluded that for acetylsalicylic acid and ibuprofen no second MS/MS-ion is detected and so only 2.5 IPs can be collected. Confirmatory analysis based on two MS/MS ions cannot be used for this compound. For some compounds increasing the cone voltage of the MS resulted in an increase of the fragmentation process and an additional MS/MS ion is monitored. One additional MS/MS ion resulted in 1.5 additional IPs. A second approach for collecting additional IPs is by using a different ionisation technique or the use of a derivatisation technique, resulting in different fragmentation, different MS/MS ions and additional IPs.

Conclusions

Confirmation of the identity, especially of prohibited compounds, must be performed in a reliable and transparent way. Fixed criteria therefore are essential. These criteria must be established in such a way that they are applicable at relevant levels. On the other hand, they must result in a truly reliable identification, meaning that they must yield adequate structure information. From the three examples shown above, it is concluded that the use of the concept of IP gives good results. The concept is purely experimentally derived and based on experiences with GC-IRMS.

For the past two decades it has been mandatory to detect at least 4 ions when applied in official residue control within the EU. Until now, no hard evidence was obtained that, in combination with proper evaluation of the results, false positive results can be obtained in this way. The use of IPs is a new approach to set up quality criteria for the identification of organic residues and contaminants, and however the first application were in the field of the control of illegal growth promoting agents in biological matrices, the concept proved also applicable to the analysis of contaminants in environmental samples. When the criteria are applied to a broad field of (residue) analysis, their practicability and usefulness will become clear.

This manuscript demonstrates the use of the criteria in different analytical methods. Even though a fundamental chemometric basis is lacking these preliminary experiences already indicate the usefulness of the approach. The concept of IPs currently can be considered as a reliable and practical approach for the verification of the identity of residues detected in biological and environmental matrices.

References

1. Geerdink, R.B.; Niessen, W.M.A. and Brinkman, U.A.Th. *J. of Chromatogr. A*. **2001**, 910, 291-300.
2. Official Journal of the European Communities L125, 3-9. Council Directive 96/22/EC of 29 April 1996 concerning the prohibition on the use in stockfarming of certain substances, having a hormonal or thyrostatic action and of beta-agonists, and repealing Directives 81/602/EEC, 88/146/EEC and 88/299/EEC. Brussels, Belgium, **1996**.
3. Official Journal of the European Communities L125, 10-31. Council Directive 96/23/EC of 29 April 1996 on measures to monitor certain substances and residues thereof in live animals and animal products and

repealing Directives 85/358/EEC and 86/469/EEC and Decision 89/187/EEC and 91/664/EEC. Brussels, Belgium, **1996**.

4. Official Journal of the European Communities L221, 8-36. Commission Decision (2002/657/EC) of 12 August 2002 concerning the performance of analytical methods and the interpretation of results. Brussels, Belgium, **2002**.

5. André, F.; De Wasch, K.K.G.; De Brabander, H.F.; Impens, S.R.; Stolker, A.A.M.; Ginkel, van L.A.; Stephany, R.W.; Schilt, R.; Courtheyn, D.; Bonnaire, Y.; Fürst, P.; Gowik, P.; Kennedy, G.; Kuhn, T.; Moretain, J.and Sauer M. *Trends in Analytical Chemistr.* **2001**, 20, 435-445.

6. Stolker, A.A.M.; Schwillens, P.L.W.J.; Ginkel, van L.A. and Brinkman, U.A.Th. *J. of Chromatogr. A.* **2000**, 893 55-67.

7. Verebey, K.G.; DePace, A.; Mule, S.J.; Kanzler M. and Jaffe J.H. *J. of Analytical Toxicology.* **1982**, 6, 294-296.

8. Öllens, S.; Singer, H.P.; Fässler, P. and Müller, S.R. *J. of Chromatogr. A.* **2001**, 911, 225-234.

9. Ahrer, W.; Scherwenk, E. and Buchberger, W. *J. of Chromatogr. A.* **2001** 910, 69-78.

Combining LC/MS/MS and TOF-MS: Identification of Unknowns

Chapter 4

Identification of Biologically Active Compounds from Nature Using Liquid Chromatography/Mass Spectrometry

Jeffrey R. Gilbert[1], Paul Lewer[1], Dennis O. Duebelbeis[1], Andrew W. Carr[1], Carl E. Snipes[1], and R. Thomas Williamson[2]

[1]Dow Agrosciences, 9330 Zionville Road, Indianapolis, IN 46208
[2]Pharmacy Department, Oregon State University, Corvallis, OR 97331

Nature can serve as an excellent source of novel biologically active compounds. Unfortunately, a large number of known active compounds are often encountered during the search for novel chemistries. Therefore, early stage identification, and determination of novelty is of critical concern to natural product chemists. We have designed a screening process utilizing a single quadrupole LC/MS system to efficiently distinguish novel compounds from known compounds in crude biologically active extracts. This "dereplication" process is accomplished by chromatographing an extract on a LC column, diverting 95% of the eluent to a 96 well plate, and directing the remainder of the eluent to the MS for spectral analysis. Bioassay of the 96 well plate contents localizes the biologically active compounds, which are then structurally interrogated using MS, accurate MS, accurate MS/MS, and MS^n. These data are often sufficient to allow identification of the active compounds from the initial small scale extract, avoiding costly sample recollection or refermentation.

Nature has proven to be an excellent source of novel chemical compounds, possessing a significant diversity of biological activities beneficial to humanity. Many of these naturally occurring compounds have been developed into, or served as synthesis templates for, commercial products. Natural products have been successfully marketed in both the pharmaceutical and agrochemical arenas. Recent agrochemical examples include the synthetic strobilurins, structurally derived from the fungicidal natural products strobilurin A (1) and oudemansin A. (2) The most commercially successful strobilurin to date, azoxystrobin, exceeded $400MM in sales in 2000. (3)

Natural products often exhibit both diverse biological activities and unique biochemical modes of action (MOAs). (4) New MOAs are of particular interest in agriculture, where they may be used to address resistance issues resulting from overuse of agrochemicals with identical MOAs. (5) Additionally, novel MOAs allow for the development of products directed against new molecular target sites. Fortunately, the search for novel MOAs in the agricultural sector is facilitated by the use of direct *in vivo* screening on the organisms of interest. For example, crude extracts can be screened directly on plants, insects, and fungi. In this way, a single assay can expose multiple biochemical target sites to numerous compounds (i.e. ligands) with which they might interact, thereby increasing the chances of finding novel and structurally interesting molecules. While known compounds with previously undiscovered biological activity can be valuable, the discovery of structurally novel active molecules offers more potential value, and for this reason will be the focus of this discussion.

Although direct *in vivo* screening facilitates the discovery of novel MOA compounds, the molecular diversity which makes natural products so appealing also creates significant technical challenges. Nature makes thousands of known, and usually commercially uninteresting, active compounds which may be detected in a broad-based *in vivo* screen. Since extracts often contain both novel as well as known active compounds, the determination of whether an active is novel or known is a critical step in assessing its value as a potential lead. This illustrates the first role mass spectrometry can play in this process: the ability to identify and discard, i.e. dereplicate, the many known compounds which will be encountered. (6,7) Once the uninteresting known compounds have been eliminated, structures of the novel active compounds must be elucidated, establishing a second critical role for mass spectrometry. Thus, an efficient natural products program must combine: the best approaches to bioprospecting (source choice, extract preparation, screen design, etc.), the ability to efficiently dereplicate large numbers of known compounds, and the ability to rapidly elucidate the structures of novel active compounds.

The Dow AgroSciences natural products screening program utilizes a diverse set of organisms from nature. Extracts of plants, fungi, bacteria, marine organisms, etc., are collected from around the world and tested in proprietary bioassays. These bioassays were developed to address our three major agrochemical product areas; insecticides, herbicides, and fungicides. High-throughput plate assays available in each of these product areas lend themselves to direct interfacing with liquid chromatography/mass spectrometry (LC/MS). The resulting combination of biological activity, UV, and mass spectral data is then utilized for compound identification.

In brief, all samples are initially dereplicated using LC/MS with concurrent UV detection. Those which are likely, on that basis, to produce novel compounds are then examined in more detail using LC combined with accurate MS, accurate MS/MS, and MS^n. This chapter focuses on the dereplication and structure elucidation system we have developed to interface with our *in vivo* screens for novel agrochemical discovery. Various aspects of molecular structure determination via mass spectrometry are illustrated using examples of molecules we have found from natural sources.

Experimental

LC/MS Dereplication

Initial screening of extracts is conducted on a Micromass (Manchester, UK) Platform LCZ single quadrupole MS coupled to a Hewlett-Packard (Palo Alto, CA) series 1100 LC equipped with a UV (photodiode array) detector. Reverse phase separations are performed on a Hypersil (Alltech Assoc. Inc, Deerfield, IL) BDS C-8 (4.6 x 250mm, 5um) column under gradient elution conditions. Mobile phase A consists of 10 mM ammonium acetate (pH 6.0) and mobile phase B consists of acetonitrile. Linear gradient elution is performed at 1.0 mL/min from 100% A to 100% B over 20 minutes. Hydrophilic interaction chromatography (HILIC) separations are performed using a PolyLC (Columbia, MD) Polyhydroxyethyl Aspartamide (200 x 4.6 mm 5 um 100 Å) column at 1 mL/min via. linear gradient elution from 95% B to 100% A over 20 minutes. The post-column eluent is split, with approximately 95% of the stream directed to a Gilson (Middleton, WI) model 215 fraction collector for collection into 96-well plates which are bioassayed to localize biological activity. The remaining 5% of the stream is directed to the Platform MS and ionized using electrospray ionization. During the run, UV, positive electrospray (+ESI), and negative electrospray (-ESI) mass spectra are collected on a scan to scan basis. The mass spectra are collected at low cone voltages in an attempt to observe molecular adduct ions for the compounds. In addition, high cone voltage spectra are acquired, providing non-selective ion fragmentation data to aid in identification within a compound class. Following acquisition, data files are processed using the Micromass QuanLynx™ software to search within defined retention time windows for UV and mass spectral features that match known compounds.

Matches are confirmed by comparing the resulting mass spec
spectra generated in-house from previously isolated and
natural products.

Accurate MS and MS/MS

Accurate MS and accurate MS/MS analyses are performed on a Micromass
Quadrupole – Time of Flight (Q-TOF) MS system coupled to a Hewlett-Packard
series 1100 LC equipped with a UV (photodiode array) detector. Samples are
analyzed using either a Hypersil BDS C-18 (4.6 x 250mm, 5um) reverse-phase
column, or a PolyLC polyhydroxyethyl A (200 x 4.6 mm, 5 um, 100 Å) HILIC
column. Column selection is made based on the active(s) retention on the
dereplication system. If the active(s) are well retained on the C-8 dereplication
assay, a shallow linear gradient is performed on the C-18 column to optimize
their retention and separation. Conversely, polar active(s) that are poorly
retained on the initial C-8 separation are analyzed on the HILIC column via a
linear gradient to optimize their retention and separation.

The post-column eluent is split, with approximately 75% of the stream
directed to a Gilson model 204 fraction collector for collection into 96-well
plates. These plates provide a source of semi-purified material for subsequent
MS^n analyses. The remaining 25% of the stream is directed to the Q-TOF MS
which is operated in either +ESI or -ESI ionization modes using data-dependent
triggering between accurate MS and accurate MS/MS acquisition. A Hamilton
(South Natick, MA) syringe pump is used to infuse an internal standard mixture
containing 10 ug/mL spinosad and 20 ug/mL caffeine post column at rates from
1-10 uL/min. Accurate mass measurements are made using spinosyn A as the
reference lock mass (theoretical $[M+H]^+$ m/z = 732.46867), with the TOF
analyzer operated at a resolution of approximately 7,000 (FWHM). This system
typically produces accurate MS data to within ± 2-5 mDA, and accurate MS/MS
data within ± 5-10 mDa over a mass range of m/z 200-1000.

LC/MS^n

Detailed mapping of the fragmentation pathways are accomplished using a
Finnigan (Finnigan MAT, San Jose, CA) LCQ ion trap MS system coupled to a
Hewlett-Packard model 1100 LC. Samples are introduced either as crude
extracts using the LC/MS conditions described above, or via infusion of semi-
purified 96-well plate contents. Analyses are performed using electrospray or
nanospray ionization, collecting a combination of MS, MS/MS, and MS^n (where
n = 3-5) spectra in data-dependent mode.

Results & Discussion

Dereplication

The dereplication process utilizes a combination of LC separation with UV and MS detection as illustrated in Figure 1. In each LC run, mass spectral data are collected using alternating positive and negative ionization with and without non-selective source induced dissociation (SID). The resulting data files provide five distinct modes of detection [UV, +ESI-MS, -ESI-MS, +ESI-MS (SID), and -ESI-MS (SID)], enhancing the ability to identify active compounds. Following each LC run, a peak search program is used to screen for the presence of known compounds or chemical families. For example, the active compound cycloheximide was detected in the extract shown in Figure 1. The confidence of this assignment was increased through the observation of peaks in three separate ion traces which had proven characteristic for cycloheximide. Use of this data

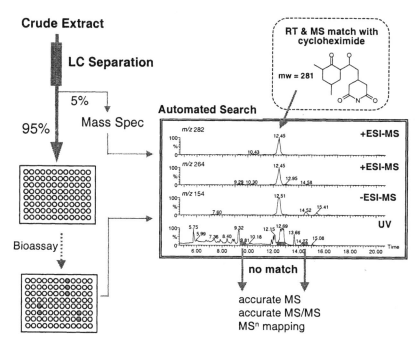

Figure 1. Flow Diagram of the Process Used for Natural Product Screening and Dereplication.

searching approach speeds interpretation of results and allo\
spent finding novel chemistries

Although the majority of compounds in our derepli
compatible with C8 reverse-phase chromatographic separati
polar active molecules have been found which require an al
technique. For example, gougerotin, a very polar and highly
molecule produced by Streptomyces, is poorly retained under C-8 separations,
making its identification in the complex extracts difficult (Figure 2A). Recent
literature on the analysis of polar compounds includes the development of a new
separation technique, hydrophilic interaction chromatography (HILIC). (8,9) A
HILIC separation of the same fermentation extract is shown in Figure 2B. Under
these conditions, gougerotin is well retained and produces useful UV and MS
spectra which could not be obtained a using reverse phase separation.

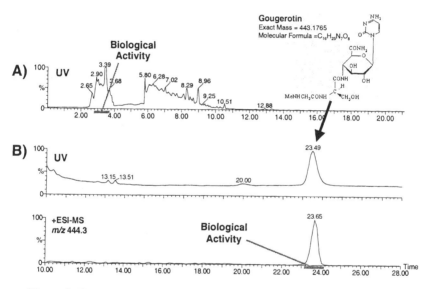

Figure 2. Streptomyces fungicidal extract A) UV chromatogram (254 nm) of
extract separated under C-8 gradient conditions, and B) UV and mass spectral
traces for the same sample analyzed using a HILIC gradient separation.

Identification of an unknown active is initiated by determining the retention
time range of the active compound(s), and then examining the spectral features
in this region. Mass spectral interpretation leads to a molecular weight
assignment which, combined with UV data, is used to search the Chapman &
Hall (C&H) database. (10) This database contains over 120,000 entries,
summarizing the majority of known natural products. Recent updates to the

H database allow MW searching using the monoisotopic data measured in MS and accurate MS experiments. These search results can be used to propose matches for known active compounds not previously encountered in our lab, which are subsequently investigated using accurate MS/MS and MS^n analyses, as described below.

Accurate Mass Based Searching

Accurate MS and accurate MS/MS data have proven to be extremely powerful inputs for database searching and structure confirmation. In one example, a marine algae sample collected in the Fiji Islands displayed activity in *in vivo* insect screens. (11) The extract was analyzed on the dereplication system, and produced no matches to our existing library spectra. Bioassay localized the *in vivo* insect activity to a 16-18 minute region of the chromatogram, as shown in Figure 3A. An intense peak at 17.7 min was observed at *m/z* 799.9 in the +ESI-MS trace, the MS and UV spectra for this peak are shown in Figure 3B. In addition to *m/z* 799.9, *m/z* 400.6 was also observed in the +ESI mass spectrum, which was consistent with assignment of these ions as the $[M+H]^+$ and $[M+2H]^{2+}$ adduct ions. Observation of the $[M+2H]^{2+}$ ion was particularly interesting, since doubly charged ions are commonly observed in peptides. Additionally, a weak signal was observed in the -ESI mass spectrum at *m/z* 798. This was designated as the $[M-H]^-$ ion, further confirming the assignment of the molecular weight of the unknown as 798.9 Da. Finally, the compound was found to have a UV maximum of 238 ± 10 nm.

Using the C&H database, a MW search for compounds with a monoisotopic MW of 798.9 ± 0.5 Da produced a total of 52 matches. When this search was refined to include only compounds with an UV maximum at 238 ± 10 nm, only seven database matches were obtained. One of these described a depsipeptide, symplostatin 1, which had also been isolated from marine algae (12), making it a likely match for the unknown. To confirm this proposal, the extract was analyzed using a shallow C-18 gradient on the Q-TOF to obtain both accurate MS and accurate MS/MS spectra of the active. Under these conditions, the active eluted at 9.3 minutes, with the UV (238 nm) and *m/z* 799.9 traces superimposed, confirming that both were attributable to a single component. The $[M+H]^+$, $[M+2H]^{2+}$, and $[M+Na]^+$ adduct ions of the active were observed in the +ESI mass spectrum. These were averaged, producing an accurate mass measurement of 798.508 ± 0.005 Da. A monoisotopic MW search of the C&H database for 798.508 ± 0.010 Da reduced the number of matches from 52 to 4 which, when combined with an UV maximum search for 238 ± 10 nm, gave a single match, symplostatin 1.

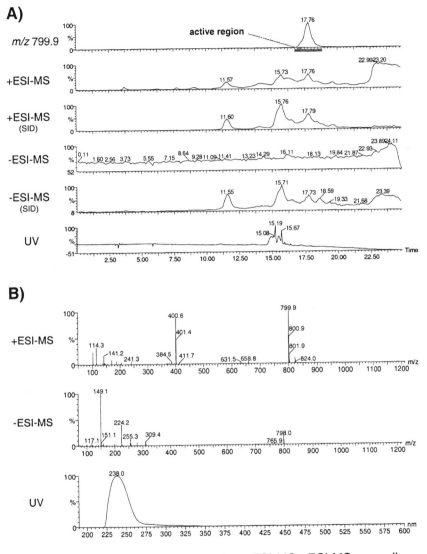

Figure 3. A) LC chromatogram showing +ESI-MS, -ESI-MS, as well as UV (254 nm) traces, and B) extracted UV and mass spectra of an insecticidal marine algae extract

The Q-TOF was operated in data-dependent mode using an inclusion list containing m/z 799.9, enabling the acquisition of both accurate MS and accurate MS/MS spectra from a single LC run. The resulting accurate MS/MS spectrum of m/z 799.9 is shown in Figure 4A. The structure of symplostatin 1 with proposed assignments for the observed fragment ions and neutral losses is shown in Figure 4B. Important losses include loss of methanol (m/z 767), as well as generation of a series of fragment ions produced via cleavage along the peptide backbone. The proposed assignments agree within ± 5 mDa of the measured values, and are summarized in Figure 4C. One exception was the m/z 496 fragment (I), whose initial assignment produced an error of > 100 mDa. Upon further examination, an alternative assignment for this fragment ion was proposed (II), involving loss of methanol in combination with cleavage along the peptide backbone. This assignment agreed to within ± 5 mDa of the measured value.

This example illustrates the power of accurate MS/MS for searching external databases, and confirming potential search matches. Based on the combination of UV, MS, and accurate MS data produced in two LC runs, a putative match was proposed for the active, symplostatin 1. This proposal was confirmed by comparison of the proposed structure with the fragment ions and neutral losses generated in the accurate MS/MS experiment. In this case, the accuracy of this data also afforded discriminated between two proposed fragmentation pathways resulting in isobaric product ions. Thus, all major ions observed in the MS of the active compound were consistent with those expected for symplostatin 1 and this extract was deprioritized from isolation.

Structure Elucidation of Unknowns Within a Family of Natural Products

The combination of UV and MS data described above can also be used to propose new structures within a family of natural products. These proposals are often supported by addition of detailed fragmentation data provided by MS^n mapping on an ion trap instrument. In one example, this approach was applied to an extract from the marine cyanobacterium *Lyngbya majuscula* collected in Curacao which had shown *in vivo* insect activity. (13) Analysis of this extract using the dereplication system did not produce a spectral match with our library compounds.

Bioassay localized the *in vivo* insect activity to a 20-23 minute region of the chromatogram. Examination of this region indicated the presence of a series of potentially related components, as shown in Figure 5. The major component of this series was observed at 20.5 minutes. This active produced a strong +ESI-MS response at m/z 748.4, and a weak -ESI-MS response at m/z 746.4, indicating that the molecular weight of the unknown was 747.4 Da. The UV

A)

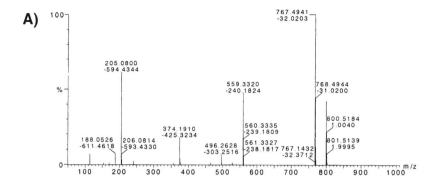

B) **Symplostatin 1**
$C_{43}H_{70}N_6O_6S_1 = 798.5071$

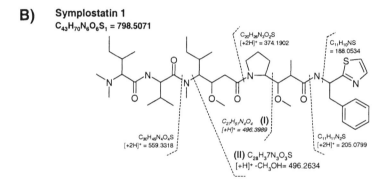

C)

Fragment Ions				Neutral Losses			
	proposed				Neutral Loss		
Measured	formula	Theo	delta	Measured	formula	Theo	delta
799.5144	C43H71N6O6S	799.5156	0.0012	-------------	-------------	-------------	-------------
767.4941	C42H67N6O5S	767.4894	0.0047	32.0203	CH3OH	32.0262	0.0059
559.3320	C30H47N4O4S	559.3318	0.0002	240.1824	C13H24N2O2	240.1838	0.0014
496.2628	*(I) C27H52N4O4*	*496.3989*	*0.1361*	*303.2516*	*C16H19N2O2S*	*303.1167*	*0.1349*
496.2628	(II) C28H38N3O3S	496.2634	0.0006	303.2516	C15H33N3O3	303.2522	0.0006
374.1910	C20H28N3O2S	374.1902	0.0008	425.3234	C23H43N3O4	425.3254	0.0020
205.0800	C11H13N2S	205.0799	0.0001	594.4344	C32H58N4O6	594.4356	0.0012
188.0542	C11H11NS	188.0534	0.0008	611.4602	C32H61N5O6	611.4622	0.0020

Figure 4. Symplostatin 1 showing A) MS/MS spectrum, B) structure and proposed MS/MS fragmentation, and C) summary of proposed accurate MS/MS fragment ions and neutral losses.

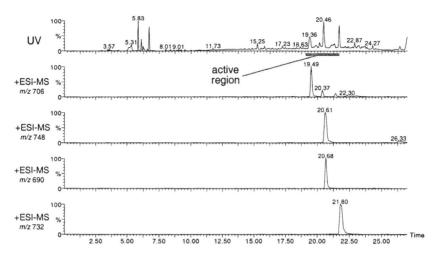

Figure 5. LC/MS Chromatogram of Insecticidal Marine Algal Extract

spectrum of the unknown did not provide a useful UV maxima in the range of 210-400 nm. A search of the C&H database for the monoisotopic MW of 747.4 ± 0.5 Da produced 15 matches.

Analysis of the unknown on the Q-TOF provided both accurate MS and accurate MS/MS spectra of the active. The $[M+H]^+$, $[M+NH_4]^+$, and $[M+Na]^+$ adduct ions from the unknown were averaged to produce an accurate mass measurement of 747.483 ± 0.001 Da. A monoisotopic MW search of the C&H database for 747.483 ± 0.010 Da produced only 2 matches. One of these, the lipopeptide microcolin A, had been previously identified from similar biological material (14) and was a likely match for the unknown. The accurate MS/MS spectrum of the unknown, shown in Figure 6A, was used to confirm this proposal. All of the major fragment ions and neutral losses in the spectrum were consistent with the structure of microcolin A. These assignments are summarized in Figure 6B (solid lines).

In addition, several potentially new members of this class of compounds were observed in the 20-23 minute region of the chromatogram. In an effort to identify these minor compounds, a Finnigan LCQ ion trap was used to characterize the MS^n (n = 2-5) fragmentation of microcolin A. These data allowed the mapping of relationships between each of the fragment ions, as well as the procession of neutral losses observed in the MS/MS spectrum of microcolin A. These data are summarized on Figure 6B (dashed lines). Using this detailed understanding of the fragmentation of this class of compounds, we were able to detect three additional known members of the microcolin family in this extract. In addition, several new members of this chemical class were identified in this sample. The structures of these new members were proposed based upon the accurate MS/MS and MS^n fragmentation mapping of their mass spectra.

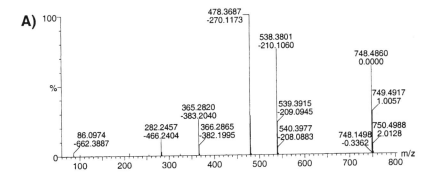

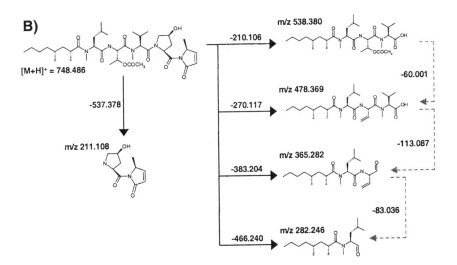

Figure 6. A) Accurate MS/MS spectrum and B) complete fragmentation pathway elucidated using a combination of accurate MS/MS and MSⁿ

Conclusions

Mass spectrometry is a crucial part of a modern natural products based discovery program. A system such as the one we describe greatly enhances the efficiency of novel compound discovery by prioritizing initial screening samples for further action. This is achieved by providing as much relevant spectral data as rapidly as possible (usually in a matter of hours) from the initial crude extract. The mass spectral data generated are either used to deprioritize an extract from isolation or, ultimately, form the initial basis for novel compound structure elucidation. Further, knowledge of the mass spectral properties of an active compound or family, even before its identity is known, can be an invaluable aid during its purification, providing a bioassay-independent means of compound tracking. Thus mass spectrometry, and particularly the combination of LC and MS, is a critical enabling technology in directing and shortening the path from weakly active crude extract to high-potency pure novel compound. As such, it has contributed significantly to our efforts a identifying novel and commercially successful compounds from nature.

Acknowledgements

The authors would like to thank Paul Graupner, Cliff Gerwick, Eleanor Chapin, John Banks, and Pat McKamey for their various contributions to this research. In addition, we thank Professor W.H. Gerwick for the Lyngbya extract, which was collected as part of his project number NIH CA52955.

References

1. Anke, T.; Hecht, H.J.; Schramm, G.; and Steglich, W. *J. Antibiot.* **1979**, 32 (11), 1112-1117.
2. Clough, J.M.; De Fraine, P.J.; Fraser, T.E.M.; Godfrey, C.R.A. *ACS Symp Ser.* **1992**, 504, 372-383.
3. Godwin, J.R., Bartlett, D.W., Clough, J.M., Godfrey, C.R.A., Harrison, E.G. and Maund, S. *Proceedings of the Brighton Crop Protection Conference – Pests and Diseases* **2000**, 533.
4. Lange, L. and Lopez, C. S. Microorganisms as a Source of Biologically Active Secondary Metabolites In *Crop Protection Agents from Nature: Natural Products and Analogues*; Copping L. G. Ed.; Critical Reports on Applied Chemistry Series 35, Royal Society of Chemistry, London; **1996**; pp 3-6.

5. Lange, L. and Lopez, C. S. Microorganisms as a Source of Biologically Active Secondary Metabolites In *Crop Protection Agents from Nature: Natural Products and Analogues*; Copping L. G. Ed.; Critical Reports on Applied Chemistry Series 35, Royal Society of Chemistry, London; **1996**; pp 18.

6. Julian, R. K. Tandem Mass Spectrometry in Natural Products Discovery, Presented at the 44th session of the American Society for Mass Spectrometry, Portland, OR, May **1996**.

7. Ackermann, B. L.; Regg, B. T.; Colombo, L.; Stella, S.; Coutant, J. E. *J. Am. Soc. Mass Spectrom.* **1996**, 7 (12), 1227-1237.

8. Alpert, A. J. *J Chromatogr.* **1990**, 499, 177-196.

9. Strege, M. A. *J. Chromatogr. B* **1999**, 725 (1), 67-78.

10. Chapman & Hall/CRC Chemical Dictionaries **2003**, HDS Software Copyright, Hampden Data Services Ltd.

11. Gilbert, J. R.; Carr, A. W.; Williamson, R. T.; Lewer, P.; Hahn, D.; Duebelbeis, D. O.; Chapin, E. L.; W. Gerwick, W. Multidimensional LC/MS for the Identification of Natural Product Lead Compounds. *Proceedings of the 48th ASMS Conference on Mass Spectrometry* [CD-ROM]; Long Beach, CA, **2000**; Oral ThOD 11:15.

12. Pettit, G.R. et al. *Tetrahedron* **1993**, 49, 9151.

13. Gilbert, J. R.; Lewer, P.; Carr, A. W.; Snipes, C. E.; Balcer, J. P. Natural Product Dereplication and Structural Elucidation Using LC/MS[n] Combined with Accurate Mass LC/MS and LC/MS/MS. *Proceedings of the 47th ASMS Conference on Mass Spectrometry* [CD-ROM]; Dallas, TX, **1999**; Oral MOF 10:15.

14. Koehn, F.E. et al. *J. Nat. Prod.* **1992**, 55, 613.

Chapter 5

LC/MS and LC/MS/MS Strategies for the Evaluation of Pesticide Intermediates Formed by Degradative Processes: Photo-Fenton Degradation of Diuron

S. Malato[1], T. Albanis[2], L. Piedra[3], A. Agüera[3], D. Hernando[3], and Amadeo Fernández-Alba[3,*]

[1]Plataforma Solar de Almeria-CIEMAT. Ctra. Senés Km. 4, 04200-Tabernas, Almeria, Spain
[2]Department of Chemistry, University of Ioannina, Ioannina 45110, Greece
[3]Pesticide Residue Research Group, University of Almeria, 04071 Almeria, Spain
*Corresponding author: amadeo@ual.es

Transformation products (TPs) obtained from diuron under photo-Fenton pilot scale degradation process were evaluated using various high performance liquid chromatography systems with electrospray ionization and provided by different analyzers such as quadrupole, ion trap and time of flight. The base peaks obtained in all cases were the protonated molecules $(M+H)^+$ and very similar fragmentation patterns were obtained in all cases by the different systems. The use of single quadrupole lead to the determination of structures based on the molecular weight and one additional fragment ion obtained by increase of the fragmentor voltage. The structures proposed in this way were related with suspected oxidative reactions as a consequence of the attack of the OH radicals generated. Ion trap allowed the confirmation of the structures proposed in this way were related with suspected oxidative reactions as a consequence of the attack of the OH radicals generated. Ion trap allowed the confirmation of the structures of suspected and unknowns compounds by applying multiple mass spectrometric analyses (MS^2 and MS^3). The application of TOF was relevant, in the determination of open ring structures that were only possible to identify by using accurate mass information. Furthermore additional parameters such as total organic carbon, organic ions and inorganic species were also evaluated to assess the proposed degradation pathway.

Introduction

Since the first European directive in 1975 related to water pollution, much progress has been made in tackling point source contamination of European waters (1). But severe pressure remains on Priority Hazardous Substances, PHS (2) and Persistent Organic Pollutants, POPs (3). Usually the major sources of those contaminants are industrial wastewaters or effluents, which may contain PHS at levels as high as few hundred mg/L. In this context, the Integrated Pollution Prevention and Control (IPPC) Directive (4) requests the development of technologies and management practices for specific industrial sectors for the minimization of pollution and recycling of water. Due to the lack of available on-site treatment technologies, a large number of the industrial activities included in Annex I of the IPPC Directive are not treating these wastewaters appropriately. As a consequence, simple, low cost, available and evaluated technologies are strongly required (5). Typically the treatment of this wastewater is based upon various mechanical, biological, physical and chemical processes. After filtration and elimination of particles in suspension, biological treatment is the ideal process. Unfortunately, a certain class of products labeled bio-recalcitrant (non biodegradable) and almost all PHS are in this category. In such cases, it is necessary to adopt much more effective reactive and abiotic systems (6) such as air stripping, adsorption on granulated activated carbon, incineration, ozone and oxidation. Their inconveniences are summarized in Figure 1.

Advanced Oxidation Processes (AOPs) have been proposed as an alternative for the treatment of this type of wastewater (7-9). All AOPs are

68

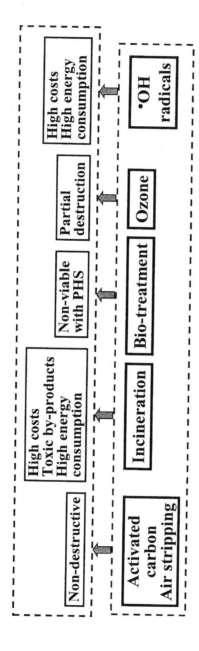

Figure 1. Inconveniences of PHS-wastewaters treatment methods

characterized by the same chemical feature: production of OH radicals ($^\bullet$OH). These radicals are extraordinarily reactive (oxidation potential 2.8 V) and attack most organic molecules with rate constants usually in the order of 10^6- 10^9 M^{-1} s^{-1}. They are characterized by their not very selective attack, which is a useful attribute for an oxidant used in pollution problems. The versatility of AOPs is also enhanced by the fact that they offer different possibilities for OH radical production, such as heterogeneous TiO_2 photocatalysis ((TiO_2 /UV) and homogeneous Photo-Fenton (Fe^{2+}/H_2O_2/UV), ozonolysis (O_3/H_2O_2) etc. Thus allowing them to conform to specific treatment requirements (8-10).

Therefore, AOPs alone are usually focused on the treatment of a specific industrial wastewater (e.g. agrochemicals) with a considerable amount of recalcitrant pollutants that cannot be eliminated by conventional biological treatment systems. But, obviously the performance of these treatments depends on the nature of the compounds, as well as on the selected operational parameters. As a consequence a wide range of transformation products and/or partial mineralization may be produced in route to complete compound destruction, which can remain after treatment. This is of great importance, considering that the transformation products generated during treatment may be more toxic than the primary ones or can generate synergistic effects on effluent toxicity. Therefore evaluation of intermediates or transformation products (TPs) evaluation is the key to optimize in each case and maximize the overall process.

Various approaches may be applied for TPs evaluation. Among them the most effective are based on GC-MS and LC-MS techniques (11). In spite of GC-MS widespread use, in the majority of the cases there are serious drawbacks due to difficulties in direct water injection, the formation of thermal artifacts, losses of polar compounds etc. (11,12). With regard to LC-MS, atmospheric pressure ionization interfaces, API can overcome many of the characteristic GC-MS limitations and provide fast and detailed information about the appearance and evolution of the TPs formed during the water treatment (11-13). However, there are several shortcomings yet to be overcome when LC-API-MS systems are used such as (i) insufficient structural information obtained, (ii) difficulties in predicting collision-induced dissociation, (CID) operational parameters when analyzing unknown compounds and (ii) the lack of universal MS libraries available for these techniques enabling unknown intermediates to be identified, meaning that a considerable effort is required to obtain conclusive information in terms of their identification. In this context the selection of appropriate LC-API-MS strategies, based on the capabilities of the different analyzers employed, as studied in this work, can at least partially overcome this lack. A good example, which encapsulates the problems above, is the evaluation of the degradation of diuron. Diuron is a very commonly used herbicide and included in the European list of POP and PHS. It is highly persistent with half-lifes over

300 days (14). Biodegradation (15) and direct photolysis (16) have been reported recently. However there are very few reports dealing with its photocatalytic oxidation by photo-Fenton (17) or TiO_2 photocatalysis (18,19). And, even in these cases, no major efforts in detailed evaluation of the degradation reaction, which is essential for the proper design of a treatment plant, has been made.

The aim of this work is to evaluate various LC-MS based techniques (LC-quadrupole-MS, LC-ion trap-MS and LC-time of flight-MS), used to evaluate the degradative process of diuron by homogeneous photocatalysis (photo-Fenton). After complementary assessment based on such parameters as total organic carbon (TOC), both the evolution of organic and inorganic ions are reported, which allow better understanding of the overall process. The use of a solar pilot plant is also relevant in this study, in which evaluation in a medium-scale field trial gives a clear idea about its industrial scale performance.

Materials and Methods

Photoreactor

All the experiments were carried out under sunlight in compound parabolic collectors (CPC) at the Plataforma Solar de Almería (PSA, latitude 37°N, longitude 2.4°W). The pilot plant (20) has three collectors, one tank and one pump. Each collector (1.03 m^2 each) consists of eight Pyrex tubes connected in series and mounted on a fixed platform tilted 37° (local latitude). The water flows at 20 L/min (see Figure 2) directly from one module to another and finally into a tank. The total volume (V_T) of the reactor (40 L) is separated into two parts: 22 liters (Pyrex tubes) total irradiated volume (V_i) and the dead reactor volume (tank + HDPE tubes). At the beginning of the experiments, with collectors covered, all the chemicals are added to the tank and mixed until constant concentration is achieved throughout the system. Then the cover is removed and samples are collected at predetermined times (t). Solar ultraviolet radiation (UV) was measured by a global UV radiometer (KIPP&ZONEN, model CUV3), mounted on a platform tilted 37° (the same angle as the CPCs), which provides data in terms of incident W_{UV} m^{-2}. This gives an idea of the energy reaching any surface in the same position with regard to the sun. With

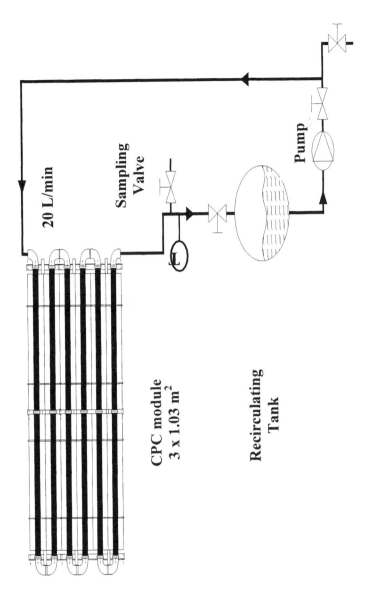

20 L/min

Sampling Valve

Pump

CPC module 3 x 1.03 m²

Recirculating Tank

Figure 2. Scheme of the solar pilot plant

Eq. 1, combination of the data from several days' experiments and their comparison with other photocatalytic experiments is possible:

$$t_{30W,n} = t_{30W,n-1} + \Delta t_n \frac{UV}{30} \frac{V_i}{V_T}; \quad \Delta t_n = t_n - t_{n-1} \qquad (1)$$

where t_n is the experimental time for each sample, UV is the average solar ultraviolet radiation measured during Δt_n, and t_{30W} is a "normalized illumination time". In this case, time refers to a constant solar UV power of 30 W m^{-2} (typical solar UV power on a perfectly sunny day around noon). As the CPCs do not concentrate light inside the photoreactor, the system is outdoors and, as it is not thermally insulated, the maximum temperature achieved inside the reactor has been 40°C. All "blank tests" (hydrolysis and photolysis) were performed in three-liter Pyrex beakers (UV transmission>80% between 320-400 nm, internal diameter 15 cm) and covered with a Pyrex top to avoid sample contamination and evaporation. During hydrolysis tests the beakers were kept in the dark. The maximum temperature inside the beakers was 35°C.

Chemicals and Reagents

Technical-grade diuron (98.5%) was supplied by Aragonesas Agro S.A. (Madrid, Spain). Analytical-standard diuron was purchased from "Dr. Ehrenstorfer GmbH" (Augsburg Germany). Analytical-grade organic solvents and inorganic salts were used for LC-MS. The photo-Fenton experiments (0.05 mM iron), the following chemicals were used: iron sulphate ($FeSO_4$-$7H_2O$), hydrogen peroxide reagent grade (30%) and sulphuric acid for pH adjustment (around 2.7-2.8). 2500 U/mg bovine liver catalase acquired from Fluka Chemie AG (Buchs, Switzerland) was used to eliminate the remaining H_2O_2 after sampling in photo-Fenton experiments. The water used in the experiments was obtained from the PSA Distillation Plant (conductivity < 10 μS cm^{-1}, Cl$^-$ = 0.2 mg/L, NO_3^- = 0.5 mg/L, organic carbon < 0.5 mg L^{-1}).

For LC-API-MS analysis, stock standard solution of diuron (100 mg·L^{-1}) was prepared in water. Diluted solution containing 10 mg·L^{-1} was prepared in order to obtain the best analysis parameters in Quadrupole, Ion-trap and Time-of-Flight analyzers.

Photo-Fenton Experiments

All experiments were performed at the highest diuron concentration (around 22 mg/L) attained dissolving it in water at around 10 °C (ambient temperature of the Plataforma Solar de Almería in November). The intention was to attain the highest possible initial concentration for evaluation of intermediates. The water used in the experiments was obtained from the PSA Distillation Plant (conductivity < 10 μS cm^{-1}, Cl$^-$ = 0.2 mg/L, NO$_3^-$= 0.5 mg/L and organic carbon < 0.5 mg/L)

LC-ESI-MS Analyses

Analyses of the samples were made using three different LC-ESI-MS systems provided with Quadrupole, Ion-Trap and Time-of-Flight analyzers.

Quadrupole system

HP Series 1100 MSD G1946A (Palo Alto, California, USA) equipped with Electrospray (ESI) ionization source and Hewlett Packard LC/MSD Chemstation Rev A.07.01 software. ESI-positive ionization mode parameters were as follow: Nebulizer pressure: 40 psi, drying gas flow: 10 mL·min^{-1}, drying gas temperature: 350 °C, capillary voltage: 2500V. To obtain information about the structure of diuron and its by-products different fragmentor voltages were used to produce mass fragmentation of the molecules, being these values of 60, 90 and 120V. Calibration of the system was automatically made by Chemstation software.

Ion Trap system

Agilent Series 1100 MSD G2445A (Palo Alto, California, USA) equipped with ESI interface, and controlled by Bruker Daltonics software. System calibration was made by MSD Trap control software. The ESI polarity ionization was set at positive mode and the conditions were as follow: Nebulizer pressure: 50 psi, Drying gas flow: 10 mL·min^{-1}, drying gas temperature: 325 °C and capillary voltage: 3500V. As in the Quadrupole system, different fragmentor voltage values were used to obtain structural information of the molecules. These values were the same that in the quadrupole system, but low fragmentation was achieved in this case. Therefore, MS^n was used to get more fragmentation and detailed information about the possible structures. Table II show the parameters for MS^n analysis.

Time-of-Flight system

MSD Mariner Workstation (Applied Biosystems, Framingham, Massachusetts) Time-of-flight system was equipped with TurboIonSpray™ (TIS) source and it was controlled by Mariner™ software. The Mariner software also made system calibration. TIS-positive ionization mode was selected; turbo probe temperature was set at 250°C and Nozzle temperature at 140 °C. In the Mariner TOF analyzer the fragmentor voltage is a parameter that doesn't exist because an "in-source" CID fragmentation occurs. The potential is controlled by the parameter called potential nozzle and it was set at 80 and 180V. Moreover, the system conducts in-source fragmentation with high-low nozzle voltage switching on alternating spectra. The instrument will subsequently deconvolute the data streams to display both high and low nozzle voltage data, providing fragmentation information from the same run.

Data Explorer (Applied Biosystems, Framingham, Massachusetts) software was used to work with the spectrum generated. With this software, we could compare the spectra with a theoretical spectra calculated by imposing the expected number and kind of atoms presents in the molecule. The software thus uses theoretical isotope scoring in conjunction with an elemental composition calculator to score comparisons of experimental and theoretical isotope patterns. Depending on the confidence, the software generates a list of about five, for instance, from as many as hundred of possible compounds. In this, the number and kinds of expected atoms was set and a list of around 15 compounds was generated per molecule analyzed: Chlorines ≤ 4; Carbons ≤ 20; Nitrogens ≤ 4; Oxygens ≤ 6.

Chromatography

Chromatographic separation was performed in all cases by using an Agilent Series 1100 liquid chromatograph (Palo Alto, California, USA) equipped with a binary solvent delivery system, autosampler and column heater with the Ion-Trap and the Time-of-flight analyzers and the same system without the column heater with the Quadrupole analyzer. The column used was a XTerra MS C_8 column (100 x 2.1 mm – 3.5 µm) purchased from Waters (Milford, MA, U.S.A.). A gradient elution was made using binary gradient of LC solvent A (acetonitrile) and LC solvent B (ammonium formate 50 mM, acetonitrile 5%, acidified with formic acid to pH 3.5) as follows: Linear gradient from 10 to 50% of solvent B in 10 min, then 100% B in 10 min. After 1 min at 100 % acetonitrile, the mobile phase was returned to initial conditions. The flow rate was kept at 0.25 mL·min^{-1} and UV monitored elution at 254 nm.

Other analyses

LC-UV

Diuron was analyzed by LC using reverse-phase liquid chromatography (0.5 mL/min of H_2O/MeOH at 40/60 ratio) with UV detection (254 nm) using an LC-UV (Hewlett-Packard, series 1100) with C-18 column (LUNA 5micron-C18, 3 x 150 mm from Phenomenex). Diuron standard solutions were daily prepared in water in a 0.2-50 mg/L concentration range and used as external standard to quantify diuron in samples.

LC-IC

Formation of inorganic anions was followed by LC-IC (Dionex-120, anions column IonPAc AS14, 250 mm long). The eluent for inorganic anions was Na_2CO_3/$NaHCO_3$ (1 mM/3.5 mM). Nessler spectrophotometric methods were used for ammonium.

TOC

Total Organic Carbon (TOC) was analyzed by direct injection of the filtered samples into a Shimadzu-5050A TOC analyzer calibrated with standard solutions of hydrogen potassium phthalate.

Sampling

Twelve 250 mL treated water samples were collected every 10 minutes from the sampling valve (see Figure 1). Each sample was treated following the experimental procedure and filtered through (0.25 μm filter) before injection in the LC systems.

Results and Discussion

Disappearance and Mineralization of Diuron

The pilot plant used (see Figure 2) in the present paper is a plug flow photoreactor (22 L) in a recirculation loop (10 L) with a well-stirred, non-reacting mixing tank (8 L). Since this a dynamic system, the photoreactor outlet concentration is not exactly the same as the mixing tank outlet concentration. In a large field system, the amount of conversion each time the mixture passes through the reactor is noticeable. But in the present case, flow rate is very high (1200 L/h) compared with the total volume of the photoreactor (22 L), which means very little conversion per run. Only if substantial conversion per run is obtained is it necessary to precisely account for mixing in the dark tank, as discussed in more detail elsewhere (21).

Several blank experiments were performed at the same initial concentrations as the photocatalytic experiments to guarantee that the results obtained during the photocatalytic tests were consistent and not due to hydrolysis and/or photolysis. Hydrolysis experiments were done at different pH, but without repeating results found in the literature (14). In all cases, hydrolysis was performed at pH 2.7, because the photo-Fenton tests were carried out at this pH. Diuron is stable into water under the experimental conditions, and no loss due to chemical hydrolysis was observed after 48 h in solutions kept in the dark at different pHs. Diuron and solar UV (latitude 37°N, longitude 2.4°W) spectra slightly overlap in the 300-to-330 nm region showing that absorption of solar photons can produce photoalteration after exposure to the environment. But such natural photodegradation has been very slow (44% of diuron has disappeared after 48 h under sunlight) under well-illuminated aerobic conditions (transparent glass, 15-cm ID) and mineralisation never occurs (no TOC disappearance have been detected). The effect must be very similar when the pesticide is disposed of in natural waters. In surface water, degradation is extremely slow and in ground water almost negligible. So disposal into the environment could be very risky. A wide range of intermediate photoproducts generated under natural sunlight by diuron has been already determined (16).

Figure 3 (left) shows the kinetics of disappearance of diuron (23 mg/L) in water during the photocatalytic treatment with Fenton reagent, with the evolution of TOC as a function of the irradiation time. The TOC values measured at the beginning of the experiment showed a good agreement with initial concentration of diuron. Due to the dark Fenton reaction 25% of initial diuron was degraded before illumination started. Although no change in TOC

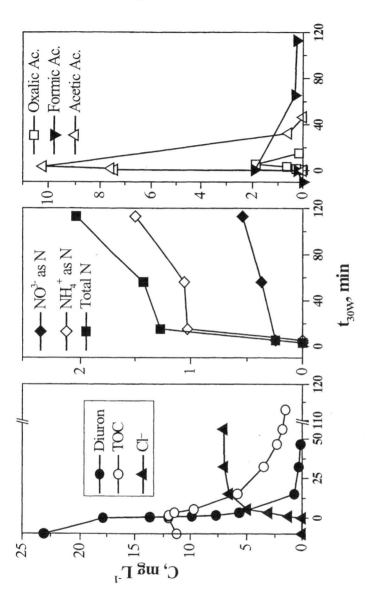

Figure 3. Degradation of diuron in solar pilot plant (by photo-Fenton treatment) as a function of t30w (illumination time). Left: Disappearance of diuron, mineralisation and chloride release. Center: Nitrogen balance during diuron degradation. Right: Evolution of organic ions

was observed during this period, appearance of a substantial number of intermediates was detected (See Figure 3). This disappearance of pesticide in the dark has been described by a multitude of authors, but in this case, it is very slight because of the small quantity of Fe^{2+} used (0.05mM), which is rapidly (less than 10 minutes) converted to Fe^{3+} by hydrogen peroxide. The total disappearance of diuron was obtained at 45 min (t_{30w}).

Almost 90% of mineralization (1.1 mg/L) was attained after approx. 110 min of photocatalytic treatment with Fenton. During the dark Fenton and the first 4 min of illumination, no appreciable decrease in initial TOC values is observed. From this time to 30 min, TOC abatement is very pronounced (near zero order kinetics), with a disappearance of more than 75% of the initial TOC. Note for a t_{30w} = 15 min, when practically no presence of diuron is detected, TOC is still high (5-6 mg/L), suggesting the presence of a considerable amount of TPs generated during the process. After 30 min at t_{30w}, TOC diminishes more slowly, and at the end of the experiment TOC was 1.1 mg/L. During photo-Fenton treatment, diuron is degraded very quickly, but TPs are also formed very quickly also. Mineralization is therefore observed to take place suddenly after the induction period.

Evolution of Inorganic and Organic Ions.

The release of heteroatoms as inorganic acids was confirmed by anion analyses (ionic chromatography) according to the stoichiometry proposed in reaction (Eq. 2). It should be remarked that nitrogen released has very often been measured as a combination of ammonia and nitrate, but as ammonia is oxidized to nitrate after long irradiation [22], reaction is given here only to the most oxidized state.

$$C_9H_{10}Cl_2N_2O + 13O_2 \rightarrow 2HNO_3 + 2HCl + 9CO_2 + 3H_2O \qquad (2)$$

Chloride evolves very quickly during the photo-Fenton treatment, suggesting a very fast degradation/dechlorination stage (Figure 3 left). The total amount of Cl^- produced at the end of the experiment is approximately 7 mg/L (100% conversion of the diuron chlorine content). This means that the residual TOC (around 1.1 mg/L) at the end of the experiment did not correspond to any chlorinated compound. The nitrogen mass balance of diuron is more complex (Figure 3 center). Both ammonia and nitrate have been detected in different relative concentrations and behavior. The nitrogen content of diuron was converted mainly into ammonia. Such incomplete nitrogen mass balance has frequently been observed in these processes (23,24) and indicates that other nitrogen-containing compounds must be present in the solution or evaporated during the process. The possible presence of alkyl nitrogen or cianno derivatives could explain the residual TOC, but the analytical procedures applied to the samples have not been able to detect them.

Substantial concentrations of oxalate, formate and acetate are detected with photo-Fenton during the first stage of the treatment (Figure 3 right). In this case organic ions are formed and mineralized very quickly, except formate, which remains at a concentration of less than 1 mg/L after 110 min, suggesting the difficulties in degrading this compound as fast as acetate or oxalate. The most important difference between the organic ions detected is concerning acetate, which appears at the beginning of the test at very high concentration but disappears very quickly. Carboxylic acids have been detected at mg/L level only when TOC was high and at the same time were quickly mineralized. These are therefore the last photoproducts prior to total mineralisation. At the end of the treatment, when there is little TOC remaining, it may be assumed that these linear acids are also produced, but in such a low concentration that detection by ion chromatography is not possible.

Identification of Intermediates by LC-MS and LC-MS/MS

Three series of analyses were carried out using LC-ESI-MS by direct injection of the filtered treated water samples in scan mode. The LC-MS systems were provided by different types of analyzer: quadrupole (Q), ion trap (IT) and time of flight (TOF). Water samples analyzed were prepared following the sampling procedures described above. The main spectra and chromatographic characteristic are shown in Table I.

LC-Q-MS

Quadrupole analyzer enabled tentative identification of 7 intermediates, TPs (Table I). A sequential increase in the voltage fragmentor from 60 to 120 volts (Figure 4A) yielded at least two fragment ions for each compound except 9 and 14 where only one was achieved. This information was enough to adequately identify the compounds in the majority of cases except for the compounds 1, 2, 3, 9 and 14. In the cases of compounds 9 and 14, this was due to their very low response in scan mode. Overlapping of compounds 1-3, the relatively high background noise and low responses made it difficult to differentiate the origin of these fragments or their evolution over treatment time and, therefore to assign them. This production of "dirty spectra" increases with fragmentor voltage as a consequence of the unspecific fragmentation obtained by this procedure. At

Table I. Retention times (Rt, min) and spectral characteristics of diuron and major photo-Fenton degradation products (relative ion intensity, %, is reported in parentheses)

Compound	Mw	LC-Q-MS (120V)	LC-IT-MS (60V)	LC-TOF-MS (80-160V)	Rt (min)
1	88	a	89(100)-72(38)		2.9
2	184	a	185		2.9
3	220	a	221	221.0311	2.9
4	200		201	201.0425	6.0
5	214	72(100)-215(25)-142(16)	215	215.0585(100)-142.0150(43)-106.0345(17)	8.1
6	214	72(100)-215(26)-142(13)	215	215.0592(100)-142.0154(40)-106.0331(19)	9.9
7	248		249(100)-195(51)		11.8
8	248		249(100)-195(48)		12.6
9	230	231[b]	231		13.0
10	234	178(100)-143(34)-161(10)-235(8)	235		14.4
11	218	127-161-219	219	127.0236(100)-160.9911(28)-219.0085(12)	13.9
12	248	72(100)-249(20)	249		14.1
13(Diuron)	232	72(100)- 233(23)	233	233.0240(100)-235.0208(76)	14.5
14	246	247[b]	247		15.7

[a] overlapping of ions 89-72-185221
[b] only detectable at fragmentor voltages <80V

relatively low cone voltages (60 V) the spectra obtained for all compounds showed protonated molecules [M+H]$^+$ as base peaks (Table I), thus agreeing with observations published in recent articles on the subject (13). By increasing the voltage to 120 V characteristic fragmentation peaks were 72 m/z and were assigned to the dimethyl urea moiety (Figure 4B) typical of diuron, except in compounds 10 and 11, where this group does not appear in any fragment, probably as a consequence of its modification by the oxidative process. This pronounced fragmentation at 72 m/z is greater when fragmentor voltage is increased. But increase in voltage also produced the completely loss of response for compounds 9 and 14. Other diagnostic fragment ions were obtained from the chloro-aryl moiety (compounds 5, 6 10, and 11). In these cases the 161 fragment ion corresponds to the parent structure (see Figure 4) and 178 (161 + 17) can be easily assigned to the addition of an hydroxyl group to that moiety and 143 to the same structure with the loss of a chlorine atom. Obviously, these last two modified fragments ions are a consequence of the oxidative treatment.

Furthermore, information regarding abundance of chlorine isotopes was of interest to evaluate dechlorination reactive processes as well as following up on loss of chlorine atoms in the collision-induced fragmentation (CID). Therefore, tentative assignment of the intermediates was mainly made on the basis of the quasi molecular peaks, combined with one or two diagnostic fragments obtained by repeated analysis at successive higher voltages. A further increase in the ESI extraction voltage (e.g., 150 V) did not provide useful information as a consequence of the great number of small fragments obtained with the subsequent loss of response in these new conditions. In spite of its considerable high identification potential the LC-Q-MS technique used in combination with CID had limitations due to the low rate of ion fragmentation which typically only allows a clear structure confirmation for suspected compounds but no unknown TPs. Furthermore, the related loss of sensitivity in scan mode in some cases made TPs identification difficult without application of sample handling procedures to preconcentrate the target compounds.

LC-QIT-MS and LC-QIT-MS/MS

As noted in Table I, working in scan mode the highest QIT sensitivity enabled the detection of three new intermediates with regard to LC-Q-MS (compounds 4, 7 and 8), as improving the response for all the compounds by a factor around 100, which then enabled adequate differentiation of the overlapped compounds 1-3. These facts are of great interest in avoiding tedious sample handling procedures for preconcentrating the TPs typically necessary with less sensitive techniques. Also of value was the ability to produce fragment ions by tandem MS avoiding the need of working at high fragmentor voltages to

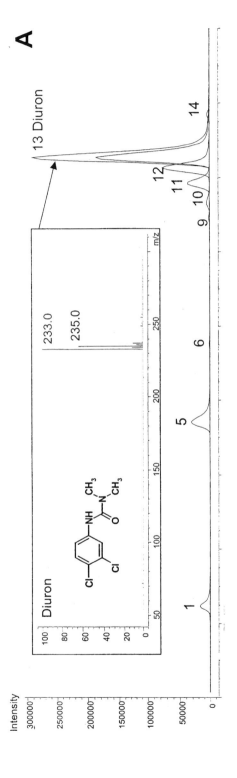

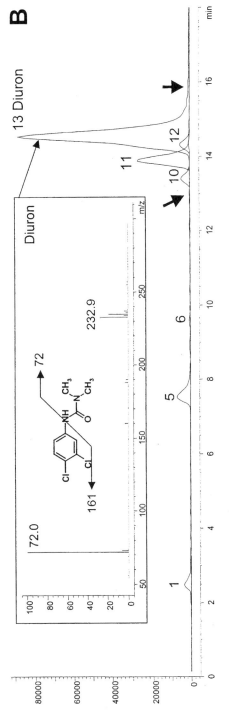

Figure 4. Typical LC-Q-MS chromatogram in scan mode obtained at two fragmentor voltages (A= 60 V and B= 120 V) corresponding to $t30_w$= 1 min of photo-Fenton treatment of diuron. Arrows in chromatogram B show the disappearance of response of compounds 9 and 14. Numbers refer to Table I

get enough information on each compound. The capability of the QIT analyzer to work in MS-MS tandem to produce product ion fragments represents a great opportunity for determining structures because the resulting fragmentation is so highly specific (see Table II). Furthermore, LC-MS/MS generates mass spectra that are less influenced (if at all) by analytical background noise than are those obtained by LC-Q-MS. Figure 5 illustrates the greater capacity of LC-MS/MS for confirmation of unknown intermediates. The full scan product spectrum for the m/z 235 precursor isolated for a retention time of 12.9 min could be assigned either to a hydrogenated compound of diuron typically present in such processes (25,26) or to the loss of a methyl group combined with the hydroxylation of the alkyl or aryl moiety. The MS^2 spectrum of ion 235 shows the fragment ion 178, which can be assigned to the cleavage of the amide (N-C) bond if the OH attack is produced in the aromatic ring. Finally, the MS^3 spectrum of fragment ion 178 produces fragment ions 161 and 143 that can be assigned to the loss of the hydroxyl and chlorine groups of the hydroxylated dichloroaryl moiety, which in turn enables confirmation of the proposed structure (see Figure 5). The loss of chlorine atoms can be also observed by the reduction in the isotope M+2 contribution. However this is not always relevant when working with LC-QIT-MS because of the isotope mass shifts that may be produced (27).

In the same way the peak corresponding to compound 14 previously detected by LC-Q-MS corresponding to a mass ion of 247 m/z at the retention time of 15.7 min could not be properly identified due to lack of information generated and the strong background noise. However, the MS^2 of this ion yielded product ions 86 and 161, which confirm the unmodified parent N-aryl moiety 161, as well as the fragment 86 (72 + 14) which can be easily assigned to the oxidation to aldehyde of a methyl group in the N-alkyl chain. Compound 1 could be assigned as to 4-isopropylaniline as predicted by the hydrolysis of diuron as stated in the literature (15). Structural elucidation of compounds 2 and 3 was not possible because they were difficult to predict by the typical reactions related to these processes such as addition of hydroxyl group, the substitution of a chlorine atom by a hydroxyl group, by the elimination of a methyl, etc.

Therefore the main advantage in the application of this technique was the high sensitivity achieved in scan mode and the possibility of generating considerably large amounts of specific product ions from the protonated molecules $[M+H]^+$ allowing at least three or more fragment ions to identify each TPs. Confirmation of the majority of TP structures generated by the help of the previous knowledge of these oxidative processes was possible.

Table II. LC-ITMS/MS operational conditions and product ions derived from multiple mass spectrometric analysis of photo-Fenton degradation products.

Compound	MS²				MS³			
	Isolation mass	Isolation width (m/z)	Amplitude (V)	Fragment ions	Isolation mass	Isolation width (m/z)	Amplitude (V)	Fragment ions
4	201	8	1.10	201(100)-144(31)	-	-	-	-
5	215	8	1.20	72(100)-215(27)	-	-	-	-
7	249	8	0.70	249(100)-72(33)	-	-	-	-
10	235	8	1.30	178(100)-180(73)-235(20)	178	6	1.20	143(100)-161(21)-178(20)
11	219	8	1.30	219(100)-161(25)	161	6	1.00	161(100)-127(64)
12	249	8	0.70	72(100)-249(46)	-	-	-	-
14	247	8	0.70	247(100)-86(24)-72(20)-162(7)	-	-	-	-

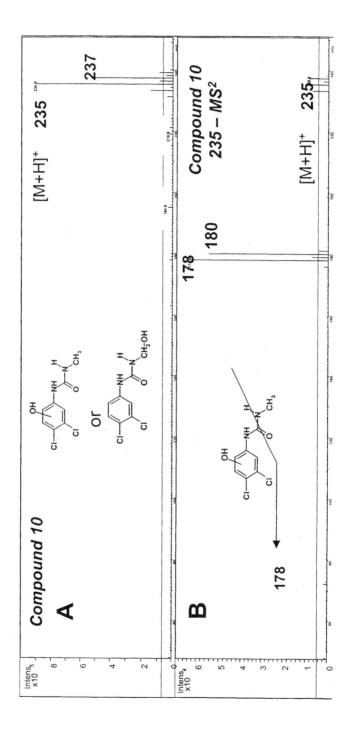

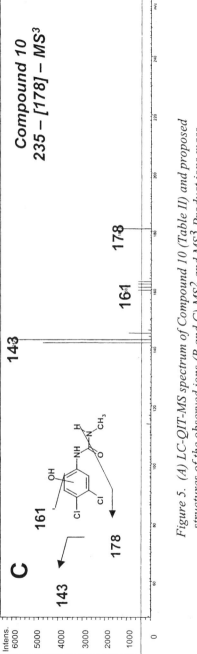

Figure 5. (A) LC-QIT-MS spectrum of Compound 10 (Table II) and proposed structures of the observed ions. (B and C) MS² and MS³ Product ions mass spectra obtained of the same compound and proposed structures

LC-TOF-MS

The possibility of alternating voltage mode with the ESI interface included in this system, not an intrinsic characteristic of the analyzer, was very useful in that multi-fragment information was obtained much faster than by repeated analysis. A good example is shown in Figure 6 where change in voltage value along the chromatogram can be observed in the millisecond range, In such a way, at least two fragment ions could be obtained in a single analysis in all cases by proper selection of two fragmentor voltage values as shown in Figure 6.

This LC-MS system only allowed the detection of five TPs (Table III). This lower number was probably affected by the unoptimized interface operating parameters (optimized with the other systems) and also by a lower response of this analyzer with respect to the IT, as it is observed elsewhere (27). Using commercially available standards, the mass spectra showed the appropriate single positively charged protonated molecule $(M+H)^+$ for diuron with a mass of 233.0240 amu and mass accuracy of 0.0003 of the expected protonated molecule (see Table III). Similar chromatographic results and mass spectra were obtained for compounds 5, 6 and 11 but in these cases, the base peak being the protonated molecule, more than one ion was detected and identified. Some fragments had less mass accuracy, even at over 5 ppm (see Table I). Despite these mass delta values observed, accuracy was high enough to confirm the structures by adding the additional information related to atomic composition mentioned above (see experimental section) to the possible list of compounds generated by the Data Explorer software. Perhaps a better choice of internal standards could improve the results of mass delta obtained.

The ability to get accurate mass of the protonated molecule and/or fragments enabled determination of the empirical formula of such unknown intermediates as compound 3. This compound shows the 221 m/z ion that could not be identified by the other systems. However, based on the exact mass of 221.031 determined (Table III) its possible elemental composition and structure could be calculated using the elemental composition program. In addition, chlorine isotope patterns and additional information on the atomic composition (see experimental) were used to limit the number of possible hits. Only one possible structure and elemental composition (see Figure 7, compound 3) was obtained. Such identification shows how the TOF analyzer can be very effective in identifying unknown compounds in these degradation studies

Proposed degradation pathway

Figure 7 shows the proposed pathway for the diuron degradation. It results from the identified photoproducts by LC-QIT-MS/MS except for compound 3

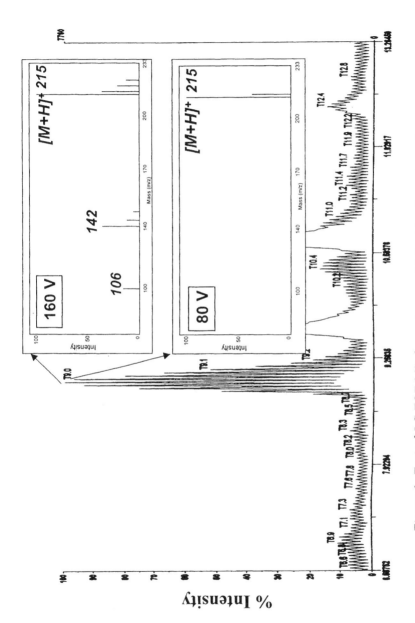

Figure 6. Typical LC-TOF-MS chromatogram obtained in alternating voltage mode (80V and 160 V) during the photo-Fenton degradation of diuron

Table III. Mass accuracy results obtained from photo-Fenton degradation products

Compound	LC-TOF-MS (real spectra)	LC-TOF-MS (calculated spectra)	Δppm
3	221.0311-223.0312	221.0324-223.0299	-5.9 ; 5.8
4	201.0425-203.0398	201.0379-203.0359	22.9 ; 19.2
5	215.0585-142.0050-106.0345	215.0582-142.0054	1.3 ; -2.8
6	215.0592-142.0056-106.0331	215.0582-142.0054	4.6 ; 1.4
11	127.0209-160.9911-219.0085	127.0183-219.0086	20.5 ; -0.5
13 (Diuron)	233.0240-235. 0208	233.0243-235.0214	-1.28 ; -2.6

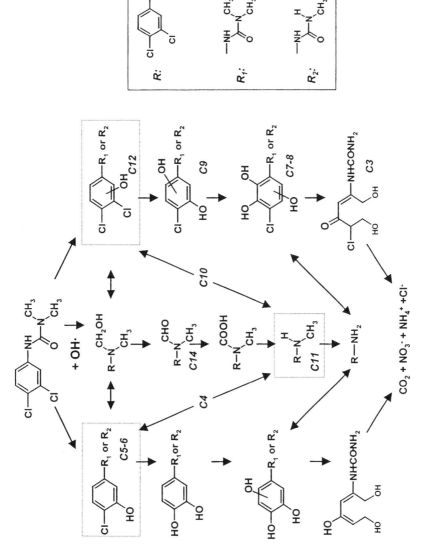

Figure 7. Proposed degradation pathway of diuron by photo-Fenton

which could only be identified by LC-TOF-MS. The first step is initiated by the attack of the OH radicals to the aromatic ring and to the alkyl chains with and without dechlorination. The next step involves a series of oxidation and decarboxylation processes that eliminates alkyl groups and chlorine atoms. The last step involves oxidative opening of the aromatic ring, leading to small organic ions and inorganic species. This degradative proposed route is consistent with previous works (13,28).

But still another question of interest in these studies are the detection of the critical oxidative steps throughout the process that is, the reaction bottlenecks that can limit the reaction kinetics. They can be estimated by considering that all the responses in the LC-QIT-MS system (without any preconcentration) may be assumed to be relatively similar and therefore comparable. Thus, compounds 5, 11 and 12 may be considered the main TPs in the process as it is shown in Figure 8. A common feature of these three compounds is that they are the addition of the OH radicals to the aromatic ring. This effective attack is carried out at two different places of the parent compound, substitution of a chlorine atom or addition to the aromatic ring before (compounds 5 and 12) and after decarboxylation of the alkyl chain (compound 11) to generate the corresponding phenols. Compounds 5, 11 and 12 are very highly concentrated compared to the other main intermediates detected but they also disappear very quickly (in less than 6 minutes of illumination time). These results imply that these compounds are the main intermediates during the first stages of degradation because they are produced in a large quantity, but are not very resistant to the degradation. It is also worth mention that the disappearance of all chlorinated TPs detected corresponds to the appearance of the stoichiometric chloride (see Figure 3) suggesting that no "resistant" chlorinated TPs are formed. It also important to remark that after 30 minutes of photo-Fenton treatment (when 25% of the initial TOC is left) no TPs have been detected apart from carboxylic acids. The degradation pathway proposed is also consistent with the N release shown in Figure 3, where it can be appreciated that 50% of the total N is detected (as ammonia and nitrate) during the first five minutes of illumination, corresponding to the mineralisation of the N-dimethyl group. A decrease of around 20% of TOC was detected at the same time, meaning that both methyl groups have been mineralized. From the proposed pathway it can be also concluded that the TPs formed do not limit dramatically the reaction kinetics. They can be easily oxidized and only the last steps (mineralisation of carboxylic acids) are slower. In any case, and for purposes of wastewater treatment, the complete oxidation of diuron by photocatalysis (a "sophisticated" treatment) is unnecessary in view of the last TPs detected, because they could be easily treated in a conventional biological treatment plant.

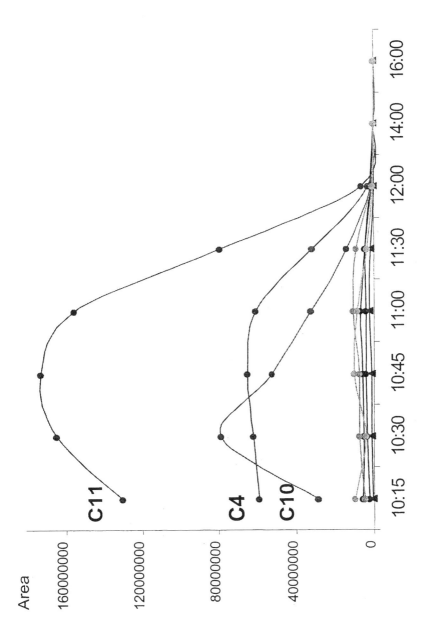

Figure 8. Intermediates (TPs) evolution obtained by LC-IT-MS during photo-Fenton degradation of diuron

Conclusions

The use of LC-ESI-MS in positive mode gave evidence the degradation pathway of diuron under photo-Fenton process. The use of quadrupole ion trap analyzer was the best strategy in terms of number of TPs evaluated and confirmation capacity of structures by using the product ion spectra. The application of quadrupole analyzer was able to detect the main TPs formed with a very simple development of the mass operation parameters but, in general the results were somewhat limited. The use of time of flight was very effective to identify unknowns difficult to predict simplifying the analysis by the use of exact mass values together with information in regard to possible atom composition. The proposed pathway was in concordance with the results obtained from total organic carbon and organic and organic ions and confirmed the efficiency of the proposed degradation treatment.

Acknowledgement

This project was supported by the Comisión Interministerial de Ciencia y Tecnología from Spain (Project nº: PPQ2000-0126-P4-05). The authors are grateful to Agilent and Applied Biosystem for instrumentation facilities.

References

1. European Commission, EU Focus on Clean Water, 1999.
2. European Commission, Socio-Economic Impacts of the Identification of Priority Hazardous Substances under the Water Framework Directive, Directorate-General Environment, 2000.
3. United Nations, Inventory of Information Sources on Chemicals. Persistent Organic Pollutants, Environment Programme, 1999.
4. European Union, Integrated Pollution Prevention and Control Directive, 96/61/EC.
5. European Commission. EU Freshwater, A Challenge for Research and Innovation, A Concerted European Response. DG XII, EUR 18098, 1998.
6. Zinkus, G.A.; Byers, W.D.; Doerr, W.W. *Chem. Eng. Prog.* May **1998**, 19.
7. Andreozzi, R.; Caprio, V.; Insola, A.; Martota, R. *Cat. Today* 1999, *53,* 51.
8. Chiron S.; Fernández-Alba, A.; Rodríguez, A.; García-Calvo, E. *Wat. Res.* **2000**, *34*, 366.
9. Malato, S.; Blanco, J.; Herrmann, J.M. (Editors). *Solar Catalysis for Water Decontamination. Cat. Today*, **1999**, *54(2-3),* complete issue.

10. Alfano, O.M.; Bahnemann, D.; Cassano, A.E.; Dillert, D.; Goslich, R. *Cat. Today* **2000**, *58*, 199.

11. Chiron, S.; Fernández-Alba, A.R.; Rodríguez, A. *Trends Anal. Chem.* **1997**, *16*, 518.

12. Fernández-Alba, A.R.; Agüera, A. *Analusis*, **1998**, *26*, 123.

13. Herrmann, J.M.; Guillard, C.; Arguello, M.; Agüera, A.; Fernández-Alba, A.R. *Cat. Today* **1999**, *54*, 353.

14. Tomlin, C.D.S. *The Pesticide Manual* (11th Edition); British Crop Protection Council: 1997.

15. Cullington, J.E.; Watker, A. *Soil Biol. Biochem.* **1999**, *31*, 677.

16. Jirkowsky, J.; Faure, V.; Boule, P. *Pest. Sci.* **1997**, *50*, 42.

17. Gallard, H.; de Laat, J. *Chemosphere* **2001**, *42*, 405.

18. Krysova, H.; Krysa, J.; Macunova, K.; Jirkovsky, J. *J. Chem. Technol. Biotechnol.* **1998**, *72*, 169.

19. Muneer, M.; Theurich, J; Bahnemann, D. *Res. Chem. Intermed.* **1999**, *25*, 667.

20. Blanco, J.; Malato, S.; Fernández, P.; Vidal, A.; Morales, A.; Trincado, P.; Oliveira, J.C.; Minero, C.; Musci, M.; Casalle, C.; Brunote, M.; Tratzky, S.; Dischinger, M.; Funken, K.-H; Sattler, C.; Vincent, M.; Collares-Pereira, M.; Mendes, J.F.; Rangel, C.M. *Sol. En.* **2000**, *67*, 317.

21. Malato, S.; Caceres, J.; Agüera, A.; Mezcua, M.; Hernando, D.; Vial, J.; Fernández-Alba., A.R. *Environ. Sci. Technol.*, **2001**, *35*, 4359.

22. Bonsen, E.M.; Schroeter, S.; Jacobs, H.; Broekaert, J. *Chemosphere*, **1997**, *35*, 1431.

23. Low, G.K.C.; McEvoy, S.R.; Matthews, R.W. *Environ. Sci. Technol.* **1991**, *25*, 460.

24. Serpone, N.; Calza, P.; Salinaro, A.; Cai, L.; Emeline, A.; Hidaka, H.; Horikoshi, S.; Pelizzetti, E. *Electrochem. Soc. Proc.* **1997**, *97*, 301.

25. Gallard, H.; de Laat, J. *Chemosphere* **2001**, *42*, 405.

26. Pelizzetti E.; Minero C. *Colloids and Surface A: Physic. Eng. Aspects* **1999**, *151*, 321.

27. Cox, K.A.; Cleven, C.D.; Cooks, R.G. *Int. J. Mass Spectrom. Ion Processes* **1995**, *144*, 47.

28. Thurman, E.M.; Ferrer, I.; Parry, R. *J. Chromatogr. A* **2002**, *957*, 3.

29. Cermenati, L.; Pichat, P.; Guillard, C.; Albini, A. *J. Phys. Chem. B* **1997**, *101*, 2650.

Chapter 6

Determination of an Unknown System Contaminant Using LC/MS/MS

Brett J. Vanderford[1], Rebecca A. Pearson[1], Robert B. Cody[2], David J. Rexing[1], and Shane A. Snyder[1]

[1]Southern Nevada Water Authority, 243 Lakeshore Road, Las Vegas, NV 89153
[2]JEOL USA Inc., 11 Dearborn Road, Peabody, MA 01960

In recent years, liquid chromatography coupled with tandem mass spectrometry (LC/MS/MS) has become an increasingly valuable tool for the study of contaminants in the environment. It has the advantage of providing structural information while simultaneously reducing background interference. The objective of the investigation presented here was to identify an unknown system contaminant using high-resolution accurate mass measurements and LC/MS/MS. In order to accomplish this, a reverse-geometry, double-focusing mass spectrometer equipped with linked scan capability was used. After an accurate mass measurement was performed, an elemental composition determination was carried out to generate a list of potential compounds. Once the list was narrowed down to the most likely molecular formula, linked scan MS/MS was used to provide enough structural information to confirm the chosen formula and identify the most probable constitutional isomer. The unknown contaminant was determined to be N-butylbenzenesulfonamide, a common plasticizer increasingly found in the environment.

Introduction

Recent reports have shown that certain contaminants at trace concentrations in surface waters can have dramatic reproductive effects on aquatic organisms (1-3). These compounds are collectively known as endocrine disrupting compounds (EDCs). Pharmaceuticals and personal care products (PPCPs) have also been detected in the aquatic environment and may act as EDCs (4-7). As more environmental contaminants such as these are discovered, the need for sensitive detection and identification methods has also increased. Traditional gas chromatography is of limited value without time-consuming derivatization because many environmental contaminants are polar, have low volatility and are thermally labile. This has led to the increased use of liquid chromatography/mass spectrometry (LC/MS) due to its ability to effectively analyze these types of molecules.

Since many of these contaminants are found at extremely low levels in the environment, extraction procedures are often used to concentrate them to detectable levels. However, these methods generally concentrate not only the compounds of interest, but also high levels of unwanted background material such as natural organic matter. This may lead to the misidentification of target compounds due to the increased chance of co-elution with compounds having the same monitored mass. In order to overcome this problem, MS/MS techniques have been developed that effectively separate the compounds of interest from background interferences through the monitoring of precursor/product ion pairs.

A second use of MS/MS is for the identification of unknown compounds. Since the analyte undergoes fragmentation to create product ions, useful structural information can be obtained by studying a product ion scan. By calculating the difference in mass between the precursor and product ions, fragment compositions can be determined which lead to structural elucidation.

In this investigation, a high-resolution mass spectrometer with MS/MS capability was used. The coupling of high-resolution and MS/MS yielded a powerful combination of structural information and accurate mass measurements. These techniques were utilized in a series of experiments to determine the structure of an unknown system contaminant that was suspected to be causing suppression of the EDCs and PCPPs of interest.

Background

During routing analytical work, it was observed that an unknown contaminant was constantly producing an extremely large background when the mass spectrometer was operated in full scan mode. This is shown in Figure 1. The signal was consistently at 100% of full scale and ion suppression was suspected. Even when the photomultiplier was attenuated by a factor of 0.25, the signal was readily observable, suggesting a complete saturation of the detector and probable negative effects on detecting any compounds of interest.

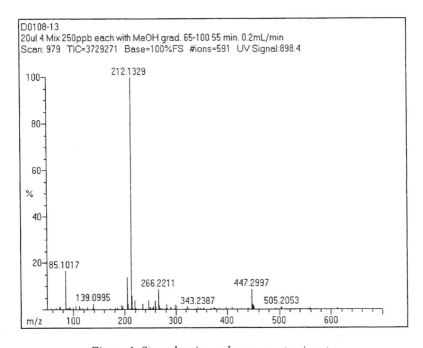

Figure 1. Scan showing unknown contaminant.

In order to determine the source of the contaminant, the components of the liquid chromatograph were systematically removed from the solvent flow path while the contaminant level was monitored. After seeing no change in contaminant signal after all of the LC components except the pump were removed, methanol was infused into the mass spectrometer using a syringe pump. As the signal was still observed, the LC pump was removed from consideration. This left only two possibilities for the contamination: the methanol or the mass spectrometer. After infusing methanol from several different manufacturers and still observing the contaminant, it was concluded that the source of the contamination was the mass spectrometer. Due to the high level of contamination, identification of the contaminant was necessary in order to determine its source with the expectation that it could possibly be removed.

Methodology

All analyses were performed on an LCMate reverse-geometry, double-focusing mass spectrometer (JEOL USA, Peabody, MA). The mass spectrometer was equipped with an electrospray ionization (ESI) source and a linked scan collision chamber. The liquid chromatography system consisted of an Agilent G1312A binary pump and an Agilent G1327A autosampler (Palo Alto, CA). Manual injections were made using a Rheodyne 7725i injection valve (Rohnert Park, CA). Infusions were performed using a single-syringe infusion pump (Cole Parmer, Vernon Hills, IL). Octylphenol (OP) and nonylphenol (NP) were provided by Dr. Carter Naylor (Huntsman Corporation, Austin, TX). HPLC grade or higher methanol was purchased from Burdick and Jackson (Muskegon, MI), Fisher (Pittsburgh, PA), and EM Science (Gibbstown, NJ). All calculations were performed using LCMate 2000 Data Reduction Version 1.9v software (Shrader Analytical Laboratories, Detroit, MI). All analyses were performed using electrospray ionization in the negative mode.

Accurate Mass Measurement

For any sector mass spectrometer, the following equation applies:

$$\frac{m}{z} = k\frac{B^2}{V}$$

where m/z is the mass/charge ratio, k is a constant, B is the magnetic field strength, and V is the accelerating voltage. This equation implies that either the magnetic field strength or the accelerating voltage may be scanned while the other is held constant in order to determine the mass of an ion that enters the mass spectrometer. Although both techniques can be used on a double-focusing mass spectrometer, an accelerating voltage scan is typically used for accurate mass measurements. This type of scan is linear and can therefore be calibrated with only two points. In addition, there is no hysteresis as with a magnetic field scan. In order to get the best mass resolution possible during an accurate mass measurement, the appropriate slits are set to their smallest widths to select for the mass of interest.

In order to calibrate an accurate mass measurement, an appropriate calibration mass set must be measured with the compound of interest. Since the unit resolution spectrum showed a mass for the contaminant of approximately 212, OP and NP were chosen due to their [M – H]⁻ exact masses of 205.1592 and 219.1749, respectively. An injection of 25μL of a 1ppm solution of OP and NP was made. A summary of the mass spectrometer conditions is shown in Table I.

After several scans were obtained, the octylphenol and nonylphenol peaks were assigned their appropriate exact masses. Using this calibration, the mass of the unknown compound was determined to be 212.0734. A scan from the accurate mass measurement is shown in Figure 2.

Table I. Accurate Mass Measurement Conditions

Parameter	Value
Orifice 1 Voltage	10 V
Ring Lens Voltage	40 V
Desolvation Plate Temperature	250°C
Orifice 1 Temperature	80°C
Main Slit	5000
Alpha Slit	1.0
Collector Slit	15μm
Multiplier	700V
Centroiding Method	Moments
Scan Speed	3 sec/scan
Magnetic Field Strength	2500
Nebulizer Gas	On
Drying Gas	Off

Elemental Composition Determination

Once an accurate mass measurement has been carried out, the value obtained can be used to determine the elemental composition of the corresponding compound. Although the mass of the compound in this study is relatively small, the number of possible elemental compositions can be quite large. In order to narrow down the list of candidates, several parameters used in the calculation algorithm must be defined.

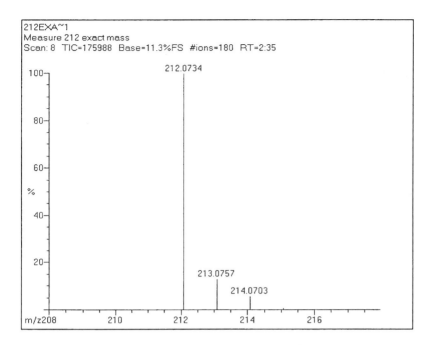

Figure 2. Scan from accurate mass measurement of unknown compound.

The first parameter to be considered is the allowable error for the calculation. Typically, a mass error of 10 mmu or less is considered reliable using an instrument with a resolving power of 5000 full-width half-maximum. Consequently, 10 mmu was used for this calculation. The range of the unsaturation value for a compound is the next parameter to be considered. The unsaturation value is a way of expressing the number of rings and sites of unsaturation on the compound of interest. Since the smallest unsaturation value for an organic compound is –0.5, this was used as the lower limit. Unless a large number of double bonds or rings is expected, a value of 20 is safely reasonable for small organic molecules. This value was used as the upper limit.

The final value to be considered is the number and type of elements allowed in the calculation. This will vary greatly depending on the particular application being considered. For the purposes of this application, the standard elements carbon, hydrogen, nitrogen and oxygen were allowed. In order to identify other possible elements, the spectrum from the accurate mass determination was closely examined. This spectrum can be seen in Figure 2. The characteristic chlorine-35/chlorine-37 and bromine-79/bromine-81 pairs were not observed, therefore they were eliminated from consideration. Since phosphorus is a relatively uncommon element of environmental contaminants, it was also removed from consideration.

Upon further review of the spectrum, an atypical [M − H + 2]⁻ peak was observed. Typically, for larger molecules, the [M − H + 2]⁻ would gradually become larger as the possibility of two carbon-13 atoms on the molecule increased. However, for a nominal mass of 213, the expected percentage of two carbon-13 atoms would be only about 0.1%. This is not in agreement with the observed [M − H + 2]⁻ peak which is approximately 5% of the base peak. Since the most abundant isotope of sulfur, sulfur-34, has a natural abundance of 4.5%, it was considered a viable candidate for the [M − H + 2]⁻ peak. Therefore sulfur was included in the calculation.

When the calculation was complete, a table showing the possible formulae, their unsaturation value, and their error was generated. This is shown in Table II.

In order to narrow down the possible formulae, several techniques may be used. For this application, the nitrogen rule and the even/odd electron rule were of the most value. The nitrogen rule states that a compound with an even nominal mass will have an even number of nitrogen atoms and a compound with an odd nominal mass will have an odd number of nitrogen atoms. Since the compound of interest was shown to have a nominal mass of 213, the nitrogen rule states that the compound must have an odd number of nitrogen atoms. When the nitrogen rule is applied to Table II, the list can be narrowed down to two formulae, $C_{13}H_{10}NO_2$ and $C_{10}H_{14}NO_2S$.

The even/odd electron rule applies to the formation of ions in a mass spectrometer. Ions can be formed with either an even or an odd number of electrons. During soft ionization processes, such as electrospray ionization, only even electron ions are formed. Thus, due to the nature of the unsaturation calculation, the unsaturation value for an ion formed using electrospray ionization will have a remainder of 0.5. When applied to Table II, the list can be narrowed down to two formulae, $C_{13}H_{10}NO_2$ and $C_{10}H_{14}NO_2S$. These results match those discussed above using the nitrogen rule.

Table II. Elemental Composition Results

Formula	Unsaturation	Error (ppm)
$C_6H_{16}N_2O_2S_2$	0	8.1
$C_6H_{16}N_2O_4S$	0	-9.7
$C_9H_{12}N_2O_2S$	5	11
$C_9H_{12}N_2O_4$	5	-6.3
$C_{10}H_{14}NO_2S$	4.5	-1.1
$C_{11}H_{16}O_2S$	4	-14
$C_{11}H_{16}S_2$	4	4
$C_{12}H_8N_2O_2$	10	15
$C_{13}H_{10}NO_2$	9.5	2.2
$C_{14}H_{12}O_2$	9	-10
$C_{14}H_{12}S$	9	7.4

In order to eliminate one of the remaining two formulae, the accurate mass spectrum was again considered. Since the formula $C_{13}H_{10}NO_2$ does not explain the unusual $[M - H + 2]^-$ peak, it was discarded in favor of $C_{10}H_{14}NO_2S$ which includes a sulfur atom. This leaves a formula of $C_{10}H_{15}NO_2S$ for the uncharged, parent compound.

Linked Scan MS/MS

Although the number of molecular formulae has now been narrowed down to one, the possibility of several isomers still exists. The structural possibilities of the most commonly used isomers are shown in Figure 3. In order to determine which of these structures is correct, linked scan MS/MS was performed on the compound. This was done because of the ability of linked scan MS/MS to provide structural information and an accurate fragment mass.

High energy, linked scan MS/MS derives its name from the high energy to which the ions of interest are accelerated and the "linking" of the two sectors of the mass spectrometer during scanning.

When a precursor ion is accelerated to a kinetic energy of approximately 1 keV or higher, the collision that occurs with the target mass is considered high energy. At this energy, excited electronic states within the precursor ion are produced. These types of collisions produce a very broad internal energy distribution that leaves virtually all structurally possible fragments with some probability of occurring. Since the center-of-mass energy of the ion is such a small percentage of the much larger kinetic energy, the target mass does not have nearly as much effect on the MS/MS spectrum as would a low energy collision. This means that parameters such as collision energy, temperature, and pressure do not have a profound effect on the MS/MS spectrum, leading to more reproducible results.

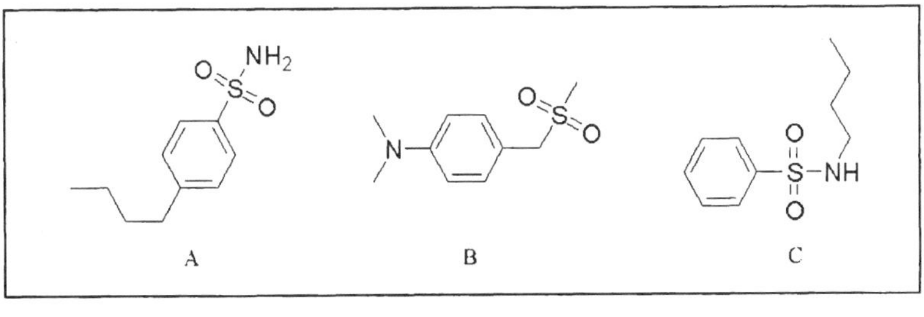

Figure 3. Structures of potential isomers. A) p-butylbenzenesulfonamide B) 4-dimethylaminobenzyl methyl sulfone C) N-butylbenzenesulfonamide

During a linked scan of a double-focusing mass spectrometer, the magnetic and electric sectors are scanned simultaneously while being held at a constant ratio. Since the magnetic sector separates ions according to their momentum (mv) and the electric sector separates ions according to their kinetic energy ($1/2mv^2$), the ratio of the two is inversely proportional to their velocity. This is shown in the formula below:

$$\text{Ratio of Sectors} = \left(\frac{2mv}{mv^2} \right) = \frac{2}{v}$$

All ions leaving the ion source are accelerated to the same kinetic energy. Since the kinetic energy of the ions is $1/2mv^2$, ions with different masses will have different velocities. In the field-free region after the ion source, the accelerated ions are introduced into a collision cell filled with a collision gas, in this case helium. The precursor ions collide with the helium to produce fragment ions. Assuming the velocity of the ion does not change when it fragments, the velocity of the product ion will be the same as the precursor. As was shown above, performing a linked scan of the two sectors selects for velocity. Thus, linked scan MS/MS selects only product ions that have the same velocity as their precursor ions.

This process was carried out for the unknown compound. A summary of the experimental conditions is shown in Table III.

The resulting linked scan MS/MS spectrum for the compound of interest is shown in Figure 4. The measured mass of the fragment, 140.9968, corresponds very well (<10 mmu difference) with the exact mass, 141.0040, of the sulfonylbenzene fragment shown in Figure 5.

N-butylbenzenesulfonamide

By using the information from the MS/MS spectrum, it can be concluded that the contaminant is the N-butylbenzenesulfonamide constitutional isomer of the molecular formula $C_{10}H_{15}NO_2S$. This compound is used as a plasticizer to increase the pliability of plastic polymers to prevent the final product from being brittle and unusable (8). It is used in the manufacture of polyamides (nylons) (9), paint films (8), and agricultural herbicides (10). It is lipid soluble, hydrophobic and is increasingly being found in the environment. It has been identified in surface water (10), ground water (11), and drinking water (12) and has been shown to have neurotoxic effects (8). Due to the nylon content of the electrospray source, it is suspected that this was the cause of the massive contamination.

Table III. MS/MS Experimental Conditions

Parameter	Value
Orifice 1 Voltage	10 V
Ring Lens Voltage	40 V
CID Gas	95%
Desolvation Plate Temperature	250°C
Orifice 1 Temperature	80°C
Main Slit	750
Alpha Slit	4.0
Collector Slit	180μm
Multiplier	500 V
Centroiding Method	Profile
Scan Speed	3 sec/scan
Nebulizer Gas	On
Drying Gas	Off

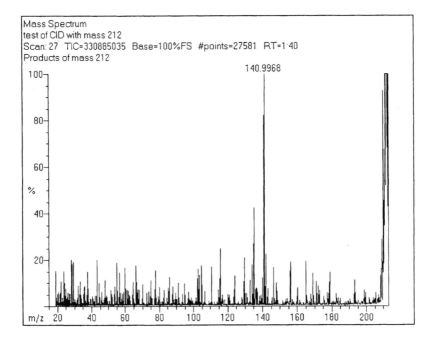

Figure 4. MS/MS scan of unknown compound.

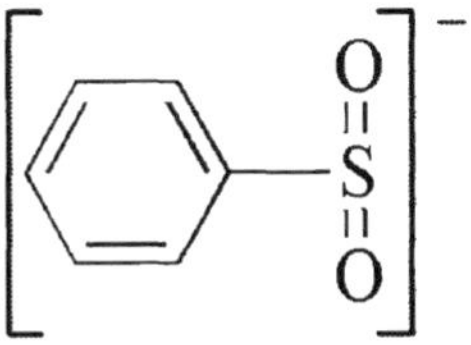

Figure 5. Depiction of observed sulfonylbenzene MS/MS fragment

Conclusion

Using a sequence of accurate mass measurements and LC/MS/MS, the structure of an unknown system contaminant was identified as N-butylbenzenesulfonamide. High-resolution mass spectrometry provided accurate mass measurements to narrow down the list of potential compounds. A combination of traditional mass spectral identification techniques with high energy, linked scan MS/MS led to the selection and confirmation of a structural isomer. These techniques proved to be invaluable in the investigation presented here and are currently being applied to other research projects involving environmental contaminants. The coupling of high resolution mass spectrometry and tandem mass spectrometry is a valuable tool for researchers studying unknown, organic contaminants in water due to its ability to simultaneously reduce background interference and provide structural information.

References

1. Jobling, S.; Noylan, M.; Tyler, C. R.; Brighty, G.; Sumpter, J. P. *Environ. Sci. Technol.* **1998**, 32, 2498-2506.
2. Snyder, S. A.; Keith, T. L.; Verbrugge, D. A.; Snyder, E. M.; Gross, T. S.; Kannan, K.; Giesy, J. P. *Environ. Sci. Technol.* **1999,** 33, 2814-2820.
3. Snyder, S. A.; Snyder, E.; Villeneuve, D.; Kurunthachalam, K.; Villalobos, A.; Blankenship, A.; Giesy, J. In *Analysis of Environmental Endocrine Disruptors;* Keith, L. H.; Jones-Lepp, T. L.; Needham, L. L.; Eds.; ACS Symposium Series 747; American Chemical Society: Washington, D. C., 2000; pp 73-95.
4. Daughton, C. G.; Ternes, T. A. *Environ. Health Perspect.* **1999,** 107, 907-938.
5. Snyder, S. A.; Villeneuve, D. L.; Snyder, E. M.; Giesy, J. P. *Environ. Sci. Technol.* **2001,** 35, 3620-3625.
6. Snyder, S. A.; Kelly, K. L.; Grange, A. H.; Sovocool, G. W.; Snyder, E. M.; Giesy, J. P. In *Pharmaceuticals and Personal Care Products in the Environment: Scientific and Regulatory Issues;* Daughton, C. G.; Jones-Lepp, T. L., Eds.; ACS Symposium Series 791; American Chemical Society: Washington, D. C., 2001; pp. 116-140.
7. Kolpin, D. W.; Furlong, E. T.; Meyer, M. T.; Thurman, E. M.; Zaugg, S. D.; Barber, L. B.; Buxton, H. T. *Environ. Sci. Technol.* **2002,** 36, 1202-1211.

8. Samiayah, G. K. Ph.D. thesis, University of New South Wales, Sydney, Australia, 1997.
9. Takeuchi, T.; Suzuki, H. Japanese Patent 7,434,947, 1979.
10. Sheldon, L. S.; Hites, R. A. *Environ. Sci. Technol.* **1979,** 13, 574-579.
11. Albaiges, J.; Casado, F.; Ventura, F. *Water Res.* **1986,** 20, 1153-1159.
12. Brambilla, A.; Broglia, L.; Nidasio, G. *Boll. Chim. Ig.* **1991,** 42, 779-785.

Chapter 7

HPLC/TOF-MS: An Alternative to LC/MS/MS for Sensitive and Selective Determination of Polar Organic Contaminants in the Aquatic Environment

M. J. Benotti[1], P. Lee Ferguson[1], R. A. Rieger[2], C. R. Iden[2], C. E. Heine[3], and B. J. Brownawell[1]

[1]Marine Sciences Research Center, Stony Brook University, Stony Brook, NY 11794–5000
[2]Department of Pharmacology, Stony Brook University, Stony Brook, NY 11794–8651
[3]Micromass, Inc., 100 Cummings Center, Beverly, MA 01915–6101

Isobaric interferences in environmental samples can compromise trace LC-MS analysis of polar organic chemicals when using single quadrupole instruments in selected ion monitoring (SIM) mode. HPLC-MS/MS with triple quadrupole instruments in multiple reaction monitoring (MRM) mode is the conventional approach for increasing selectivity and improving sensitivity. Time-of-flight (ToF) MS offers an alternative approach with higher mass resolving power and full spectral sensitivity. Extracts from STP effluent were evaluated for PPCPs using three different types of Micromass, Inc. mass spectrometers: a single quadrupole (LCZ™), a triple quadrupole (Quattro LC™), and an orthogonal acceleration ToF instrument (LCT™). HPLC-ToF-MS provided significant S/N improvement over SIM analysis for each of the analytes detected and approached S/N values afforded by MRM analysis. In addition, the ToF was also able to identify non-target analytes based on accurate mass measurements and associated elemental composition calculation as illustrated by the detection and confirmation of a polyethylene glycol (PEG) homologous series in STP influent. Limited dynamic range on the ToF as compared to the quadrupole was partially corrected for using digital dead time correction (DDTC).

Introduction

Over the past decade, the combination of high performance liquid chromatography and with mass spectrometry (HPLC-MS) has proven to be an important tool for trace analysis of polar organic contaminants in the aquatic environment. A recent development within the field of HPLC-MS has been the emergence of time-of-flight (ToF) mass analyzers with atmospheric pressure ionization interfaces (including electrospray and atmospheric pressure chemical ionization). ToF mass spectrometers operate by accelerating ions into a field-free drift region of a fixed path length. The time required to traverse the flight tube and strike a detector is precisely measured and related to the mass-to-charge ratio (m/z). The square of the flight time is proportional to m/z of a particular ion. Modern ToF instruments display superior resolving power than quadrupoles and ion traps along with faster data acquisition rates than quadrupoles and sector instruments. These attributes facilitate the routine measurement of accurate masses, even from narrow chromatographic peaks. Advantages inherent to ToF mass spectrometers also include a wide mass range and full spectral sensitivity. These features make HPLC-ToF-MS instruments attractive for environmental analyses as analytes are often detected at trace levels in a complex matrix. The high sensitivity of HPLC-ToF-MS allows for trace-level quantitation (ppt) and the enhanced resolution in comparison to quadrupole instruments make it better suited for resolving isobaric spectral interferences. Also, since ToF instruments continually acquire data for all ions across a given mass range, non-target analytes can be investigated without compromising sensitivity.

In the past, detection and quantification of polar organic contaminants in environmental samples have been most often performed with scanning MS instruments such as quadrupoles. Single quadrupole HPLC-MS systems operated in single ion monitoring (SIM) mode are robust and sensitive for targeted analyses in relatively clean matrices. This instrumental configuration has been the workhorse for many environmental applications but is susceptible to isobaric interferences (i.e. the low resolving power of the instrument cannot distinguish analytes from interferences with nominally identical m/z). This lack of specificity can lead to decreased signal-to-noise (S/N) ratio, (thereby degrading sensitivity) and an incorrect estimation of analyte concentration (*1*). Despite these limitations, robust environmental HPLC-MS applications have been developed using single quadrupole mass spectrometers. For example, LC-MS has been used in the investigation of antimicrobials in groundwater and surface water (*2*), polyethylene glycol (PEG) in environmental waters (*3*), as well as alkylphenol ethoxylates and steroid hormones in wastewater impacted water (*4,5,6*). Other applications of HPLC-MS include investigation of pesticides and metabolites in surface and groundwater (*7,8*) and characterization of alcohol ethoxylate biodegradation intermediates (*9*). Triple quadrupole (MS/MS) instruments operated in multiple reaction monitoring (MRM) mode are less affected by isobaric interferences and have been the analytical gold standard for quantitative environmental analysis of polar organic contaminants. For example, HPLC-MS/MS has been used in the study of neutral pharmaceuticals (*10,11*) antibiotics (*11,12*) and X-ray contrast agents (*13*) present in aqueous media (e.g. groundwater, river water, wastewater, etc.). Although HPLC-MS/MS affords superior selectivity and sensitivity, considerable method development is required (e.g. determining MRM transitions for each analyte and optimizing instrument parameters for each separate transition). More importantly, quadrupole instruments operating in SIM or MRM mode will only detect target ions, and increasing the number of target analytes or transitions will reduce instrument sensitivity.

HPLC-ToF-MS provides a powerful alternative or valuable compliment to traditional HPLC-MS or HPLC-MS/MS approaches in environmental contaminant analyses. Initial studies include accurate mass confirmation and analysis of pesticides and metabolites in groundwater (14,*15*), and surface water (*16*), as well as screening and identification of unknown contaminants in surface water (*17*). However, to date there have been no direct comparisons of HPLC-ToF-MS to quadrupole-based MS for the analysis of trace contaminants in environmental samples.

Scope of Study

In the present study, three mass analyzers were compared for the analysis of pharmaceuticals and personal care products (PPCPs) in wastewater treatment plant (WWTP) effluent: an HPLC-ToF-MS instrument, a single quadrupole MS operated in SIM mode, and a triple quadrupole MS operated in MRM mode. A set of 22 frequently detected PPCPs was targeted for this work. Analyses were performed using nearly identical chromatography and electrospray ionization conditions (identical LC columns, MS ion sources, etc.) so that the mass spectrometric analyses could be directly compared. The relative performance and utility of the three instruments with respect to useful sensitivity, selectivity, and dynamic range were assessed. Furthermore, the ability of HPLC-ToF-MS to perform trace-level identification of unknown peaks and confirmation of target and non-target analytes was evaluated.

Methods

Sample Collection and Preparation

Wastewater samples (100mL influent and 4L effluent) were collected on March 22, 2002 from a tertiary WWTP using solvent rinsed glass bottles. At the same time, a field blank (4L of Milli-Q purified water) was prepared in an identical sampling container. Effluent was immediately filtered under vacuum upon return to lab using Whatman glass fiber filters (0.7μm particle retention). Two 500mL aliquots of filtered effluent and 500mL of the field blank were each transferred to separate 1L solvent-rinsed Boston Round bottle and 50 ng of surrogate standard was added (10μL of a 5ng/μL standard of ^{13}C-phenacetin in H_2O). Sample extraction was based on a recently published method for the extraction of PPCPs from surface and groundwater (11). Briefly, samples were extracted at approximately 15 mL/min using Oasis HLB SPE cartridges (6cc, 500mg sorbent) that had been conditioned with 2x3mL aliquots of methanol, followed by 2x2mL aliquots of MilliQ H_2O. Upon completion of the extraction, SPE cartridges were washed with 1mL of 5% methanol in MilliQ H_2O and dried under vacuum for 1 hour.

PPCPs were eluted under vacuum using 2x3mL aliquots of methanol followed by 2x2mL aliquots of 0.1% TFA in methanol. Eluent was collected in glass test tubes, and the two separate extracts were combined, mixed, and again split in half to ensure an identical matrix. The samples were then evaporated to dryness under a gentle N_2 stream, reconstituted with 1 mL of HPLC starting mobile phase containing 50 ng internal standard (^{13}C-caffeine), and ultrasonicated for 15 min to ensure complete solubilization and homogenization.

Samples were then filtered using Millipore Ultrafree-MC 0.45μm centrifugal filters and transferred to 1 mL HPLC injection vials.

The two 100 mL influent samples were vacuum filtered using Whatman glass fiber filters (0.7μm particle retention) and transferred to 100mL round bottom flasks which had been baked at 400°C and solvent rinsed. No preconcentration or clean-up step was employed prior to analysis.

Instruments and LC Conditions

Three Micromass, Inc. mass spectrometers were used in the present study: a single quadrupole MS (LCZ™) with an Hewlett Packard 1100 LC, a triple quadrupole MS (Quattro LC™) with a Hewlett Packard 1100 LC, and a ToF-MS (LCT™, equipped with a 4.6 GHz time-to-digital converter) with a Waters 2695 LC. LC conditions (mobile phases, gradient, etc...) for each instrument were similar as follows: mobile phase "A" was 5 mM formic acid/5 mM ammonium formate buffer (pH = 3.7) in MilliQ water and mobile phase "B" was acetonitrile. 10 μL of sample was injected and separated using a Betasil C_{18} (Keystone Scientific) 150 x 2.1mm, 3 μm analytical column with a similar guard column at a flow rate of 0.2 mL/min. Initial solvent composition was 95:5 (A:B) and a linear gradient was employed with a final solvent composition of 10:90 (A:B) at 28 min. Following the gradient, the initial solvent conditions were restored over a 5 minute ramp and the column was allowed to re-equilibrate for an additional 12 minutes. For quantitation, a six-point calibration curve (from 1 ng/mL to 500 ng/mL) was constructed for each analyte on each instrument with analyte response normalized to the internal standard.

Single-quadrupole MS conditions

Analytes were detected in SIM mode as $(M+H)^+$ ions. The sample cone voltage in the electrospray source was held at 31 volts in order to limit fragmentation. Eleven SIM masses were monitored during the first 17 minutes. Ten SIM masses were monitored for the last 11 minutes. Table 1 includes a list of analytes and their corresponding SIM ion m/z values.

Triple-quadrupole MS Conditions

Detection of analytes was performed using MRM mode. The sample cone voltage in the electrospray source was optimized for minimal fragmentation and maximum overall sensitivity (34 volts). Eleven MRM transitions were monitored for the first 16 minutes, followed by an additional 11 MRM transitions over the last 12 minutes. Table 1 includes precursor-product MS/MS transitions as well as CID collision energies for each analyte.

LCT Conditions

The ToF analysis was conducted in ESP+ mode with a 1.050 second cycle time (1 sec acquisition time and 0.05 sec post-acquisition delay). The selected m/z range was 100 to 1000 Da. The instrument was operated at a mass resolution of 6800 (FWHM), and was externally mass calibrated using polyalanine (Sigma #P9003). Five ng/mL Leucine enkephalin ((M+H)$^+$ = 556.2771) was used as a lock mass (to compensate for drift of the external calibration) and was added post-column at a flow rate of 1 µL/min. This delivery rate was optimized to ensure that the leucine enkephalin signal would not saturate the detector (ion counts per scan) when the mobile phase was at its highest organic content and ionization/desolvation efficiencies were highest. Data files were internally mass calibrated (vs. the leucine enkephalin lock mass) using the manufacturer's all file accurate mass measure (AFAMM) software process. The resulting files were used in the successive data analyses. Accurate mass measurements were used in conjunction with chromatographic retention time for analyte confirmation. Table 1 lists calculated exact masses for each of the analytes detected.

Results and Discussion

Chromatography

Figure 1 shows a base peak intensity (BPI) chromatogram resulting from an HPLC-ToF-MS analysis of a 50 ng/mL calibration. A base peak intensity chromatogram plots the intense signal from each mass spectrum on a chromatographic time scale. Peaks were generally well separated with minimal tailing, and in general, chromatography was reproducible among instruments. Relative retention times did not vary significantly within or among the

Table I: Mass analyzer parameters for sample analysis

Analyte	LCZ^{TM}	Quattro II^{TM}		LCT^{TM}
	SIM mass	MRM transition	Collision energy (v)	Calculated M+1 (Da)
13C-caffeine	197.8	197.8>139.7	22	198.0984
13C-phenacetin	180.8	180.9>109.8	21	181.1058
Acetaminophen	151.8	151.7>109.7	15	152.0711
Antipyrine	188.8	188.9>55.8	26	189.1028
Caffeine	194.8	195.0>137.7	22	195.0882
Carbamazepine	236.8	236.8>193.8	19	237.1028
Cimetidine	252.8	253.0>158.7	19	253.1235
Cotinine	176.8	176.7>79.7	24	177.1028
Diltiazem	414.9	414.8>177.7	27	415.1691
Diphenhydramine	255.9	255.7>166.7	11	256.1701
Erythromycin	734.1	734.5>576.2	25	734.4690
Fenofibrate	361.1	361.1>232.8	18	361.1206
Fluoxetine	309.9	310.0>147.7	10	310.1418
Metformin	129.8	129.8>70.8	21	130.1092
Nifedipine	347.0	347.1>315.0	10	347.1243
Paraxanthine	180.8	180.7>123.8	23	181.0725
Ranitidine	314.9	315.0>175.7	27	315.1491
Salbutamol	239.9	239.9>147.7	16	240.1599
Sulfamethoxazole	253.8	253.8>155.7	21	254.0599
Trimethoprim	290.9	291.0>229.9	27	291.1457
Warfarin	308.9	309.2>162.7	20	309.1127

116

instruments. Absolute retention times varied only slightly between the single quadrupole and triple quadrupole instruments as they used nearly identical HPLC systems. However, absolute retention times in the HPLC-ToF-MS analysis (Waters HPLC system) were shorter. This was likely the result of differences between the Waters and Hewlett Packard HPLC systems, as well as slight differences in column temperature. Relative analyte responses among instruments were similar, indicating that electrospray conditions were indeed relatively constant for the three instruments considered here.

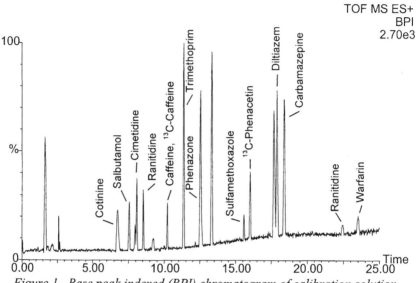

Figure 1. Base peak indexed (BPI) chromatogram of calibration solution containing 50 ng/mL of each PPCP

Comparison of Instrument Sensitivity and Selectivity

For individual PPCP measured in wastewater effluent extract, the useful sensitivity and selectivity of analysis depended upon several factors. The nature of the analyte (m/z, ionization efficiency, chromatographic behavior including peak width and retention time), properties of the sample matrix, and ion transmission/detection properties of each instrument all influenced instrumental analysis to varying degrees. Under the present conditions, the single quadrupole instrument was the least sensitive for nearly all of the PPCPs due to greater isobaric interferences and baseline noise in complex sample matrices. In general, the triple quadrupole was the most sensitive due to the selectivity and noise reduction afforded by MS/MS. The overall sensitivity of the HPLC-ToF-

MS operated in accurate mass mode was much better than that afforded by the single quadrupole in complex mixture analysis and often approached that obtained by the triple quadrupole. This performance was a result of the increased selectivity (reduction of isobaric interferences) afforded by the higher resolving power of the ToF analyzer.

The results obtained for diltiazem in wastewater effluent extract illustrate the relative analytical performance of the three instruments (Figure 2). Peak-to-peak signal to noise (S/N) ratios were calculated for the SIM and MRM analyses in addition to the ToF accurate mass chromatogram. These chromatograms are shown in Figure 2 for diltiazem in wastewater effluent, including a section of expanded baseline, as well as corresponding S/N values.

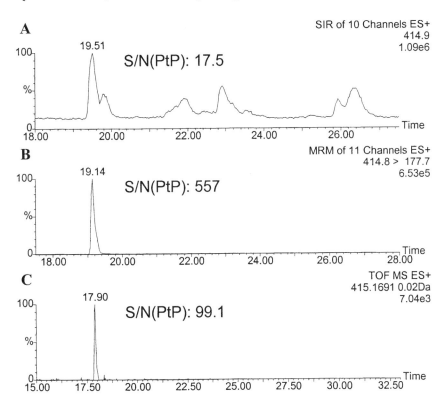

Figure 2. Diltiazem response on single quadrupole (A), triple quadrupole (B), and ToF (C): note sectional baseline enhancement scales

The complex matrix present in the wastewater effluent extract caused greatly enhanced signal background and isobaric noise in the single quadrupole

HPLC-MS analysis. In this case, the baseline accounted for a significant portion (about 15%) of the instrument response throughout the illustrated portion of the chromatogram. Accordingly, the lowest signal S/N was obtained from the single quadrupole SIM analysis. Several distinct peaks other than that produced by diltiazem peak were also present. These peaks were likely due to the presence of partially or completely chromatographically resolved compounds sharing the same nominal mass with diltiazem, which were inherently monitored by the relatively wide (~1Da) SIM window for this compound. Thus, SIM was found to be neither sufficiently selective nor sensitive for the present analysis of PPCP compounds in wastewater effluent.

The triple quadrupole MS/MS analysis was the most selective and sensitive. It produced the highest S/N ratio and completely resolved diltiazem from other isobaric interferences. In the case of the triple quadrupole, the noise reduction was due to the inherent specificity of monitoring specific precursor to product ion transitions during the MRM process. Isobaric compounds are unlikely to have identical CID MS/MS transitions and therefore do not produce an instrument response in MRM mode.

The accurate mass chromatogram of diltiazem obtained from HPLC-ToF-MS analysis was centered about the calculated $(M+H)^+$ m/z with a 20 mDa window. The resulting S/N was a marked improvement over the single quadrupole SIM S/N and begins to approach that obtained by MS/MS. The improvement over the SIM analysis can be attributed to the high mass accuracy of the ToF analyzer. Where any analyte within approximately 0.5 Da produced a signal in SIM, only analytes within 0.010 Da registered a signal in the ToF accurate mass chromatogram. In addition, most of the isobaric interferences observed during SIM were not apparent in the ToF analysis. This was a result of considerably higher mass resolution obtained by the ToF, which was in most cases capable of eliminating nominally isobaric spectral mass contaminants. In general, although the ToF was usually less sensitive than the triple quadrupole instrument for analysis of targeted PPCPs in wastewater, it outperformed the single quadrupole in complex mixtures while providing an additional level of confidence in analyte identification via accurate mass measurement.

Figure 3 further illustrates the specificity of accurate mass analysis using ToF-MS. All five reconstructed ion chromatograms are from the same ToF analysis of diltiazem in the wastewater effluent sample, however each is centered on the calculated exact mass with mass windows of varying size. Thus, when a small mass window is used, the instrument is selective and produced a higher S/N (Figure 3A). As the mass window is widened, the instrument detects isobaric interferences for diltiazem, resulting in a noisier baseline and decreasing S/N values. The smallest window size achievable is practically limited by the average mass measurement error of the instrument. In the case of the LCT™ presently investigated, it was found that a 0.20 mDa window was optimal. Figure 3E represents 1 Da mass window and can be regarded as a simulation of an instrument operating at unit mass resolution.

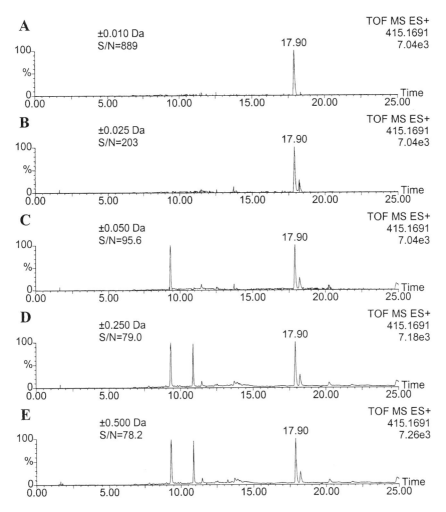

Figure 3. Selectivity and sensitivity improvements in HPLC-ToF-MS analysis of diltiazem in wastewater effluent with narrowing mass window (note the improvement in S/N with decreasing mass window)

Quantitation of PPCPs in wastewater effluent

The analytical performance of the three MS instruments compared in the present work was observed to significantly impact their relative abilities to perform reliable quantitation of PPCPs in wastewater effluent. Table II lists the concentrations determined for the PPCPs detected in the wastewater effluent extract as well as accurate mass measurements and corresponding errors obtained from the ToF analysis. Concentrations obtained from SIM analysis using the single quadrupole instrument were consistently more than a factor of two different than concentrations estimated from MS/MS or ToF analyses. In the cases where a higher concentration was observed from the single quadrupole analysis (as compared to the MS/MS or ToF values), the instrument response was likely influenced by co-eluting isobaric interferences leading to an overestimation of analyte concentration. For analytes detected using the single quadrupole in lower concentrations than estimated by MS/MS or ToF, it is possible that poor chromatographic resolution from a noisy baseline contributed to observed inaccuracies. It is unlikely that these discrepancies were a result of matrix enhancement or suppression as nearly identical chromatography and ionization sources were used for each instrument.

The ability of the ToF mass analyzer to resolve interferences leads to more reliable determinations of analyte concentrations (closer to the results achieved using MS/MS) in comparison to the results obtained from the single quadrupole MS. With few exceptions, concentrations determined from HPLC-ToF-MS were within a factor of two of those obtained using the triple quadrupole. In the case of sulfamethoxazole, it is not clear why the ToF seems to have underestimated the concentration relative to the MS/MS results. In the cases of caffeine and paraxanthine, however, the ^{13}C isotope peak from a co-eluting compound with a mass 1 amu lower than the analytes caused an overestimation of analyte concentration (Figure 4). Although the interferences are not resolved for either example, the ratios of the 194 to 195 and 180 to 181 peaks suggest a significant contribution from the ^{13}C isotope from the lower mass compounds. Consequently, the accurate mass calculations for the target peaks reflect the influence of the suspected isobaric interferences (Table II). The associated mass error is significantly higher than the 2.0 mDa accuracy specified by the manufacturer. Thus, the accurate mass measurement capability of the ToF instrument can sometimes provide valuable evidence for an isobaric interference. Such information was not available using either single or triple quadrupole instruments.

Table II: PPCP Quantification using different mass analyzers

Compound*	PPCP concentrations in effluent (ng/L)			ToF accurate mass	
	LCZ^{TM}	Quattro LC^{TM}	LCT^{TM}	Accurate mass (Da)	Mass error (mDa)
Caffeine	210	43.6	109	195.1082	-20.0
Carbamazepine	888	194	119	237.1027	0.1
Cimetidine	88.6	193	240	253.1230	0.5
Cotinine	nd**	10.2	22.0	177.1016	1.2
Diltiazem	21.8	72.4	52.4	415.1685	0.6
Diphenhydramine	117	361	314	256.1698	0.3
Erythromycin	nd	307	nd	-	-
Fenofibrate	nd	19.9	nd	-	-
Metformin	nd	243	261	130.1087	0.5
Paraxanthine	478	35.3	154	181.0956	-23.1
Ranitidine	51.4	184	91.0	315.1528	-3.6
Salbutamol	9.44	nd	35.6	240.1597	0.2
Sulfamethoxazole	2420	1570	458	254.0604	-0.5
Trimethoprim	19.2	130	105	291.1452	0.5
Warfarin	nd	9.78	nd	-	-

*Only those PPCPs that were detected in the STP effluent extract are listed.

**Not detected

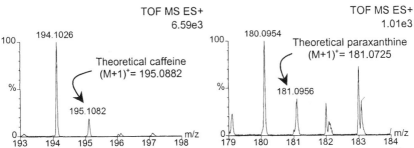

Figure 4. Caffeine and paraxanthine mass spectra illustrating interfering ^{13}C isotope peaks derived from intense spectra peaks ~1 Da smaller than the analytes

Target and non-target analysis of environmental contaminants in complex samples using accurate mass HPLC-ToF-MS

Three distinct advantages offered by ToF-MS include the ability to collect data across a wide mass range without sacrificing sensitivity, the ability to resolve interferences away from signals of interest, and mass measurement accuracy adequate for the estimation of elemental composition. The accurate mass measurements for the target analytes in the wastewater effluent involved a single point correction of the base calibration (to compensate for slight drift of the calibration due to temperature fluctuations in the flight tube and instabilities of the power supplies) utilizing a reference compound, or lock mass. Table II includes the measured masses for those compounds in addition to their mass error relative to calculated exact masses (listed in Table 1). Mass measurement error was determined as the difference between the calculated and measured $(M+H)^+$. For most of the target analytes, this error was very low, usually below 1 mDa. In these cases, elemental composition calculated by the instrument software provided an additional means of analyte confirmation. As previously mentioned, in the cases where the mass error was highest, it appeared that an unresolved peak had interfered with the mass measurement of the target compound (Figure 4).

The resolving power, accurate mass measurement capability, and full spectral sensitivity also make LC-ToF-MS attractive as a tool for identifying non-target compounds in complex environmental matrices. In principle, if masses can be measured with sufficient accuracy, it is possible to assign unique elemental compositions to peaks observed during the course of an analysis. In practice, a mass measurement within 2 mDa gives rise to a short list of elemental compositions to consider. In combination with other information such as calculated isotope ratios and chromatographic retention times of standards, it is obvious that HPLC-MS employing accurate mass measurement holds great promise for rapid qualitative analysis of "unknown" environmental mixtures. As an example, a filtered wastewater influent sample was analyzed using the HPLC-ToF-MS system directly without an extraction/cleanup step. Utilizing a presentation of peak intensity as a function of m/z and retention time, patterns of sequentially eluting homologous series separated by 44 Da were identified. One series displayed m/z values consistent with protonated polyethylene glycol (PEG) molecules. Extracted ion chromatograms were created for representative members of the series, and accurate mass measurements confirmed elemental compositions of the PEGs. Figure 5 shows narrow window the accurate mass chromatograms of selected members of the observed PEG homologous series as well as the mass measurement error. It should be noted that identification of the PEG series in the present example was greatly facilitated by the polymeric nature of the material (i.e. the presence of a distinct pattern of repeating peaks in the observed mass spectrum). This feature made it relatively easy to find the series visually in the display of the HPLC-MS dataset. General "non-target" analyses will likely require sophisticated data processing software tools for rapidly and reliably interpreting data from accurate mass HPLC-ToF-MS runs.

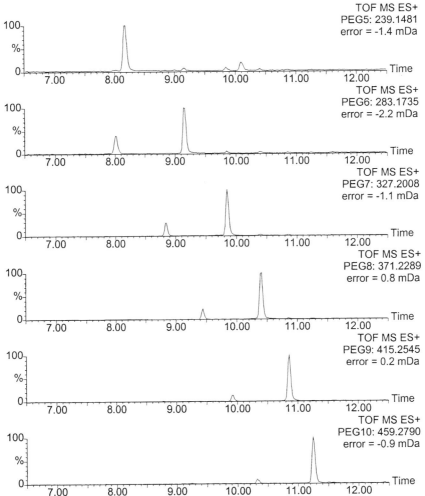

Figure 5. Accurate mass chromatograms of selected PEG oligomers in wastewater influent: PEG ethoxy-chain length and corresponding m/z are shown at right, top; mass error (mDa) is shown at right bottom

Lock mass as suppression indicator

One problem associated with electrospray ionization is suppression or enhancement of signal in the presence of co-eluting sample matrix. Often, components co-eluting with an analyte can preferentially occupy surface sites of the electrosprayed droplets, or compete for available charge, thereby decreasing analyte signal intensities (*18,19*). In the present case, the signal obtained from the lock-mass used to determine accurate mass (leucine enkephalin) inadvertently, but advantageously, provided a useful record of matrix effects. Because the lock mass was added post-column (just prior to the electrospray source), comparison of the response from a "clean" sample (e.g. a calibration solution) to that of a matrix-rich sample provided an indication of the presence and degree of analyte signal suppression. Figure 6 shows two overlaid leucine enkephalin response plots; one from a calibration solution, and the other from a wastewater effluent sample. Aside from the area between 12 and 17 minutes, there are few differences in the quantitative responses of the lock mass in the two analyses. During the period between ~12 and 17 minutes, the lock mass signal from the effluent analysis appears to be substantially suppressed relative to that measured in the analysis of the calibration solution. Consequently, it is likely that analytes eluting in this window may also have been subject to signal suppression. This added information provided a valuable internal check on the efficiency of electrospray ionization in the analysis of a complex mixture. Thus, post-column lock mass infusion (which is commonly employed for accurate mass studies in HPLC-ToF-MS) can serve as a useful secondary purpose as an indicator of matrix-induced signal suppression events often associated with HPLC-ESI-MS analyses of environmental samples.

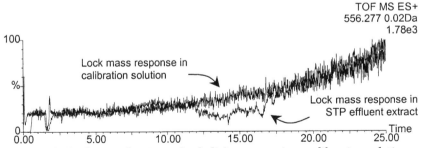

Figure 6. Lock mass (leucine enkephalin) response in a calibration solution and in the wastewater effluent extract showing evidence of matrix-induced signal suppression.

The digital dead time correction algorithm and its effect on ToF-MS dynamic range

The majority of commercially available HPLC-ToF-MS systems employ a time-to-digital (TDC) converter to process signals associated with individual ion arrival times at a multichannel plate detector. There is a finite time interval (4 nsec on the present instrument) required between ion detection events, which limits the linear dynamic range of the instrument. As the TDC systems become saturated (arrivals of nearly the same m/z ions at less than 4 nsec intervals), there occurs both a loss of linearity in detector response (Figure 7), and a shift in the signal to lower m/z, resulting in poor mass accuracy. In the present work, it was found that the effective linear dynamic range of the ToF-MS was between two and three orders of magnitude. This is significantly lower than the dynamic range typically observed on quadrupole instruments (typically >4 orders of magnitude). In actuality, this limitation is one of the most important drawbacks of using HPLC-ToF-MS for quantitative measurements of environmental contaminants. Software corrects moderate levels of detector saturation and the associated mass shifts with an algorithm called digital dead time correction (DDTC). This technique allows limited correction of the mass centroid (and peak intensity) of acquired spectra by a back-calculation algorithm triggered when the number of ion counts per second exceeds a user-defined "dead-time saturation" value. Figure 7 illustrates the effect of DDTC processing on the quantitative calibration dynamic range of cotinine acquired with the HPLC-ToF-MS system.

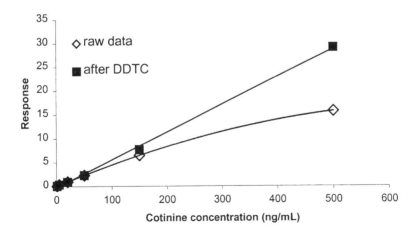

Figure 7. Effect of digital dead-time correction (DDTC) on ToF dynamic range for cotinine

Conclusion

The present work illustrates that HPLC-ToF-MS provides powerful and complimentary information to that obtained by LC-MS/MS in the analysis of polar organic contaminants in a complex environmental matrix. Figure 8 provides a comparison of the relative advantages and disadvantages of the three types of instruments used when applied to environmental analysis. Due to higher mass resolving power, ToF-MS can provide S/N improvements when compared to single quadrupole analyses of complex samples. An important benefit of accurate mass determinations using ToF-MS is useful information about elemental compositions, which can confirm or rule out potential molecular formulas. The full spectra sensitivity of ToF-MS provides an important advantage for conducting survey or discovery based analyses in comparison to scanning MS instruments where sensitivity can be limited even when scanning over a narrow mass range. Consequently, the strengths of HPLC-ToF-MS appear well suited for challenges inherent to the analysis of polar organic contaminants and their metabolites in the aquatic environment.

	Sensitivity	Selectivity	Identity confirmation	Dynamic range	Analyte discovery	cost
LC-ToF-MS	++	+++ (accurate mass)	+++ (elemental composition)	+ $(10^2\text{-}10^3)$	+++	$$
LC-MS	+	+ (SIM)	+ (nominal mass)	+++ $(>10^4)$	+	$
LC-MS/MS	+++	++ (MRM)	++ (product spectra)	+++ $(>10^4)$	++	$$$

Figure 8. Relative comparison of evaluated HPLC-MS systems

Acknowledgements

This work was supported by an EPA STAR Grant (#R-82900701-0). Although the research described in this article has been funded wholly or in part by the United States Environmental Protection Agency through grant/cooperative agreement (#R-82900701-0) to Bruce Brownawell, it has not been subjected to the Agency's required peer and policy review and therefore does not necessarily reflect the views of the Agency and no official endorsement should be inferred.

References

[1]Ternes, T. A., *TrAC, Trends Anal. Chem.* **2001**, *20*, 419-434

[2]Lindsey, M. E., Meyer, M. and Thurman, E. M.; *Anal. Chem.* **2001**, *76*, 4640-4646

[3]Crescenzi, C., Di Corcia, A., Marcomini, A. and Samperi, R.; *Environ. Sci. Technol.* **1997**, *31*, 2679-2685

[4]Crescenzi, C., Di Corcia, A., Samperi, R. and Marcomini, A.; *Anal, Chem.* **1995**, *67*, 1797-1804

[5]Ferguson, P. L., Iden, C. R. and Brownawell, B. J.; *Anal Chem.* **2000**, *72*, 4322-4330

[6]Ferguson, P. L., Iden, C. R., McElroy, A. E. and Brownawell, B. J.; *Anal. Chem.* **2001**, *73*, 3890-3895

[7]Crescenzi, C., Di Corcia, A., Marchese,S. and Samperi, R.; *Anal. Chem.* **1995**, *67, 1968-1975*

[8]Di Corcia, A., Crescenzi, C., Guerriero, E and Samperi, R.; *Environ. Sci. Technol.* **1997**, *31*, 1658-1663

[9]Di Corcia, A., Crescenzi, C., Marcomini, A. and Samperi, R.; *Environ. Sci. Technol.* **1998**, *32*, 711-718

[10]Ternes, T., Bonerz, M. and Schmidt, T.; *J. Chromatogr. A* **2001**, *938*, 175-185

[11]Kolpin, D. W., Furlong, E. T., Meyer, M. T., Thurman, E. M., Zaugg, S. D., Barber, L. B. and Buxton, H. T.; *Environ. Sci. Technol.* **2002**, *36*, 1202-1211

[12]Hirsch, R., Ternes, T, A., Haberer, K., Mehlich, A., Ballwanz, F and Kratz, K. L.; *Journal of Chromatography A* **1998**, *815*, 213-223

[13]Ternes, T. A. and Hirsch, R.; *Environ. Sci. Technol.* **2000**. *34*, 2741-2748

[14] Maizels, M. and Budde W. L.; *Anal. Chem.* **2001**, 73, 5436-5440

[15]Thurman, E. M., Ferrer, I. and Parry, R.; *J. Chromatogr. A* **2002**, 957, 3-9.

[16]Hogenbloom, A. C., Niessen, W. M. A., Little, D and Brinkman, U. A. T.; *Rapid Com. Mass Spectrom.* **1999**, *13*, 125-133

[17]Bobeldijk, I., Vissers, J. P. C., Kearney, G., Major, H. and van Leerdam, J. A.; *J. Chromatogr. A* **2001**, *929*, 63-74

[18]Choi, B. K., Hercules, D. M. Gusev, A. I.; *Fresenius' J. Anal. Chem.* **2001**, *369*, 370-377

[19]Choi, B. K., Hercules, D. M. and Gusev, A. I.; *J. Chromatogr. A* **2001**, *917*, 337-342

Chapter 8

TOF-MS and Quadrupole Ion-Trap MS/MS for the Discovery of Herbicide Degradates in Groundwater

E. M. Thurman[1,2], Imma Ferrer[1,3], and Edward T. Furlong[1]

[1]U.S. Geological Survey, Denver Federal Center, Box 25046, Denver, CO 80225
[2]Current address: U.S. Geological Survey, 4821 Quail Crest Place, Lawrence, KS 66049
[3]Current address: immaferrer@menta.net

Time-of-flight mass spectrometry (TOF/MS) and quadrupole ion-trap mass spectrometry/mass spectrometry (QIT/MS/MS) were combined for the identification and discovery of two new 2nd amide degradates of acetochlor, alachlor, and metolachlor in groundwater. These degradates were hypothesized to be environmentally important intermediates in the degradation of these acetanilide herbicides with over 100 million pounds applied annually to soil. The strategy involved four steps. First was the theoretical hypothesis that the secondary-amide ESA of the acetanilide herbicides were present in groundwater because of similarities of structure and likelihood of degradation and persistence (called the discovery process). Second was the QIT/MS/MS analysis of several samples for the molecular ion and the characteristic fragmentation and diagnostic ions. Third was the synthesis of a standard and verification of standard with retention time and MS/MS spectra followed by accurate mass analysis and molecular formula with TOF/MS. Finally, the last step involved the discovery of these 2nd amide ESA degradates in groundwater with HPLC and QIT/MS/MS. The analytical procedure described herein has the potential for widespread use in the discovery of degradates for many herbicides used on crops in the United States. The QIT/MS/MS is powerful for structural elucidation because of the capability of MS^n. For example, two degradates with the same exact mass could be differentiated because of differences in MS/MS fragmentation. The TOF/MS analysis was useful for the assurance of synthesis of pure standards of the secondary-amide degradates. Analysis of 82 shallow groundwater samples from the Midwestern United States showed that the secondary amide of ESA degradates of acetochlor, alachlor, and metolachlor occur at detection frequencies of 21-26%.

The advent of high performance liquid chromatography/mass spectrometry (HPLC/MS) quadrupole instruments has made analysis of polar pesticides in groundwater a common procedure (1-2). During the past 5 years many papers have been published on the analysis of pesticides and their degradation products by quadrupole HPLC/MS (3-12); however, there are several short comings yet to be overcome. For example, often polar pesticides give only a protonated or de-protonated molecule or a weak fragment ion, especially when the interface is electrospray ionization. The fragmentor or cone voltage is used to enhance collision-induced dissociation (CID) in the source and transport region of the electrospray source, and this fragmentation voltage may vary substantially among different analytes and sources, which makes fragmentation difficult to predict in an analysis of unknown compounds. Second, there are no universal libraries available for pesticide analysis by HPLC/MS, as in electron impact gas chromatography/mass spectrometry (GC/MS), this problem makes identification of unknown pesticides or their degradates nearly impossible by simple quadrupole HPLC/MS analysis.

These shortcomings may be overcome partially by the application of time-of-flight mass spectrometry (TOF/MS) (13) and high performance liquid chromatography/quadrupole ion-trap-mass spectrometry/mass spectrometry (HPLC-QIT-MS/MS) (14-15). The HPLC-QIT-MS/MS does MS/MS in time rather than in space, which means that ions are retained in a trap through a set time-period. If all the ions are ejected, then the result is a full-scan spectrum. If the protonated or de-protonated molecule is retained in the trap and all others are ejected, and this ion is fragmented, the result is MS/MS. This process may be repeated multiple times, which results in MS^n. In contrast the triple quadrupole MS/MS does the isolation and fragmentation in space, which means that the fragmentation is continuous in time, but the selected ion travels through the flight tube of the mass spectrometer to collision chamber where fragmentation occurs and then onto the third quadrupole for the mass spectrum.

Two advantages of the ion trap are that it gives excellent sensitivity while trapping ions in full scan mode, which then may be selected and fragmented to yield MS/MS spectra, and second is the ability of the ion trap is to do MS^n (14).

Typically, three or four isolations and fragmentations are possible before sensitivity is too low to record ions in unknown samples. The ability to do multiple isolation and fragmentation allows one to build a library of spectra using standard compounds, which give both characteristic fragmentations and diagnostic ions that then can be used to identify unknown pesticides or their degradates, such as the ethane-sulfonic-acid (ESA) degradates of the herbicides acetochlor and alachlor. These compounds have the same exact mass and are common ground-water contaminants (16-17). TOF/MS is useful also for identification of synthesized standards to verify the analysis of QIT/MS/MS when no commercial standards are available and new standards are synthesized, as well as the identification of degradates in actual groundwater samples (13).

Recently these two ubiquitous ground water contaminants have been analyzed for the first time by a HPLC-MS/MS method using triple quadrupole analysis (18). The triple quadrupole MS/MS method was tested on spiked water samples and environmental data are reported in several chapters in this book. These ESA degradates and their further degradation products are the subject of this paper using quadrupole ion-trap mass spectrometry/mass spectrometry (QIT/MS/MS) and time-of-flight/mass spectrometry (TOF/MS). This paper deals with (1) the process of hypothesis of pesticide degradates in ground water, (2) their chemical analysis using characteristic fragmentations and diagnostic ions, and (3 and 4) the identification and occurrence of these new degradates in water samples from the Midwestern United States using QIT/MS/MS and TOF/MS. It is hypothesized that the approach explained here using QIT/MS/MS and TOF/MS will have broad application among many areas of environmental analytical chemistry when used to generate a library of characteristic fragmentations and diagnostic ions for the unknown identification of both pesticides and pharmaceuticals in groundwater.

Experimental Methods

Reagents

HPLC-grade acetonitrile, methanol, and water, along with reagent-grade acetic acid were obtained from Fisher Scientific (Pittsburg, PA, USA). The analytical standards for acetochlor, alachlor, and metolachlor were obtained from Chem Service, Inc. (West Chester, PA). The analytical standard for acetochlor ESA was obtained from Zeneca Agrochemicals (Fernhurst, Haslemere Surrey, UK), and the standard for alachlor ESA was obtained from the U.S. Environmental Protection Agency Repository (Cincinnati, OH, USA). Metolachlor ESA was synthesized in the U.S. Geological Survey laboratory in Lawrence, KS in a previous study (19). Standard solutions were prepared in methanol. The acetochlor, alachlor, and metolachlor secondary amide ESAs were prepared by the method of Thurman et al. (*20*).

HPLC-QIT-MS/MS

Liquid chromatography electrospray ion-trap tandem mass spectrometry (HPLC-QIT-MS/MS), in negative ion mode of operation, was used to separate and identify the ESA analytes. The analytes were separated by using a series 1100 Hewlett Packard liquid chromatograph (Palo Alto, CA) equipped with a reverse-phase C_{18} analytical column (Phenomenex RP18, Torrance, CA) of 250 x 3 mm and 5-μm particle diameter. Column temperature was maintained at 60° C. The mobile phase used for eluting the analytes from the SPE and HPLC columns consisted of acetonitrile and 10-mM ammonium formate buffer at a flow rate of 0.3 mL/min. This HPLC system was connected to an ion trap mass spectrometer, an Esquire HPLC-QIT-MS/MS (Bruker Daltonics, Bellerica, MA) system equipped with an electrospray ionization (ESI) source. Operating conditions of the MS system were optimized in full-scan mode (m/z scan range: 50 to 400) by flow-injection analysis of selected compounds at 10-μg/mL concentration. The maximum accumulation time value was set at 200 ms.

TOF/MS

A ToF-MS (LCT^{TM}) Micromass instrument (Manchester, UK) was equipped with a 4.6 GHz time-to-digital converter and was used in ESI negative mode. The sample was added with flow injection (10 μL injection) with a mobile phase of acetonitrile:water (50:50) with 0.1% formic acid at a flow rate of 300 μL/min. The instrument was operated with a cone voltage of 25V. The resolution was 6000 FWHM and was externally calibrated with polyalanine and a leucine enkephalin solution (554.2615 m/z) was added as a lock mass to compensate for drift of the external calibration, which was added post column at a flow rate of ~1 μL/min. Molecular weight formulae were calculated using Micromass software package for composition.

Sample Collection and Analysis

Ground-water samples were collected from 82 wells in Indiana and Minnesota as part of several surveys of shallow ground-water quality (*16-17*). The water was filtered through 0.7-μm glass-fiber filters (Whatman GF/F, Maidstone, England) and stored on ice and shipped to our laboratory in Lawrence, KS. Samples were processed according to the method just described for HPLC/MS analysis. The solid-phase extraction procedure was performed using an automated Millipore Workstation (Waters, Milford, MA, USA) as described by Ferrer et al. (12). The SPE cartridges (Sep-Pak) were obtained from Waters-Millipore (Milford, MA, USA). They contained 360 mg of 40-µm C_{18} bonded silica. Each C_{18} cartridge was preconditioned as follows: 2 mL methanol, 2 mL ethyl acetate, 2 mL water, followed by 2 mL distilled water. A 100-mL sample was passed through the cartridge at a flow rate of 10 mL/min, and the cartridge was purged with air to remove excess water. The cartridge was eluted with 3 mL of ethyl acetate, followed by 3 mL of methanol. The ethyl acetate removed the parent pesticide and secondary amide of the parent compound. The methanol eluted the secondary amide of the ESA, which is the ionic degradate of the parent pesticide and was analyzed by HPLC/MS.

For routine analysis of acetochlor, alachlor, and metolachlor ESA and secondary degradates in ground water a single quadrupole HPLC/MS system was used. The compounds were separated on a Hewlett Packard 1100 HPLC coupled to a Hewlett Packard 1100 mass selective detector (MSD) operating in negative-ion electrospray mode. The mobile phase consisted of 0.3% acetic acid, 24% methanol, 35.7% water, and 40% acetonitrile with a flow rate of 0.3 mL/min. The analytical columns consisted of two Phenomenex 5-µm, 250- x 3- mm C_{18} columns coupled to one Phenomenex 3-µm 150- x 2.0-mm C_{18} column. Column temperatures were set at 70° C to achieve better separation and peak shapes. The drying gas flow was set at 6 L/min, the nebulizer pressure was 25 psi, the drying gas temperature was 300° C, the capillary voltage was 3100 V and the fragmentor voltage was 70 V (21). The following mass spectral ions were monitored: acetochlor ESA (314 m/z), alachlor ESA (314 m/z), metolachlor ESA (328 m/z), the secondary amide of acetochlor (256 and 121 m/z), the secondary amide of alachlor (270 and 121 m/z) and metolachlor secondary amide (256 and 121 m/z).

Results and Discussion

The Discovery Process: Step 1

Figure 1 shows the degradation pathway for acetochlor, alachlor, and metolachlor in soil via glutathione conjugation and subsequent oxidation to the sulfonic acid as previously reported (13, 22). The degradation proceeds by first conjugation to glutathione (22) followed by cleavage to a thiol group and its subsequent oxidation in soil to the sulfonic acid. These negatively-charged ESA degradates of acetochlor (314 m/z), alachlor (314 m/z), and metolachlor (328 m/z) have been reported in surface and ground water (16-17, 23-24) throughout the United States by the authors. However, the fate of these compounds has not been reported in the literature; therefore, an important research question is: What is the degradation and fate of these compounds in the environment? Figure 1

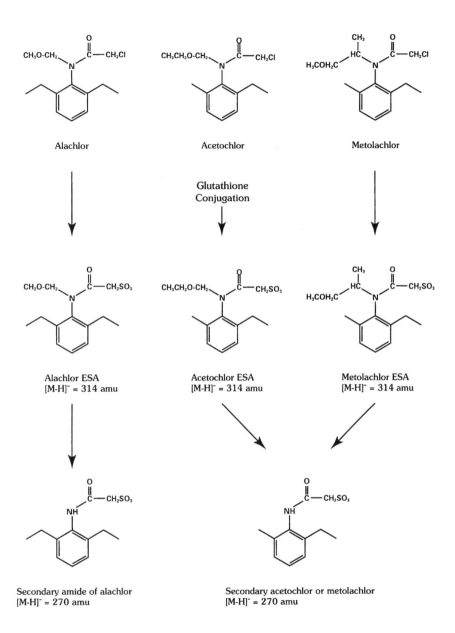

Figure 1. Proposed pathway of degradation for chloroacetamides of alachlor, acetochlor, and metolachlor to their secondary amides.

also shows the hypothesized pathway for further degradation of the ESA degradates and is the first step in the discovery process. It is known from the general weed-science literature that the half lives in soil for these compounds are 14 days for acetochlor, 21 days for alachlor, and 30 days for metolachlor. The major difference in half lives may be attributed to the side chain at the nitrogen because the remainder of the molecule is nearly identical.

This result leads to the conclusion that degradation down the side chain should be the avenue for future research on the identification of degradates for this class of herbicides. This intellectual process is step one in discovery. The second step is to gather data on the fragmentation of members of the family and their characteristic fragmentation and any diagnostic ions that may also form. The third step is to look for the new candidate degradates in real water samples that may contain high concentrations of the parent or other related degradate compounds. If this is successful, the final steps are the synthesis of the degradate standard and the identification and discovery of the compound in groundwater. This process is outlined in the following sections of this paper for the discovery of the secondary amide ESA degradates of acetochlor, alachlor, and metolachlor.

Characteristic Fragmentation and Diagnostic Ions: Step 2

The concept of using diagnostic ions and deprotonated molecules to generate a library spectrum and to identify unknowns was introduced by Ferrer et al. (*25*) in the study of unknown surfactants by HPLC-QIT-MS/MS (also see Ferrer et al. Chapter 22 in this book). The diagnostic ion is simply a fragment ion found in all members of a family of compounds, which alerts the analyst that a possible family member is present. For example, the chromatogram in Figure 2 shows the MS/MS spectra for acetochlor ESA, alachlor ESA, and metolachlor ESA. The only ion present in all three compounds is the 121 m/z ion, which is diagnostic of the sulfonic acid—methylene—carbonyl structure shown in Figure 2.

The fragmentation that leads to this ion is therefore characteristic of this class of degradates and is an important fragmentation to look for in the new secondary amide degradates. Another possible diagnostic ion that we observed was the 80 m/z fragment ion. This ion was seen only in collision-induced dissociation (CID) in the source. It was not seen during QIT-MS/MS experiments of any of the ions. However, the 80 m/z ion, SO_3^-, was detected by MS/MS experiments with the triple quadrupole (*18*). Apparently, either the fragmentation of the deprotonated molecule in the ion trap with He as a collision gas is not capable of fragmentation to the 80 m/z ion, or more likely, the trap is not able to trap the 80 m/z ion in the MS/MS experiment when the fragmented ion is 121 m/z and quite close in mass to the ion being trapped, in this case 80 m/z (personal communication with Bruker Instruments). This is an interesting difference between fragmentation in these two regions of the instrument when using the ion trap, and this result identifies a possible limitation to fragmentation or ion collection in the ion trap. Nonetheless, we used the 80 m/z ion in CID fragmentation as a diagnostic ion in order to help discover the secondary amides of the ESAs (see next section).

The 80 m/z fragment ion is a rare type of ion to be seen in HPLC-MS/MS in that a typical HPLC-MS/MS spectrum is dominated by even-electron ions. The 80 m/z ion is an odd-electron ion, and in some sense is a perfect odd-electron ion, in that the odd electron is stabilized across the 3 oxygen atoms that make up the SO_3^- ion, where all oxygen atoms have equal sharing of the odd electron! Thus, it is an especially useful diagnostic ion for the discovery of the secondary amide degradates.

Another powerful feature of QIT-MS/MS is its ability to distinguish compounds of the same exact mass by different product ions. For example, both acetochlor ESA and alachlor ESA have the same exact mass of 314.0678 (Figure 1). Thus, the measurement of [M-H]⁻ by LC/MS does not distinguish between these two compounds (*12*), which is a problem that has been reported previously (*18*). However, it is possible to measure unique fragments of 162 and 146 m/z for acetochlor ESA and 176 and 160 for alachlor ESA by ion trap MS/MS as found by LC/MS/MS triple quadrupole (*18*) as well as the diagnostic ion of 121 (See Figure 2). The structures of the 162 and 176 ions have been assigned to two structures that maintain the sulfonic-acid group (Figure 2). The two ions differ by 14 mass units because of an extra CH_2 group attached to the amide nitrogen in acetochlor. Assignments are not made for the 146 and 160 ions, which are 16 m/z less than the characteristic ions for acetochlor and alachlor ESA. The fragment ions of 162 and 176 m/z for alachlor and 146 and 160 m/z for acetochlor are what is being called a characteristic fragmentation of the family of ESA degradates and, because of this fragmentation, may be used to identify these compounds and others in this family of compounds. Thus, it is possible to accomplish the unequivocal identification of both acetochlor and alachlor ESA using HPLC-QIT-MS/MS similar to the identification of both ESAs using HPLC-MS/MS with a triple quadrupole (*18*). Finally, metolachlor ESA is not separated completely from acetochlor ESA and alachlor ESA but is distinguished with the isolation of the [M–H]⁻ ion at 328 m/z and its subsequent fragmentation to 256, 192, and 121 m/z ions (Figure 2).

The differentiation by mass spectrometry among the three ESAs is quite important because it is difficult to separate these two compounds completely by liquid chromatography (*12, 21*) and both compounds have been reported frequently in ground water of the Midwestern United States where the parent compound, metolachlor, is used on corn and soybeans with a total usage of over 25 million kg/yr (*16-17*).

Identification of Secondary-Amide ESAs: Step 3

It is important that the preliminary identification just described be tested on actual ground-water samples that may contain the secondary amides of

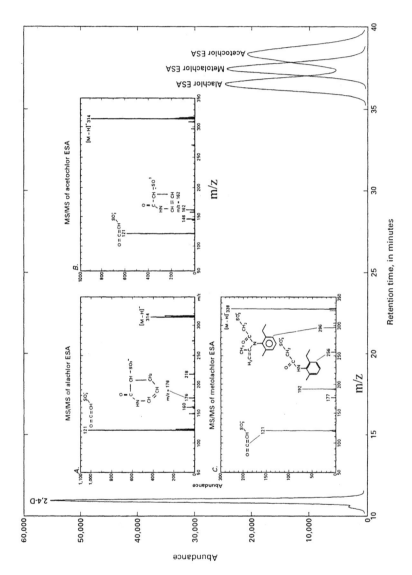

Figure 2. Chromatogram, MS/MS spectra, and diagnostic ions for identification of acetochlor, alachlor, and metolachlor ESAs.

acetochlor, alachlor and, metolachlor ESA. The test samples for this step were collected from a Minnesota groundwater site that was known to contain high concentrations of alachlor ESA and metolachlor ESA degradates (300-1000 μg/L). This site represents a point source spill of alachlor and metolachlor that had degraded in soil to their ESA degradates and generated concentrations that were about 100 times higher than is typically found in groundwater (*16-17*). Thus, this sample was an excellent one to check for the new secondary-amide ESA degradates of acetochlor, alachlor and metolachlor. The sample was first processed through a C_{18} cartridge according to the method described in the Experimental Section. This method is known to isolate the alachlor ESA degradates (*12*). Thus, it seemed likely that the methanol eluate would contain the new degradates of the secondary amide of acetochlor, alachlor, and metolachlor ESA.

The final identification procedure involves the following steps for the discovery of the secondary amides of the ESA metabolites (Figure 1) in the methanol extract of the Minnesota groundwater sample. First, the ion trap was operated in the MS scan mode to look for the deprotonated molecule [M-H]⁻ of 256 and 270 m/z and then the subsequent diagnostic ions of 121 and 80 m/z. A slow chromatographic gradient was used to separate the various compounds present and to help separate the peaks that gave rise to the 256 and 270 m/z ions and their diagnostic ions. The deprotonated molecule and diagnostic ions were discovered in the reconstructed ion chromatogram at 256 and 270 m/z. Then the sample was re-chromatographed, and the deprotonated molecule was isolated and fragmented in the trap (MS/MS) to give the diagnostic ion of 121 m/z, thus providing direct evidence that the suspected deprotonated molecule of acetochlor and alachlor ESA secondary amides actually contained the diagnostic ions of 121 m/z. In this way a preliminary identification was made. Thus, this step 3 provides us a 99.9% probability of correct identification and leads us to the synthesis and purification of a standard, which is published elsewhere (*20*).

The purity of the synthesized standards was checked by TOF/MS. The standard was analyzed by ESI negative with a low cone voltage to minimize any fragmentation and to give molecular ions. The deprotonated molecule of the alachlor secondary amide gave a major peak at 270.0790 m/z (90%) and a minor peak at 314.1059 m/z (10%). The major peak was with -0.8 millimass units (μ) of the calculated mass of 270.0798 for the deprotonated molecule of the secondary amide ESA of alachlor, which is an error of -0.8 μ or –3.0 ppm. The best match formula also matched the secondary amide alachlor ESA ($C_{12}H_{16}N_1O_4S$). The minor peak at 314.1059 m/z was alachlor ESA (accuracy was 0.1 μ), which was inadvertently synthesized in the process of making the secondary degradate. Thus, the sample was re-processed by preparative chromatography (20) until greater than 95% purity was obtained and the 314.1059 m/z ion was not present.

Likewise the deprotonated molecule of acetochlor secondary amide ESA was checked for purity by TOF/MS and the only peak obtained was the 256.0656 m/z, which is only -0.1μ from the calculated mass. The best match formula gave $C_{11}H_{14}N_1O_4S$, which is the correct formula for the acetochlor secondary amide ESA standard. Thus, TOF/MS assured us of pure standards for the QIT/MS/MS analysis and the LC/MS analysis (21) of groundwater samples. The excellent sensitivity of the TOF/MS was further assurance of pure standards for identification purposes.

Groundwater Results: Step 4

The final step then in the identification of the secondary amide of the acetochlor, alachlor, and metolachlor ESA was the matching of chromatographic and mass spectra of the new degradate standards to unknowns from the environment. Figure 3 shows the exact match for the synthesized standards with the unknown degradates in the ground-water samples from Minnesota along with the chromatography match of the standards with the unknowns in the water sample. The ion-trap MS/MS mass spectra of the synthesized standards for both compounds and follow the general use pattern of these herbicides in the United States (metolachlor 30 million kg > acetochlor 15 million kg > alachlor at 3 million kg (26). The increased detections of alachlor ESA over acetochlor ESA reflect the fact that, although acetochlor parent compound has recently exceeded alachlor use, the long-term use of alachlor (since 1972 compared to 1995 for acetochlor) has left a signature of alachlor ESA in the soil and aquatic environment. The apparently longer half life and increased water solubility of the alachlor ESA is responsible for this remaining pesticide in the aquatic environment. The frequency of detection of the secondary amide ESA degradates of acetochlor or metolachlor (256 m/z ion) and alachlor (270 m/z ion) show similar frequencies of detection of 26 and 21%, respectively (Table 1A). Thus, a first conclusion is that the first ESA degradate (Figure 1) of the acetochlor and alachlor ESA secondary amides gives the same retention time and mass spectra as the unknown. This completes the procedure of identification of two new degradates of acetochlor, alachlor, and metolachlor ESA.

Finally, eighty-two groundwater samples were analyzed from the Minnesota and Indiana (areas of intense use of these herbicides) for acetochlor, alachlor, and metolachlor, their ESA, and their secondary-amide ESA degradates. Table IA shows the frequency of detection for each of the degradates compared to their parent herbicides at two detection levels (the detection limit of the method is 0.05 μg/L and the European Health Standard for pesticide degradates of 0.1 μg/L--the U.S. has no health standards for these compounds). Parent herbicides in ground-water samples showed that the frequency of detection was metolachlor (2%) > acetochlor (1%) = alachlor (1%). Next, the frequency of detection of the ESA degradates indicates metolachlor ESA (72%) > alachlor ESA (65%) > acetochlor ESA (17%). These detections greatly exceed parent

Table I. Frequency of detection (A) and median and highest concentration (B) for 82 groundwater samples taken for chloroacetanilide herbicides and their degradates.

A. Frequency of Detection in Percent at 0.05 and 0.1 µg/L

	Total Samples	Acetochlor (%)	Acetochlor ESA (%)	2nd Amide ESA of Acetochlor or Metolachlor (%)	Alachlor (%)	Alachlor ESA (%)	2nd Amide Alachlor ESA (%)	Metolachlor (%)	Metolachlor ESA (%)
0.05 µg/L	82	1	17	26	1	65	21	2	72
0.1 µg/L	82	1	13	10	0	50	11	1	65

B. Median and Highest Concentration in µg/L

	Total Samples	Acetochlor (µg/L)	Acetochlor ESA (µg/L)	2nd Amide ESA of Acetochlor or Metolachlor (µg/L)	Alachlor (µg/L)	Alachlor ESA (µg/L)	2nd Amide Alachlor ESA (µg/L)	Metolachlor (µg/L)	Metolachlor ESA (µg/L)
Median	82	<0.05	<0.05	<0.05	<0.05	0.1	<0.05	<0.05	0.26
Highest	82	0.41	0.74	0.66	0.07	4.9	2.4	0.86	40

140

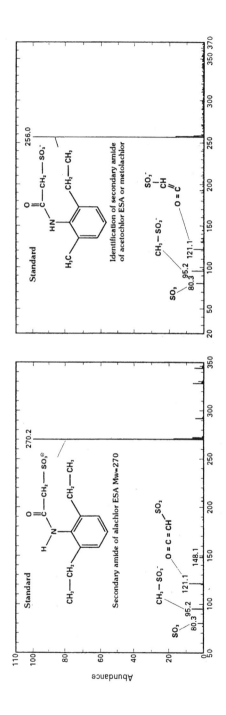

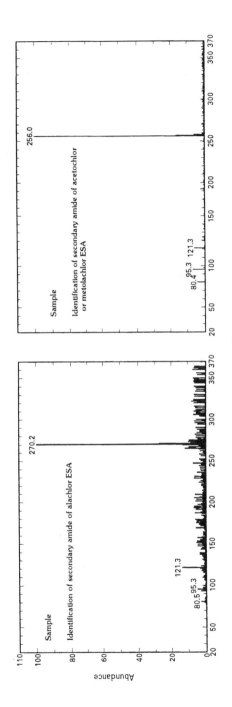

Figure 3. Fragmentation assignments for acetochlor, alachlor, and metolachlor secondary amide ESAs matching with synthesized standards. The spectra are CID spectra at 70V for the fragmentor. The MS/MS spectra only give the 121 m/z ion.

chloroacetanilide herbicides is the most frequently detected in groundwater, followed by its degradate, the secondary amide ESA and lastly by the parent compound, which is detected a least one order of magnitude less frequently (Table IA). These results corroborate and extend previous work done in Iowa on groundwater samples where the primary ESA degradates greatly exceeded the parent compounds (16).

The median concentrations of parent herbicides found for these groundwater samples (Table IB) were identical in that all values were less than 0.05 µg/L. The highest concentrations found were metolachlor (0.86 µg/L) > acetochlor (0.41 µg/L) > alachlor (0.07 µg/L). None of the samples exceeded the United States EPA health standard for metolachlor of 100 µg/L, acetochlor of 2 µg/L, or alachlor of 2 µg/L. The median concentrations for the first ESA degradate of all three herbicides were metolachlor ESA (0.26 µg/L) > alachlor ESA (0.1 µg/L) > acetochlor ESA (<0.05 µg/L). Thus, this first ESA degradate was greater in concentration than the parent compound and followed the same trend as the frequency of detection. This is an interesting result and shows how the first ESA degradate (Figure 1) is rapidly formed and transported to groundwater. The highest concentrations for the first ESA degradate of all three herbicides were metolachlor ESA (40 µg/L) > alachlor ESA (4.9 µg/L) > acetochlor ESA (0.74 µg/L). Finally, the secondary amide ESA degradates are much lower in concentration than the acetochlor ESA, alachlor ESA, and metolachlor ESA with all median concentrations <0.05 µg/L. The highest concentrations of these compounds was also less than the primary ESA degradates. They were 0.66 µg/L for the acetochlor/metolachlor secondary amide ESA and 2.4 µg/L for the secondary amide ESA of alachlor. These data suggest that the secondary amide ESA degradates are probably not the main sink for the first ESA degradates or that they degrade more rapidly and do not accumulate in the environment. Rather the first ESA degradates (Figure 1) with the highest concentrations and frequencies of detection are the major degradates yet discovered for the chloroacetanilide herbicides and this result is a major water-quality finding for states in the Midwestern United States that use large amounts of these herbicides on corn and soybeans.

A last consideration concerning the secondary amides of ESA degradates is their origin from either acetochlor ESA or metolachlor ESA. There is the question of how to distinguish whether the 256 m/z ion was from acetochlor or metolachlor ESA (see Figure 1). Because metolachlor has a much longer half life than acetochlor (30 days versus 10 days) and because the chemical structure differs between these two compounds at the nitrogen atom, it was suspected from this simple model that acetochlor ESA might be the major contributor to the secondary amide ESA. A second line of evidence would be a correlation of the secondary amide ESA with the primary ESA degradate. Unfortunately, there was essentially no correlation with either the acetochlor ESA or the metolachlor ESA in these groundwater samples (Table I); thus, the question of origin of the 256 m/z ion from either acetochlor ESA or metolachlor ESA is unresolved. However, another line of evidence is demonstrated using surface-water samples from New York State where only metolachlor is applied. In this data set (not shown) the correlation coefficient was 0.49 for an N of 44 samples (metolachlor ESA versus the 256 m/z ion), which is significant at the 0.05 level and does strongly suggest that metolachlor is an important source of the 256 m/z ion of the secondary amide ESA. The surface-water samples represent a much more recent snapshot in time than groundwater samples (days versus years in residence time); thus, the result makes intuitive sense that the 256 m/z ion may originate from both acetochlor and metolachlor ESA.

Finally, preliminary toxicological studies of alachlor suggest that the primary ESA degradate is nontoxic to mammals (27-28) and also does not have herbicidal effects. Nothing is known concerning the secondary amides, but it seems reasonable to assume that their action would be similar to their related ESA degradates. Regardless of the health effects of these compounds, the discovery of these new ESA degradates in groundwater samples is important for the fate and transport of these residual pesticides residues in the environment. Measurements of these compounds do give a useful tool for hydrologists to understand shallow groundwater movement and the fate of agricultural chemicals in the environment.

Acknowledgments

The use of brand names is for identification purposes only and does not constitute endorsement by the U.S. Geological Survey. Special thanks to Herb Buxton of the U.S. Geological Survey Toxic Substances Program for financial support of this research. Thanks also to Curt Heine of Micromass for TOF analysis of the secondary-amide ESA standards.

References

1. Barcelo, D. Applications of LC-MS in Environmental Chemistry; Elsevier; Amsterdam, **1996**.
2. Ferrer, I.; Barceló, D. *Analusis* **1998**, *26*, M118.
3. DiCorcia, A.; Nazzari, M.; Rao, R.; Samperi, R.; Sebastiani, E. *J. Chromatogr. A.* **2000**, *878*, 87.
4. Curini, R.; Gentili, A.; Marchese, S.; Marino, A.; Perret, D. *J. Chromatogr. A.* **2000**, *874*, 187.
5. Geerdink, R.B.; Kooistra-Sijpersma, A.; Tiesnitsch, J.; Kienhuis, P.G.M.; Brinkman, U.A.Th. *J. Chromatogr. A.* **1999**, *857*, 147.
6. Steen, R.J.C.A.; Hobenboom, A.C.; Leonards, P.E.G.; Peerboom, R.A.L.; Cofino, W.P.; Brinkman, U.A.Th. *J. Chromatogr. A.* **1999**, *857*, 157.
7. DiCorcia, A.; Crescenzi, C.; Samperi, R.; Scappaticcio, L. *Anal. Chem.* **1997**, *69*, 2819.
8. Thurman, E.M.; Ferrer, Imma; Barcelo, Damia. *Anal. Chem.* **2001**, *73*, 5441-5449.

9. Cai, Z.; Cerny, R.L.; Spalding, R.F. *J. Chromatogr. A.* **1996**, *753*, 243.

10. Lacorte, S.; Barceló, D. *Anal. Chem.* **1996**, *68*, 2464.

11. Ferrer, I.; Hennion, M-C.; Barceló, D. *Anal. Chem.* **1997**, *69*, 4508.

12. Ferrer, I.; Thurman, E.M.; Barcelo, D. *Anal. Chem*. **1997**, *69*, 4547.

13. Thurman, E.M.; Ferrer, I.; Parry, R.*J. Chromatogr.* **2002**, *957*, 3-9.

14. Busch, K.L., Glish, G.L., McLuckey, S.A. (Mass Spectrometry/Mass Spectrometry: Techniques and Applications of Tandem Mass Spectrometry. VCH Publ., New York, 1988; 333 p.

15. March, R.E., Quadrupole Ion Trap Mass Spectrometer, 1998, Encyclopedia of Analytical Chemistry, Meyers, R.A., John Wiley & Sons Ltd., Chichesterr, UK, pp. 1-25.

16. Kolpin, D.W.; Thurman, E.M.; Linhart, S.M. *Arch. Environ. Contam. Tox.* **1998**, *35*, 385.

17. Kolpin, D.W.; Thurman, E.M.; Linhart, S.M. *Sci. Tot. Environ.* **2000**, *248*, 115.

18. Vargo, J.D. *Anal. Chem.* **1998**, *70*, 2699.

19. Aga, D.S.; Thurman, E.M.; Yockel, M.E.; Zimmerman, L.R.; Williams, T.D. *Environ. Sci. Technol.* **1996**, *30*, 592.

20. Thurman, E.M., Ferrer, Imma, and Lee, E. *Anal. Chem.* **2003** Submitted.

21. Lee, E.A., Kish, J.L., Zimmerman, L.R., and Thurman, E.M., *U.S. Geological Survey Open-File Report 01-10*, 2001,17 p.

22. Field, J.A.; Thurman, E.M. *Environ. Sci. Technol.* **1996**, *30*, 1413.

23. Thurman, E.M.; Goolsby, D.A.; Aga, D.S.; Pomes, M.L.; Meyer, M.T. *Environ. Sci. Technol.* **1996**, *30*, 569.

24. Heberle, S.A.; Aga, D.S., Hany, Roland, Muller, S.R. *Anal. Chem.* **2000**, *72*, 840.

25. Ferrer, Imma,; Furlong, E., Thurman, E.M. Abstr. *ASMS Meeting,* Chicago, 2001.

26. Gianessi and Puffer, 2001, Herbicide Use in the United States, Resources for the Future:Quality of the Environment Division, Washington, DC.

27. Heydens, W.F.; Siglin, J.C.; Holson, J.F.; Stegeman, S.D. *Fundam. Appl. Toxicol.* **1996**, *33*, 173.

28. Heydens, W.F.; Wilson, A.G.E.; Kraus, L.J.; Hopkins II, W.E.; Hotz, K.J.; *Tox. Sci* **2000**, *55*, 36.

Emerging Contaminants

Pharmaceuticals

Chapter 9

LC/MS/MS and LC/NMR for the Structure Elucidation of Ciprofloxacin Transformation Products in Pond Water Solution

Laurie A. Cardoza, Todd D. Williams, Bob Drake, and Cynthia K. Larive[*]

Department of Chemistry, University of Kansas, 1251 Wescoe Hall Drive, Lawrence, KS 66045–7582

LC/MS/MS and LC/NMR have been utilized to investigate the transformation of the fluoroquinolone antibiotic, ciprofloxacin, in a natural water sample. Antibiotics, such as fluoroquinolones, are often prescribed in relatively high doses of which only a percentage of the active drug is actually effectively metabolized. The unmetabolized drug has the potential to make its way into the environment where it can have a major impact on the natural microbial community. Although the human metabolites of these pharmaceuticals are well characterized, the identification and structure elucidation of metabolites formed in the environment presents a significant analytical challenge. LC/MS/MS and LC/NMR methods have been developed for the separation and structure elucidation of ciprofloxacin and its transformation products. Two predominant transformation products were identified using NMR to obtain chemical shift, coupling and integration information. Tandem mass spectrometry was exploited to confirm the structure elucidated by NMR through the fragmentation patterns and neutral losses observed. The transformation products identified included (1) the substitution of the fluorine on the fluoroquinolone ring with a hydroxyl group and (2) the loss of ethylene by cleavage of the piperazine ring.

Introduction

The detection of pharmaceutical contaminants in surface waters has generated a great deal of concern, however little is known about the fate and effects of these compounds in the environment (1-6). Structure elucidation of the transformation products formed in the environment is an important step in developing an understanding of the fate of pharmaceutical contaminants. Mass spectrometry (MS) and nuclear magnetic resonance spectroscopy (NMR) can be used in combination to determine the structures of new or unknown compounds, such as transformation products formed from the degradation of a parent contaminant. Because these transformation products will typically be present in a complex environmental matrix, it is necessary to couple the structure analysis with a separation technique. The structure elucidation of organic environmental contaminants has traditionally been carried out using LC/MS/MS (7-11).

LC/NMR is a relatively new tool for the analysis of environmental contaminants. However, the potential of this method for structure elucidation, especially in combination with mass spectrometry, has been aptly demonstrated (12-16). Although LC/NMR is a relatively new tool, the instrumentation has been extensively illustrated in the literature (17-19). Nuclear magnetic resonance is a relatively insensitive detection method compared with mass spectrometry. The detection limits achieved with a conventional LC/NMR flow-probes are typically in the range of hundreds of nanograms of analyte when long analysis times are utilized. However the introduction of cryoprobes and microcoil probes, improvements in flow-cell design and increasingly higher field magnets promise additional improvements in detection limits (19).

One class of pharmaceutical contaminants that has generated significant concern is the antibiotics because of the potential of these drugs to contribute to the development of antibiotic resistance in the environment. The fluoroquinolone antibiotics are of particular concern because they have broad-spectrum activity against both gram-positive and gram-negative bacteria (20-22). Unlike most other antibiotics, which are derived as natural products, the fluoroquinolones are synthetic compounds. Ciprofloxacin (Figure 1a) is one of the most commonly prescribed fluoroquinolones, used primarily for the treatment of urinary and respiratory tract infections. Ciprofloxacin is often prescribed in relatively high doses of which 40-70% of the active drug is excreted unmetabolized (22). In addition, the metabolites of ciprofloxacin have been shown to have antimicrobial activity, which are similar to first and second-generation fluoroquinolone antibiotics (22, 23). It is well recognized that many pharmaceutical compounds are not completely eliminated by wastewater treatment processes and can enter receiving waters through wastewater effluents *(4,24)*. However, the fate of ciprofloxacin and its human metabolites in the waters that receive these effluents is unknown.

Several human and fungal metabolites of ciprofloxacin resulting from reaction at the piperazine ring and the quinolone ring system have been characterized by NMR and MS *(25-28)*. Gau and coworkers isolated and identified four metabolites from human urine and used MS and NMR to determined their structures *(25)*. Volmer *et al.* used LC/MS/MS to establish the fragmentation patterns and common neutral losses for ciprofloxacin and its four major urinary metabolites *(26)*. The metabolism of ciprofloxacin by the fungus

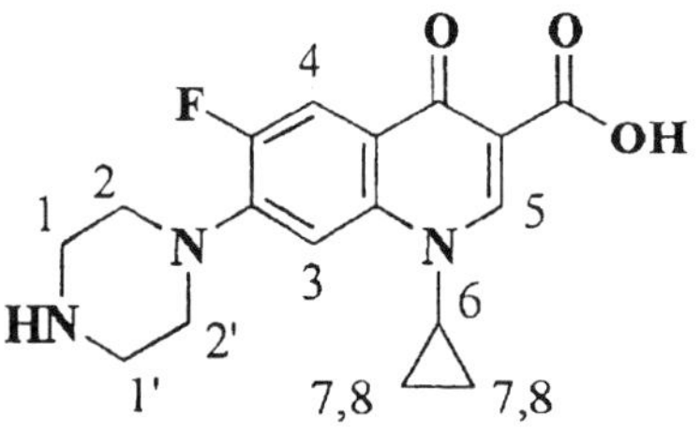

(a)

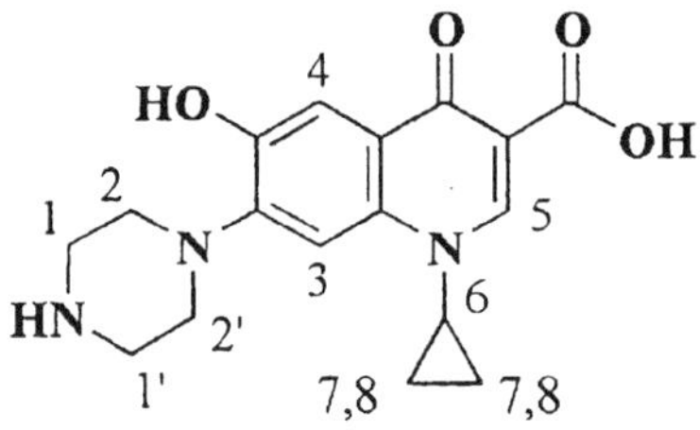

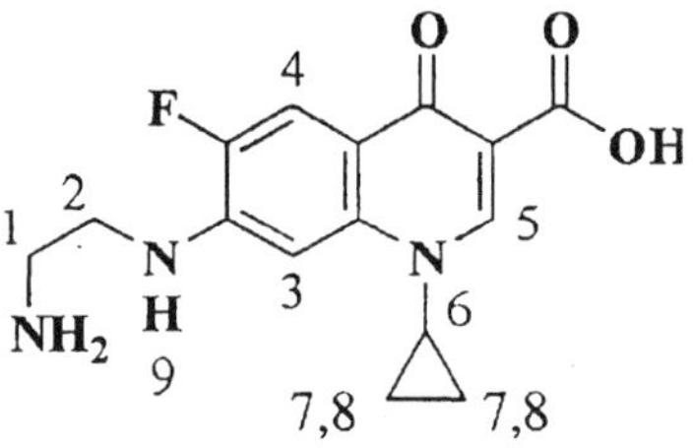

(b)

(c)

Figure 1. Structure of ciprofloxacin (a) and transformation products 1 (b) and 2 (c) with proton NMR assignments

Pestalotiopsis guepini was investigated as a model for mammalian metabolism of fluoroquinolones *(27)*. However, metabolism by the fungus *Gloeophyllum striatum* leads to the formation of a number of unique hydroxyated products not observed in human metabolism studies *(28)*. In addition to the reports of ciprofloxacin metabolism, Burhenne *et al.* studied the photochemical degradation pathways of ciprofloxacin in an aqueous medium and identified the products formed *(29)*.

In this study, LC/MS/MS and LC/NMR methods have been applied for the structure elucidation of ciprofloxacin transformation products. This preliminary study was designed primarily for the development of the analytical methods to be used in a larger and more systematic investigation of ciprofloxacin transformation in aquatic ecosystems. The types of structural information that can be obtained by the application of these two powerful and complimentary analytical techniques will be demonstrated through the structure elucidation of two ciprofloxacin transformation products.

Experimental

Reagents

Methanol (MeOH) and formic acid (H_2CO_2) used for the LC/MS and LC/MS/MS experiments were purchased from Fisher Scientific (Springfield, NJ). The Omnisolv™ acetonitrile (MeCN) used in LC/NMR experiments was purchased from EM Science (Gibbstown, NJ). The trifluoroacetic acid (TFA) was purchased from Sigma (St. Louis, MO). HPLC-grade water with resistance greater than 18 MΩ was obtained using a Labconco Water Pro PS purification system (Kansas City, MO).

Ciprofloxacin Transformation Experiment

A 250 mg ciprofloxacin tablet was dissolved in 50 mL of water obtained from a farm pond near Lawrence, KS. The sample was placed in a clear glass vial and capped. The solution was exposed to natural sunlight at ambient conditions over a 115-day period. The sample solution was decanted then passed through a J. T. Baker (Phillipsburg, NJ) brand SAX solid phase extraction cartridge to remove sample matrix components before being analyzed by LC/NMR and LC/MS/MS.

LC/NMR Stopped-Flow Experiment

The separation was performed on a Varian Pro-Star model 230 liquid chromatography system, equipped with a model 330 photodiode array detector (Varian Instruments, Walnut Creek, CA). The system was controlled with a Varian Star LC workstation. A 150 μL injection of the filtered pond water sample was introduced into a 100 μL sample loop and separated on a 4.6 mm x 15 cm Supelco Discovery HS C-18 column with 3 μm particles (Supelco, Bellefonte, PA). The solvent system utilized included (A) 10 mM TFA at pH 3 and (B) MeCN. The gradient elution method employed started with a composition of 85:15 and changed to 59:41 A:B over the course of 35 minutes at a rate of 0.5 mL/min. The separations conducted at room temperature were monitored at 285 nm with a photodiode array detector (PDA). Eluting peaks were transferred to the NMR controlled by Varian LC/NMR 2000 software.

Stopped-flow spectra were obtained on a Varian Inova 600 MHz spectrometer (Varian NMR Systems, Palo Alto, CA.). The sample was stopped in a triple resonance flow probe with a 120 μL flow cell (60 μL detection volume) and one-dimensional ^1H NMR spectra were acquired. The free induction decays (FIDs) were sampled at 32,768 data points over a spectral width of 9000.9 Hz. An acquisition time of 1 sec with no additional relaxation delay was used to collect spectra obtained in protonated solvents. The number of scans acquired ranged between 1280 and 12800 transients, depending on the concentration of the analyte of interest. Water Eliminated through Transverse gradients (WET) solvent suppression with ^{13}C decoupling was applied to suppress the MeCN and H_2O (or HOD) signals (30). The FIDs were apodized by multiplication by an exponential decay equivalent to 1 Hz line-broadening. In a later experiment, the chemical shifts were referenced to maleic acid (6.29 ppm), which was incorporated into the buffer system. The NMR spectra presented were measured using protonated solvents; however, the integrals of the NMR peaks were determined from totally relaxed spectra acquired for analytes separated using D_2O with TFA and protonated acetonitrile.

Preliminary LC/MS Experiments

Initial LC/MS experiments were conducted on pond water sample using a Quattro Ultima (Micromass Ltd. Manchester, UK) "triple" quadrupole instrument with an electrospray ionization (ESI) source. The MS was coupled to a Waters 2690 HPLC equipped with a 2487 dual wavelength detector (Waters Corp., Milford, MA). Samples were analyzed in positive ion mode using a sampling cone voltage of 35 V. Full scan acquisitions conducted by scanning

Q_1 at a rate of 10 s over the mass range 150 - 1000 amu with the analyzer resolution adjusted to 0.6 amu FWHH. The initial LC/MS experiments were conducted using the same separation conditions as the LC/NMR experiments.

LC/MS/MS Experiments

A quadrupole time-of-flight tandem mass spectrometer (Q-TOF) was employed for the identification of the transformation products by exact mass and fragmentation behavior. The Q-Tof-2™ (Micromass Ltd, Manchester UK) with an ESI source was coupled to an Ultra-Plus II microbore LC system (Micro-Tech Scientific, Sunnyvale CA). The separations were performed on a Zorbax C-18 column (Micro-Tech Scientific) with dimensions of 1 mm ID x 5 cm with 3.5 µm particles. The original sample was diluted 10-fold and 5 µL was used for the analysis. The LC solvents were: (A) 99%H_2O, 1%MeOH and 0.08% HCO_2H and (B) 99% MeOH, 1%H_2O and 0.06% HCO_2H. A gradient at 50 µL/min was developed from 5% to 41%B in 12 minutes followed by a wash 95%B. The Q-Tof-2™ was operated in highest resolution mode (10,000 RP FWHH, Ar in the collision cell with 5eV offset) and spectra were acquired at 22,727 Hz accumulating for 5.0 s/cycle. Product ion spectra were collected in positive ion mode with a cone voltage of 30 V and collision energy of 30 V. A daily time-to-mass calibration was made with CsI clusters and corrected with the known mass of the $[M+H]^+$ ion of ciprofloxacin ($C_{17}H_{19}FN_3O_3^+$) for accurate mass determinations. All subsequent measured ion masses were 15 ppm or better when compared to assigned formula.

Results and Discussion

Separation and Ciprofloxacin Analysis

The separation method was designed for the separation of ciprofloxacin from more polar transformation products (TP) and the humic and fulvic material present in natural waters. The separation of the ciprofloxacin in the pond sample produced well-resolved chromatographic peaks for both analytes and the parent compound ciprofloxacin (Figure 2). The peaks 1 and 2, which eluted at 10.5 min. and 13.5 min. were labeled TP-1 and TP-2, respectively. A significant portion of the initial ciprofloxacin precursor remained in the solution after the 115-day experiment and was observed as a large broad peak (3), which emerged at 15.5 minutes.

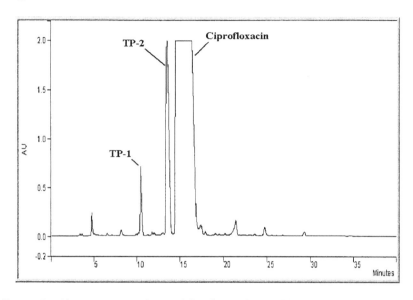

Figure 2. Chromatogram obtained for the analysis of the ciprofloxacin pond sample by LC/NMR

The analysis of ciprofloxacin was conducted on the compound separated from the pond sample rather than a solution of the pure commercial drug, since the concentration in the water sample was sufficiently high. A ¹H NMR spectrum was acquired for ciprofloxacin (Figure 3). Although the WET solvent suppression is quite effective, there are residual features due to the protonated acetonitrile (2.1 ppm) and water (4.7 ppm) used in the chromatographic separation. These solvent suppression features are present in all the NMR spectra presented.

The ¹H NMR spectrum of ciprofloxacin (Figure 3) can be examined in terms of two regions, aliphatic (1-4 ppm) and aromatic (7-9 ppm). The proton assignments (Figure 1a) for the ¹H NMR spectrum of ciprofloxacin were made based on the chemical shifts, coupling patterns and integrals, and are consistent with the literature *(29)*. The quinolone ring system contains three isolated aromatic protons 3, 4 and 5 (Figure 1a). The aliphatic region is made up of methylene protons (1, 1′, 2 and 2′) of the piperazine ring and the methyne (6) and methylene (7 and 8) protons of the cyclopropyl ring (Figure 1a). One of the

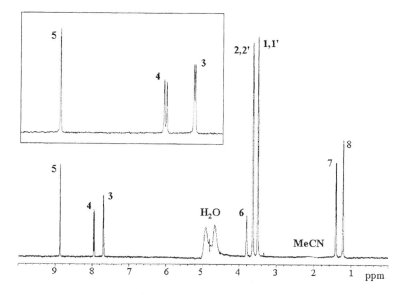

Figure 3. ¹H-NMR spectrum of ciprofloxacin

most distinct coupling patterns is observed between the fluorine atom on the quinolone ring and aromatic protons 3 and 4. The coupling constants for the doublets 3 and 4 are 11.54 Hz and 7.14 Hz, respectively, consistent with protons, which are meta and ortho to a fluorine.

Collision induced dissociation (CID) spectra were obtained using the Q-Tof-2™ instrument for the identification of ciprofloxacin and its transformation products. The CID spectrum obtained for ciprofloxacin was consistent with data reported in the literature (27). The information acquired from the CID experiment with ciprofloxacin served as a starting point for the identification of new transformation products. LC/NMR and LC/MS/MS clearly demonstrate unreacted ciprofloxacin is the major component of the water sample. To identify the transformation products, the mass assignments of the ions collected during LC/MS/MS were corrected using the exact mass of the [M+H]⁺ ion of ciprofloxacin (332.141 m/z). The CID spectrum was used to predict neutral loss and product ions for the fragmentation of ciprofloxacin transformation products (Figure 4). The product ions 314.130, 294.124, 245.109 and 231.05 m/z were consistent with those reported by Volmer *et al.* (26). Fragmentation is thought to occur primarily on the moieties attached to the quinolone ring including the fluorine, piperazine ring and the cyclopropyl group. Neutral losses of 18, 38, 89, 101 and 129 amu are observed as five major product ions. The loss of the mass of water (18 amu) is not a significantly selective neutral loss since there are many molecules that lose water upon fragmentation.

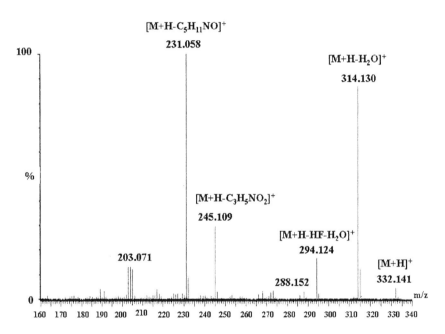

Figure 4. MS/MS spectrum of ciprofloxacin

However, the loss of 38 amu, which is a neutral loss of [H$_2$O + HF] is a more significant transformation product marker since there are very few natural organic compounds that contain fluorine. Interestingly, loss of HF alone was not observed in any of the mass spectra recorded. Additionally, the neutral losses of 89, 101 and 129 amu may be representative of the neutral losses observed in the transformation products produced by ciprofloxacin.

Structure Elucidation of Ciprofloxacin Transformation Products

The two most prominent transformation products TP-1 and TP-2 (Figure 1 b and c) were analyzed in the same fashion as the parent compound, ciprofloxacin. A ^{1}H NMR spectrum of TP-1 was obtained using 12800 transients (Figure 5). Comparing the spectrum of TP-1 (Figure 5) to that of ciprofloxacin (Figure 3) reveals a number of similarities and differences in the two spectra. In the aliphatic region, there are no significant changes in the chemical shifts of the peaks or the number of protons associated with each peak, determined by integration. This conservation of spectral information in the aliphatic region signifies that there are no structural changes associated with the piperazine or cyclopropyl rings upon the formation of TP-1. However, there are significant spectral changes in the aromatic region of TP-1 compared with the spectrum of ciprofloxacin. In the aromatic region, proton 5 showed no significant change, however, peaks 3 and 4 are both present as singlets, each of which integrate to one proton. The loss of coupling observed for protons 3 and 4 in the spectrum of TP-1, is attributed to the loss of the fluorine atom. In addition to the loss in coupling, the meta proton 3 and the ortho proton 4 are shifted up field by 0.01 and 0.3 ppm, respectively. The shifts observed for protons 3 and 4 are consistent with a substitution at the fluorine position, most likely by a hydroxyl group. The preliminary structure constructed (Figure 1b) is consistent with the information gleaned from the NMR as well as the aqueous conditions under which the compound reacted. To further ascertain the structure of TP-1 LC/MS/MS experiments were conducted to the find the exact mass and investigate the product ions generated. The Q-TOF spectrum (Figure 6) of TP-1 yielded an exact mass of 330.148 m/z for the [M+H]$^+$ ion, which is consistent with the substitution of the fluorine atom of ciprofloxacin with a hydroxyl group. In addition, the fragmentation pattern of TP-1 is consistent with that observed for ciprofloxacin. There are product ions formed due to the loss of water (312. 133 m/z) and further fragmentation of the molecule TP-1 including 229.063 m/z (loss of 101), which is consistent with the 231 m/z product ion observed for ciprofloxacin. Additionally, there was no neutral loss of 38 amu corresponding to the loss of [H$_2$O+HF], consistent with the proposed structure. The MS/MS results confirm the structure of TP-1 (Figure 1b) suggested by the NMR results.

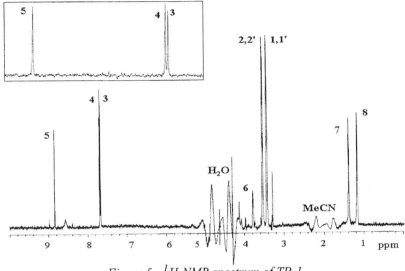

Figure 5. ^{1}H-NMR spectrum of TP-1

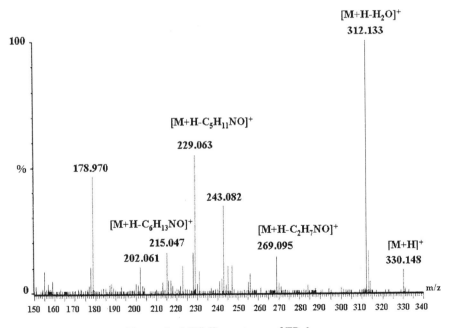

Figure 6. MS/MS spectrum of TP-1

The product TP-2 (Figure 1c) was present in a higher concentration than TP-1 so only 6400 transients were needed to acquire a satisfactory ^{1}H NMR spectrum (Figure 7). Comparing the NMR spectrum for TP-2 with the spectrum of ciprofloxacin (Figure 3) reveals no significant change in the cyclopropyl ring since the methyne proton (6) and the two-methylene protons (7 and 8) have similar chemical shifts and integrals. There are however, significant differences in the regions of the spectrum incorporating the piperazine and aromatic protons. The region encompassing aliphatic protons 6 and 2 integrates to three protons in the TP-2 spectrum versus the five protons associated with the same peaks in ciprofloxacin. This result suggests the loss of two of the piperazine protons close to the secondary amine on the piperazine ring. In the aromatic region there is no significant change in proton 5, however, protons 3 and 4 have shifted up field by 0.4 and 0.1 ppm, respectively. An additional broad peak observed at 7.8 ppm in the aromatic region in protonated water, was thought to be a proton on nitrogen bonded to an aromatic ring. The NMR spectral information suggests that the piperazine ring was cleaved with net loss of a C_2H_2 group upon formation of TP-2 (Figure 1c). The proposed structure suggested from the LC/NMR results was confirmed by the MS/MS analysis, as described below.

The Q-TOF spectrum of TP-2 (Figure 8) yielded an exact mass of 306.124 m/z, which is consistent with the structure suggested by the NMR data (Figure 1c). The fragmentation pattern of TP-2 was significantly different than that obtained from ciprofloxacin and TP-1 using the same collision conditions. The only similarities in the spectral data coincide with the neutral losses of 18 and 38, which are consistent with the losses of H_2O and $[H_2O + HF]$ that were

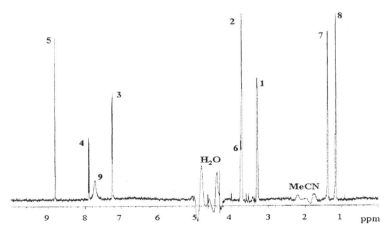

Figure 7. ^{1}*H-NMR spectrum of TP-2*

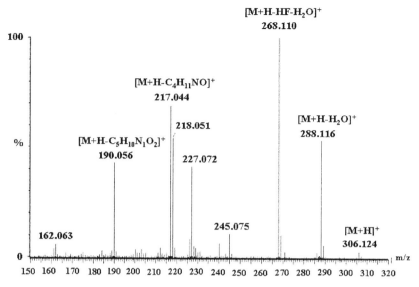

Figure 8. MS/MS spectrum of TP-2

also observed in ciprofloxacin. The product ions generated due to the fragmentation of the piperazine and cyclopropyl rings did not yield the same intense ions observed for ciprofloxacin. This may be due the presence of the primary amine in the structure of TP-2 (Figure 1c). The preliminary structure of TP-2 determined from the NMR chemical shifts and peak integrals was confirmed by the exact mass as well as similarities in neutral losses in the formation of product ions compared to the parent compound.

The two transformation products identified in this study have been observed in other degradation and metabolism studies reported in the literature. The compound TP-1 was initially identified as one of the many hydroxylated metabolites of ciprofloxacin produced by the brown rot fungus *Gloeophyllum striatum* (28). Gau and coworkers initially identified TP-2 as one of the four metabolites found in human urine (25). The compound TP-2 has also been identified as a photochemical degradation product of ciprofloxacin (29).

Conclusion

The complementary information acquired from the use of LC/MS/MS and LC/NMR has been used in a preliminary study to elucidate the structures of two major transformation products formed by the ciprofloxacin in a natural water solution. The major transformation products formed were generated by cleavage of the piperazine ring or the substitution of the fluorine atom on the quinolone group with a hydroxyl group. The information used for the structure elucidation of the transformation products included the use of chemical shift, coupling constants, integration information as well as exact mass and neutral loss correlations. The incorporation of both NMR and MS/MS for the structure elucidation of transformation products generates complimentary structural information for the unequivocal identification of new products. However, it should be noted that a more efficient means of structure elucidation would be to incorporate a coupled system such as LC/NMR/MS/MS *(30)*.

Acknowledgements

This research was supported by EPA grant R82900801-0. NSF grant DBI-0088931 and Kansas administered NSF EPSCoR partially funded the purchase of the 600 MHz NMR spectrometer used in this research. The purchase of the Q-Tof-2 was supported by Kansas administered NSF EPSCoR and the University of Kansas.

References

1. Hartmann, A.; Alder, A. C.; Koller, T.; Widmer, R. M. *Environ. Toxicol. Chem.* **1998**, *17*, 377-382.
2. Daughton, C.G.; Ternes, T. A. *Environ. Health Perspect.* **1999**, *107*, 907-938.
3. Golet, E. M.; Alder, A. C.; Hartmann, A.; Ternes, T. A.; Giger, W. *Anal. Chem.* **2001**, *73*, 3632-3638.
4. Kolpin, D. W.; Furlong, E. T.; Meyer, M. T.; Thurman, E. M.; Zaugg, S. D.; Barber, L. B.; Buxton, H. T. *Environ. Sci. Technol.* **2002**, *36*, 1202-1211.
5. Jones, K. C.; de Voogt, P. *Environ. Pollut.* **1999**, *100*, 209-221.
6. Al-Ahmad, A.; Daschner, F. D.; Kummerer, K. *Arch. Environ. Contam. Toxicol.* **1999**, *37*, 158-163.
7. Bobeldijk, I.; Vissers, J. P. C.; Kearney, G.; Major, H.; van Leerdam, J. A. *J. Chromatogr., A* **2001**, *929*, 63-74.
8. Hernandez, F.; Sancho, J. V.; Pozo, O.; Lara, A.; Pitarch, E. J. *J. Chromatogr., A* **2001**, *939*, 1-11.
9. Ternes, T. A. *Trends Anal. Chem.* **2001**, *20*, 419- 434.

160

10. Lagana, A.; Fago, G.; Marino, A.; Santarelli, D. *Anal. Lett.* **2001**, *34*, 913-926.
11. Behnke, B.; Albert, K.; Schlotterbeck, G.; Tallarek, U.; Strohschein, S.; Tseng, L.-H.; Keller, T.; Albert, K.; Bayer, E. *Anal. Chem.* **1996**, *68*, 1110-1115.
12. Preiss, A.; Sanger, U.; Karfich, N.; Levsen, K.; Mugge, C. *Anal. Chem.* **2000**, *72*, 992-998.
13. Godejohann, M.; Preiss, A.; Mugge, C. *Anal. Chem.* **1997**, *69*, 3832-3837.
14. Levsen, K.; Preiss, A.; Godejohann, M. *Trends Anal. Chem.* **2000**, *19*, 27-48.
15. Schrader, W.; Geiger, J.; Godejohann, M.; Warscheid, B.; Hoffmann, T. *Angew. Chem. Int. Ed.* **2001**, *40*, 3998-4001.
16. Benfenatia, E.; Pierucci, P.; Fanelli, R.; Preiss, A.; Godejohann, M.; Astratov, M.; Levsen, K.; Barcelo, D. *J. Chromatogr., A* **1999**, *831*, 243-256.
17. Ternes, T. A.; Hirsch, R.; Mueller, J.; Haberer, K. *Fresenius' J. Anal. Chem.* **1998**, *362*, 329-340.
18. Albert, K.; Dachtler, M.; Glaser, T.; Händel, H.; Lacker, T.; Schlotterbeck, G.; Strohschein, S.; Tseng, L.-H.; Braumann, U. *J. High Resol. Chromatogr.* **1999**, *22*, 135-143.
19. Lacey, M. E.; Subramanian, R.; Olson, D. L.; Webb, A. G.; Sweedler, J. V. *Chem. Rev.* **1999**, *99*, 3133-3152.
20. Ho, A.; Cupo, J.; Gill, C. L.; Gill, M. A. *J. Contin. Educ.* **2000**, 12
21. Bearden, D. T.; Sonthisombat, P. *Clin. Trends* **1999**, *13*, 57-64.
22. Bayer Corp. USA Pharm. Division Cipro tablet/suspension Prescribing Information. http://www.univgraph.com/bayer/INSERTS/CIPROTAB.pdf and http://www.univgraph.com/bayer/INSERTS/CIPROIV.pdf (accessed March 2002), Tablet and IV insert information.
23. Zeiler, H.-J.; Petersen, U.; Gau, W.; Ploschke, H. J. *Arzneim.-Forsch. Drug Res.* **1987**, *37*, 131-134.
24. Golet, E. M.; Alder, A. C.; Giger, W. *Environ. Sci. Technol.* **2002**, *36*, 3645-3651.
25. Gau, W.; Kurz, J.; Petersen, U.; Ploschke, H. J.; Wuensche, C. *Arzneim.-Forsch. Drug Res.* **1986**, *36*, 1545-1549.
26. Volmer, D. A.; Mansoori, B.; Locke, S. J. *Anal. Chem.* **1997**, *69*, 4143-4155
27. Parshikov, I. A.; Heinze, T. M.; Moody, J. D.; Freeman, J. P.; Williams, A. J.; Sutherland, J. B. *Appl. Microbiol. Biotechnol.* **2001**, *56*, 474-477.
28. Wetzstein, H-G.; Stadler, M.; Tichy, H-V.; Dalhoff, A.; Karl, W. *Appl. Environ. Microbiol.* **1999**, *65*, 1556-1563.
29. Burhenne, J.; Ludwig, M.; Nicoloudis, P.; Spiteller, M. *Environ. Sci. Pollut. Res. Int.* **1997**, *4*, 10-15.
30. Smallcombe; S. H.; Patt, S. L.; Keifer P. A. *J. Magn. Reson. A* **1995**, *117*, 295-303.

Chapter 10

Tetracycline and Macrolide Antibiotics

Trace Analysis in Water and Wastewater Using Solid Phase Extraction and Liquid Chromatography–Tandem Mass Spectrometry

D. D. Snow, D. A. Cassada, S. J. Monson, J. Zhu, and R. F. Spalding

Water Sciences Laboratory, University of Nebraska, 103 Natural Resources Hall, Lincoln, NE 68583–0844

Liquid chromatography-tandem mass spectrometry (LC/MS/MS) provides a sensitive, reproducible, and specific instrumental method for quantifying agricultural antibiotics in a variety of matrices. Sensitivity can be enhanced and interferences minimized by concentrating these compounds using solid phase extraction. Comparison of polymeric and reverse phase extraction cartridges suggests that either phase may provide quantitative results. Buffering and/or pH adjustment improves the recovery of these highly polar and sometimes problematic compounds. Tetracycline method detection limits determined in water are between 0.2 and 0.3 µg/L, and between 1.0 and 2.0 µg/L in buffered wastewater. Macrolides, lincomycin, and tiamulin method detection limits determined in buffered water are between 0.4 and 1.4 µg/L. Macrolides and tiamulin detection limits are between 1.0 and 3.0 µg/L in wastewater.

Introduction

Antibiotics are widely used in agriculture as growth enhancers and for disease treatment and control in animal feeding operations. Concerns for increased antibiotic resistance of microorganisms have prompted research into the environmental occurrence of these compounds. Tetracycline and macrolide antibiotics account for over 16% of all antibiotics used for animals in the United States *(1)*. Previous analytical methods developed for these compounds have focused on the determination of these compounds in food and animal tissues *(2-6)*. Liquid chromatography-tandem mass spectrometry (LC/MS/MS) provides a sensitive, reproducible, and specific instrumental method for determining antibiotics in a wide variety of matrices.

One difficulty in the analysis of tetracycline antibiotics involves interaction of these compounds with residual silanol groups and metal ions on the LC column, often resulting in severe peak tailing and variable analyte recoveries. Depending on the source configuration, compensation for this problem by the addition of nonvolatile modifiers to the mobile phase may complicate analysis using electrospray ionization (ESI). Thus, if ESI tandem mass spectrometry detection is used, the best approach may be to use a high purity deactivated reverse phase column in combination with a mobile phase containing a volatile buffer.

Solid phase extraction (SPE) is frequently used for concentrating and purifying antibiotics extracted from animal tissues and food samples, and is highly suited for extracting aqueous samples. Cheng et al. *(7)* described a method for extraction of tetracyclines from porcine serum using a newly developed macroporous copolymer (divinylbenzene-*N*-vinylpyrrolidone) sorbent. Recently published methods for concentrating and analyzing tetracycline antibiotics in aqueous samples by LC/MS *(8-10)* have used either polymeric or reversed phase C-18 solid phase extraction. The polymeric cartridges may also be suitable for extraction of other antibiotics, such as the macrolides, from water and wastewater samples. Comparison of the polymeric and reverse phase extraction cartridges *(8)* suggests that either phase may provide quantitative results, though the use of an internal standard is critical for quantitation by LC/MS. Given the very low concentrations of antibiotics detected in surface waters reported by Lindsey *et al (9)* and by Hirsch *et al (10, 13)*, there is a pressing need for sensitive and reliable levels of these compounds in environmental matrices.

The objective of this work is to develop robust multi-residue methods for trace level determination of antibiotics commonly used in concentrated animal feeding operations. The first method focuses on the tetracycline compounds and accounts for a major group of antibiotics used in animals. The second method includes macrolide antibiotics, such as tylosin and erythromycin, a related product lincomycin, and the diterpene tiamulin. These methods are used in studies focused on the occurrence of these compounds in water and wastewater to help understand their impact on the environment.

Experimental

The structures of selected tetracyclines and macrolides, as well as lincomycin, and tiamulin, are shown in Figure 1. Demeclocycline, doxycycline, roxithromycin, and josamycin are not approved for use in swine, cattle, dairy cows, or poultry in the U.S and are thus suitable for use as internal standards and surrogates at confined feeding operations. Standards for most of these compounds were obtained from Acros Organics (Fisher Scientific, St. Louis, MO), Sigma-Aldrich (St. Louis, MO), Fluka Chemical (Milwaukee, WI), and ICN Biomedicals, Inc. (Aurora, OH). Tilmicosin was generously provided by Lilly Research Laboratories (Indianapolis, IN). Solvents and other reagents were obtained from Fisher Scientific (St. Louis, MO), J.T. Baker (Phillipsburg, N.J.) and Mallinkrodt Chemical Works (St. Louis, MO). Stock solutions of standards (5 µg/µL), spiking solutions (10-50 ng/µL), and calibration solutions (0.05-5 ng/µL) were prepared in methanol. One gram trifunctional reverse phase (tC18) and 200 mg polymeric HLB Oasis solid phase extraction cartridges used for concentrating antibiotics were obtained from Waters Corporation (Milford, MA).

Instrumental Conditions

Analyses were performed on a high performance liquid chromatograph (HPLC) interfaced with a quadrupole ion trap mass spectrometer using selected reaction monitoring (SRM). The HPLC system consisted of a Waters model 717+ autosampler, model 616 pump and gradient module with 600S controller. (Waters Corporation, Milford, MA). The Finnigan LCQ quadrupole ion trap mass spectrometer (Thermoquest, San Jose, CA) was operated in electrospray ionization (ESI) positive ion mode. Tetracyclines and macrolides were analyzed separately using a narrow bore (250 x 2 mm) BetaBasic C18 reverse-phase column (Keystone Scientific, Bellfonte, PA). This packing utilizes a high purity silica bonded with a high density end-capped bonded phase, and provided the most symmetrical peak shapes for both groups of compounds.

Isocratic separation provided the most reproducible analytical results and also eliminated the need for re-equilibration after each run. A mobile phase consisting of acetonitrile:methanol: 2% formic acid in water (25:12:63)

Tetracyclines	R_1	R_2	R_3	R_4
Chlortetracycline	Cl	OH	CH_3	H
Oxytetracycline	H	OH	CH_3	OH
Tetracycline	H	OH	CH_3	H
Minocycline	$N(CH_3)_2$	H	H	H
Doxycycline	H	H	CH_3	OH
Demeclocycline	Cl	OH	H	H

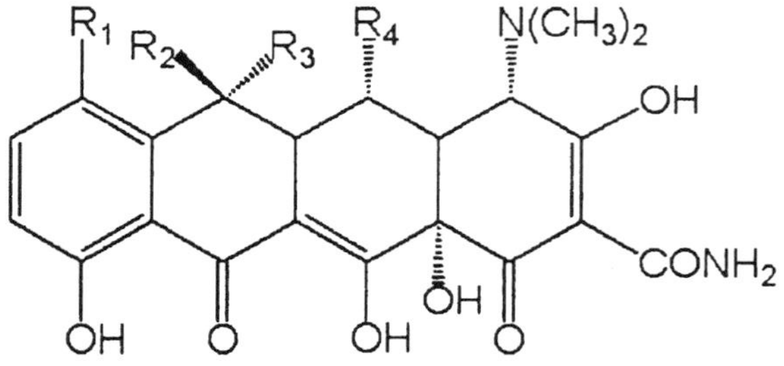

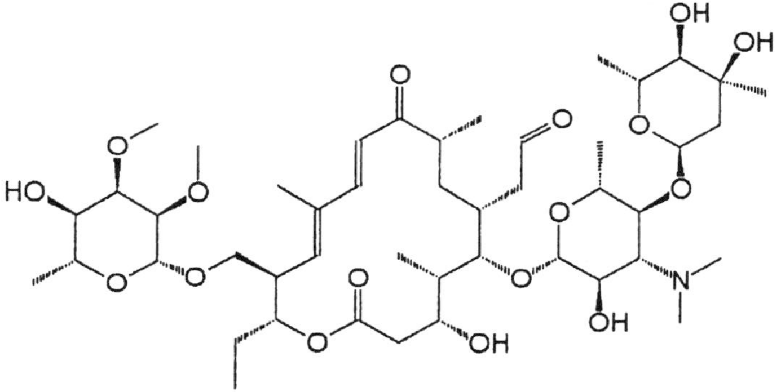

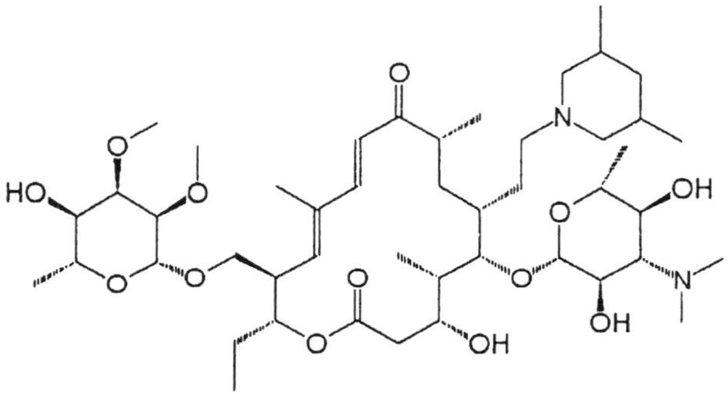

Tilmicosin

Tylosin

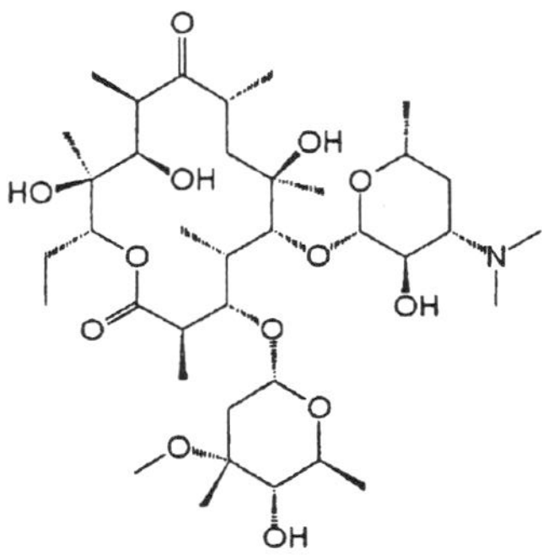

Josamycin

Roxithromycin

Oleandomycin

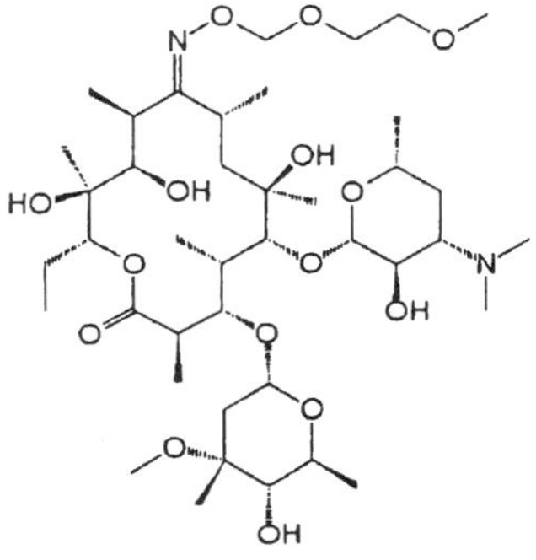

Erythromycin A

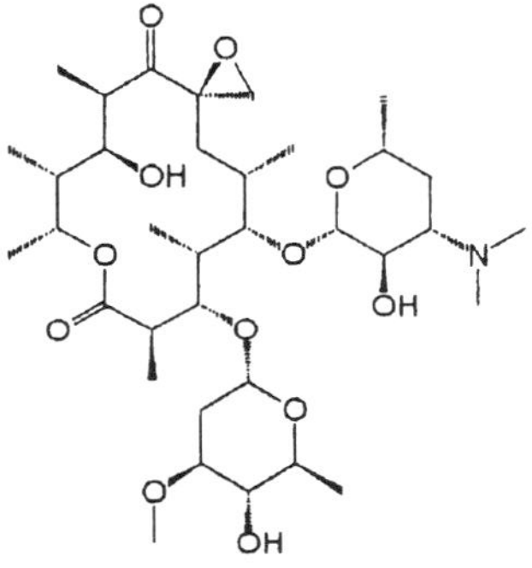

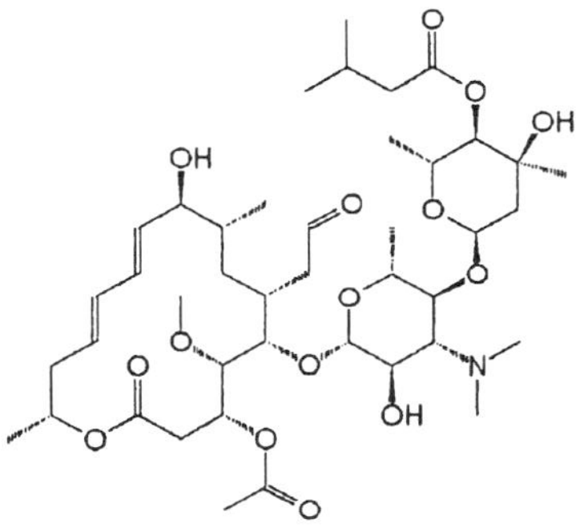

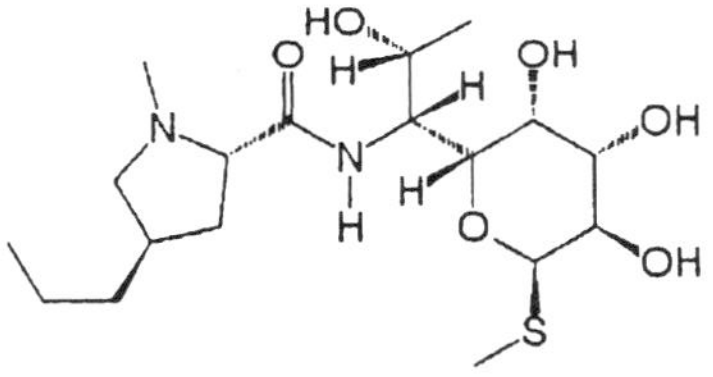

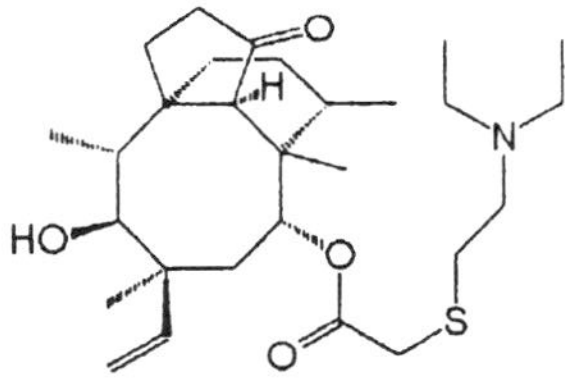

Tiamulin

Lincomycin

Figure 1. Chemical structures of antibiotics investigated

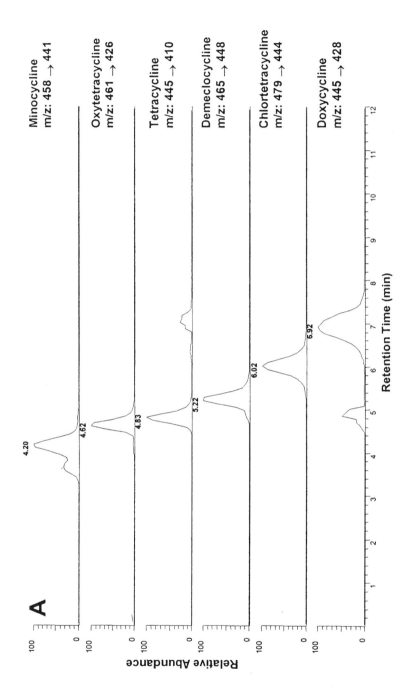

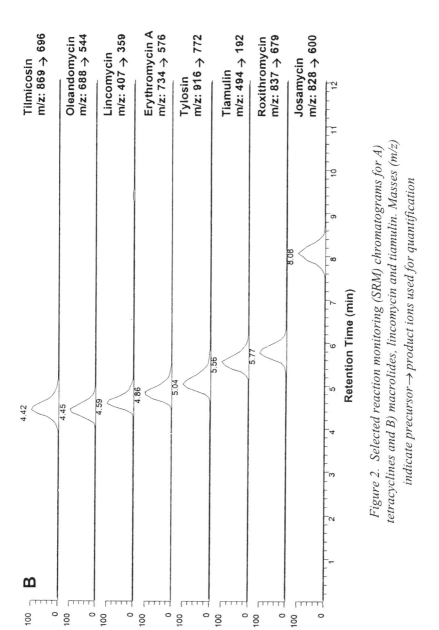

Figure 2. Selected reaction monitoring (SRM) chromatograms for A) tetracyclines and B) macrolides, lincomycin and tiamulin. Masses (m/z) indicate precursor→product ions used for quantification

produced good peak shapes and sufficient separation for tetracyclines at a flow rate of 0.2 mL/min and temperature of 50°C. Selected reaction monitoring (SRM) chromatograms using LC-ion trap tandem MS for the tetracycline compounds are shown in Figure 2A. The macrolides, lincomycin and tiamulin were separated using an isocratic mobile phase of methanol:60 mM ammonium formate in water (75:25) at a flow rate of 0.2 mL/min and a column temperature of 50°C. Liquid nitrogen was used as a source of nitrogen gas at a maintained pressure of 550 kPa (80 p.s.i.).

Instrumental conditions of the LCQ are listed in Table I. Ion source and trap conditions were adjusted to obtain the most intense and stable product ions via selected reaction monitoring in the trap. Precursor and product ion masses for both the tetracycline and macrolide methods are given in Figure 2A and 2B. Fragmentation was produced via collision-induced dissociation in the ion trap with collision energies ranging from 20-40% of the 5 volt maximum end cap potential. Fragmentation of tetracyclines corresponded to a loss of ammonia, water or both from the protonated molecule $[M+H]^+$. Fragmentation of the macrolides was consistent with previous investigations and generally corresponded to a loss of desosamine, cladinose, or mycarosal functional groups from the lactone ring *(10)*. Fragmentation of lincomycin produced a product ion at m/z=359 and probably corresponds to loss of the methyl sulfide functional group (Figure 1). Tiamulin produced a product ion at m/z=192 which is postulated to result from cleavage of the ester linkage from the diterpene moiety (Figure 1).

Table I. Source parameters used for LCQ Ion Trap Mass Spectrometer.

Parameter	Tetracycline Method	Macrolide Method
Ion Mode	Positive	Positive
Sheath Gas Flow Rate	75	70
Auxiliary Gas Flow Rate	15	15
Spray Voltage (kV)	4.00	4.00
Spray Current (μA)	18-38	4-20
Capillary Temperature (°o)	230	150
Capillary Voltage (V)	-9.00	-12.00
Tube Lens Offset (V)	0.00	0.00
Octapole 1 Offset (V)	-3.40	-3.00
Lens Voltage (V)	-30.00	-42.00
Octapole 2 Offset (V)	-8.50	-6.50
Octapole RF Amplitude (Vp/p)	720.00	720.00

Fragile Ions

Recently, McClellan et al. *(11)* reported that macrolide antibiotics exhibit a chemical mass shift in the ion trap because they fragment during application of the resonance ejection for mass analysis. Fragmentation of these "fragile ions" during isolation can cause peak broadening in the ion trap mass spectra and reduce the intensity of the product ions during selected reaction monitoring. Both precursor and product ion intensity for fragile ions can be affected by trapping time or width of the isolation waveform. A 16 millisecond trapping time, equal to the product of four 4-millisecond cycles of the isolation waveform, cannot be changed by the user. However the isolation width, determined by the width of the notched broadband waveform (m/z), can be set by the user and should be adjusted to maximize the sensitivity of the instrument. Figure 3 shows the effect of varying the isolation width on product ion intensities during selected reaction monitoring of the macrolides, lincomycin, and tiamulin.

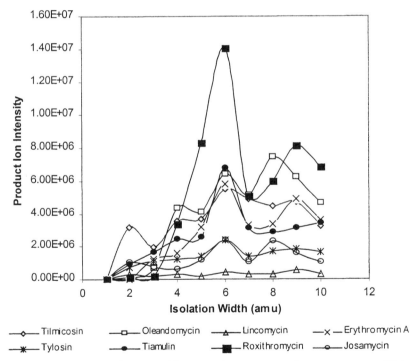

Figure 3. Effect of isolation width on product ion intensity at constant collision energy for the macrolide antibiotics, and lincomycin and tiamulin

The effect is most dramatic for the internal standard roxithromycin. The standard isolation width of 3 amu produced a relatively low product ion response suggesting a very low sensitivity in MS/MS detection of this compound. Increasing the isolation width to 6 amu increased the product ion intensity by more than 60X. For the other macrolide antibiotics, increasing the isolation width from 3 to 6 amu increased the product ion intensity by a factor of 2 to 10X, but had little effect on the product ion intensity for lincomycin (Figure 3). Increasing the isolation width also effectively decreases the kinetic energy applied to the ion in the trap, and helps to explain why product ion intensity decreases at isolation widths greater than 6 amu (Figure 3). Too large of an isolation width can also increase the potential for positive interferences by trapping compounds with masses similar to the analyte. The step-wise pattern of product ion intensities is thought to be the result of the way the isolation waveform is calculated and applied *(11)*.

Recovery and Sensitivity

Tetracyclines

Details of the cartridge conditioning, extraction, and elution of tetracyclines are reported elsewhere *(8)*. Internal standards and surrogates were added to samples after the addition of acids or buffers and before extraction to compensate for variations in analyte recovery. A series of buffering and elution experiments were performed before the optimal conditions for recovery were determined. Method detection limits were determined first in reagent water and then in waste water samples to estimate sensitivity in real matrices *(12)* Waste water lagoons from 11 sites across Nebraska were sampled and analyzed as part of a study to determine the impacts of these operations on ground water quality *(8)*. Several samples with no detectable levels of antibiotics were used to determine matrix effects. Chlortetracycline levels in waste water sampled for this study ranged from below detection to nearly 12,000 µg/L, while tetracycline and oxytetracycline concentrations ranged from below detection to 1300 µg/L. These concentration ranges are comparable to those reported by Meyer et al. *(13)* for hog waste lagoon samples screened using radioimmunoassay and confirmed by LC/MS.

Tetracyclines could be extracted from fortified reagent (deionized-organic free) water samples without modification using the tC18 cartridges. Acidification to a pH near 2.5 with concentrated phosphoric acid immediately prior to extraction improved recovery with the HLB cartridges. Combining animal waste water samples with a buffer containing 0.05M disodium phosphate (Na$_2$HPO$_4$), citric acid (C$_6$H$_8$O$_7$) and disodium ethylenediamine tetraacetate (Na$_2$EDTA) prior to extraction improved recovery with the tC18 cartridges. A 0.05M phosphate and citric acid buffer (pH~2.5) added to waste water samples prior to extraction with the HLB cartridges minimized complex formation and improved recovery. Acidified samples for tetracycline determination were extracted immediately after addition of acid to minimize the potential for degradation *(3)*.

Recoveries of tetracyclines using the HLB and tC18 SPE cartridges were compared after spiking and extracting reagent water buffered with phosphate, citric acid, and EDTA. The internal standard was added after elution to measure absolute recovery of each compound added. Elution of tetracylines from the HLB cartridges with 1% TFA in methanol produced the most consistent recoveries in fortified reagent water (Figure 4). Improved recovery of pharmaceutical compounds extracted using the HLB packing has been reported previously by eluting with methanol modified with trifluoracetic acid or triethylamine *(13)*. The addition of an acid modifier may help with solvation of

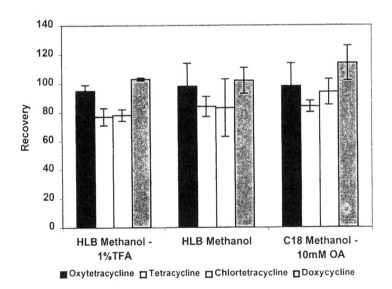

Figure 4. Average recovery of tetracyclines extracted from five samples of fortified reagent water using the Oasis HLB and tC18 SPE cartridges. Error bars equal one standard deviation

ionizable compounds. Use of the HLB cartridges permits a more aggressive wash (5% methanol in water) after extraction to help remove potential interferences from waste water samples *(8)*.

Tetracycline method detection limits were estimated from the variability in analysis of eight 100 mL aliquots of reagent water fortified with 1 μg/L oxytetracycline (OTC), tetracycline (TC), and chlortetracycline (CTC) were 0.21, 0.20, and 0.28 μg/L *(8)*. Method detection limits in 20 mL samples of waste water fortified at 10 μg/L for OTC, TC, and CTC were 1.1, 1.8, and 1.7 μg/L, respectively.

Macrolides, lincomycin and tiamulin

Extraction of macrolide antibiotics, lincomycin, and tiamulin was more problematic than for the tetracycline compounds. Highly variable recoveries were initially observed for the macrolide antibiotics extracted from reagent water with either the C18 or the HLB cartridges. Hirsch et al. *(10, 15)* noted reduced recovery of macrolide antibiotics from pure reagent water and prepared solutions for extraction in mountain spring water containing dissolved salts. Extraction and analysis of fortified tap water (pH=7.2) improved consistency in the recovery of macrolides between replicates, but also produced anomalous recoveries of erythromycin and tiamulin.

Sample pH seems critical for reproducible extraction of these compounds (pK_a between 7 and 10). Buffering reagent water and waste water samples with 0.05M disodium phosphate and potassium phosphate ($KH_2PO_4/0.05$ M Na_2HPO_4) at a pH near 7 improved both reproducibility and recovery of the macrolides, lincomycin and tiamulin using either cartridge.

Recovery experiments from acidic and basic solutions produced low recoveries of the macrolides, especially erythromycin A and tylosin. Macrolides are known to be unstable in acidic and basic solutions *(16)*. Erythromycin A has been shown to be stable at neutral pH, but rapidly decomposes to its anhydro form at pH less than 4 *(17)*. Extraction from neutral buffered waste water using the polymeric HLB cartridge produced the most reproducible and quantitative recoveries of the macrolides, lincomycin and tiamulin. Elution from the HLB cartridges with acetonitrile produced the more consistent recovery than methanol (Figure 5).

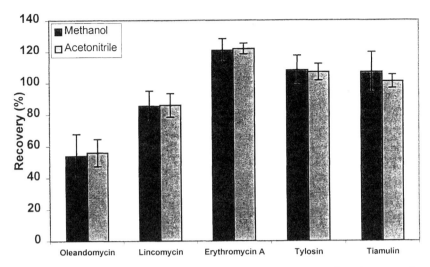

Figure 5. Recovery of macrolides, lincomycin and tiamulin after elution with either methanol or acetonitrile. Errors bars represent one standard deviation

Extraction of 100 mL aliquots of buffered reagent water fortified with 3 µg/L of the macrolide antibiotics, lincomycin and tiamulin, and eluting using both acetonitrile and methanol sequentially yielded method detection limits of 0.5 µg/L for tylosin and tiamulin and 1.5 µg/L for lincomycin and erythromycin. Extraction of 20 mL samples of buffered waste water resulted in method detection limits between 1.0 and 1.5 µg/L for tylosin, tiamulin, and erythromycin A, 2.3 µg/L for tilmicosin, and 9.0 µg/L for lincomycin. Concentrations of the macrolide antibiotics and related compounds in animal waste water have not been reported, but depending on dosage rates and persistence may be expected to be comparable to concentrations of tetracycline antibiotics. Because many of these compounds are highly sorptive *(18)* it is likely that their concentrations will also depend upon the composition of sludge and other solids contained in the waste water impoundments.

Future work will focus on the improvement of these methods and on development of methods for other antibiotics. The sensitivity of these methods may be improved by increasing the sample size while maintaining consistency in compound recovery.

Conclusions

Solid phase extraction and liquid chromatography tandem mass spectrometry provides a method for sensitive analysis of commonly used tetracycline and macrolide antibiotics in water and waste water. Buffering of the samples and modification of eluting solvents improves consistency in the recovery of these compounds. Optimal detection of the macrolide antibiotics using an ion trap mass spectrometer depends on selection of the isolation width. Method detection limits estimated from replicate analyses of fortified reagent water and waste water indicate that these antibiotics can be analyzed at environmentally relevant concentrations.

References

1. *Animal Health Institute 1999 Survey of Manufacturers,* Washington, D.C., 2001.
2. Niessen, W. M. A. *J. Chromatogr. A.* **1998,** *812,* 53-75.
3. Oka, H.; Patterson, J. In: *Chemical Analysis of Antibiotics Used in Agriculture*; Oka, H.; Nakazawa, H.; Harada, K.I; MacNeil, J.D.; Eds.; AOAC International: Arlington, VA, 1995, pp. 333-406.
4. Kennedy, D. G.; McCracken, R. J.; Cannavan, A.; Hewitt; S. A.; *J. Chromatogr. A.* **1998,** *812,* 77-98.
5. Oka, H.; Ito, Y.; Ikai, Y.; Kagami, T.; Harada, K. *J. Chromatogr. A* **1998,** *812,* 309-319.
6. Schenck, F. J.; Callery, P. S.; *J. Chromatogr. A* **1998,** *812,* 99.
7. Cheng, Y.F.; Phillips, D. J.; Neue, U.; *Chromatographia,* **1997,** *44* ,187-190.
8. Zhu, J.; Snow, D. D.; Cassada, D.A.; Monson, S.J.; Spalding, R. F.; *J. Chromatogr. A* **2001,** *928* , 177-186.
9. Lindsey, M.; Meyer, M.; Thurman, E.M. *Anal. Chem.* **2001,** *73,* 4640-4646.
10. Hirsch, R.; Ternes, T.A.; Haberer, K.; Mehlich, A. ; Ballwanz, F. ; Kratz, K.L. *J. Chromatogr. A* **1998,** *815,* 213-223.
11. McClellan, J.E.; Murphy, J.P.; Mulholland, J.J.; Yost., R.A. *Anal Chem.* **2002,** *74,* 402-412.
12. Keith, L.H. Crummet, W. Deegan, J.; Libby, R.A.; Taylor, J.K.; Wentler, G. *Anal. Chem.* **1983,** *55,* 2210-2218.
13. Meyer, M.T; Bumgarner, J.E.; Thurman, E.M.; Hostetler, K.A.; Daughtridge, J.V.; *Sci. Total Env.* **2000,** *248,* 181-187.
14. Martin, P.; Wilson, I. *J. Pharm. Biomed. Anal,* **1998,** *17,* 1093-1100.
15. Hirsch, R. Ternes, T. Haberer, K. Kratz, K.L. *Sci. Total Env.* **1999,** *225,* 109-118.
16. Horie, M. In: *Chemical Analysis of Antibiotics Used in Agriculture*; Oka, H.; Nakazawa, H.; Harada, K.I; MacNeil; J.D; Eds.; AOAC International: Arlington, VA, 1995, pp. 165-205.
17. Volmer, D.A.; Hui, J.P.M. *Rapid Commun. Mass Spectrom.* **1998,** *12,* 123-29.
18. Tolls, J. *Env. Sci. Technol.* **2001,** *35,* 3397-3406.

Chapter 11

Identification of Labile Polar Organic Contaminants by Atmospheric-Pressure Ionization Tandem Mass Spectrometry

Edward T. Furlong[1], Imma Ferrer[1,2], Paul M. Gates[1], Jeffery D. Cahill[1], and E. Michael Thurman[1,3]

[1]National Water Quality Laboratory, U.S. Geological Survey, P.O. Box 25046, Denver, CO 80225
[2]Current address: immaferrer@menta.net
[3]Current address: U.S. Geological Survey, 4821 Quail Crest Place, Lawrence, KS 66049

Mass spectrometry, in single or multiple stages, can be coupled to high-performance liquid chromatography by an atmospheric-pressure ionization interface (HPLC/MS or HPLC/MS/MS), thus providing a sensitive and selective tool for the identification and quantitation of labile polar organic contaminants in aquatic and sedimentary environments. An example is the identification of unknown compounds in a water sample showing substantial toxicity to plankton. This sample was analyzed by full-scan HPLC/MS with atmospheric-pressure chemical ionization in the positive mode. Iterative, data-dependent HPLC/MS/MS analyses identified characteristic fragment ions for these compounds and confirmed the presence of an extended series of nonylphenol ethoxylate surfactant homologues at microgram-per-liter concentrations in these agriculturally influenced samples. An additional example is provided by the use of ion-trap HPLC/MS/MS with electrospray ionization to identify pharmaceuticals and cationic surfactants in sediment extracts at concentrations ranging from 2 to 200 ng/g. The MS/MS capabilities of this instrument permitted positive identification at subnanogram-per-gram concentrations in the presence of substantial coextracted interferences.

Introduction

For several years, the combination of high-performance liquid chromatography with mass spectrometry (HPLC/MS) by using an atmospheric-pressure ionization interface has been proposed as the ideal tool for sensitive and selective identification and quantitation of labile polar organic contaminants, such as pesticides, in aquatic and sedimentary environments (1). Although numerous applications for identifying and quantifying pesticides, drinking-water disinfection byproducts, surfactants, and other contaminant classes at less than 1-μg/L concentrations have been reported (2-5), there has not been a parallel increase in use of HPLC/MS for routine environmental monitoring. Although at least one large-scale study has used routine monitoring by HPLC/MS (6), two problems seem to impede wider implementation. The first problem is limited fragmentation in the atmospheric-pressure ionization (API) source, which can preclude unambiguous identification. The second problem is coeluting sample matrix components that also can interfere with identification, or can affect analyte ionization efficiency, which result in signal suppression or enhancement, thus introducing a positive or negative bias into quantitation. Limited fragmentation and matrix effects are not limited to environmental analysis (7-14). Approaches often used in gas chromatography/mass spectrometry, such as the incorporation of isotopically labeled internal standards into the instrumental analysis, can reduce but not eliminate the impact of limited fragmentation and matrix effects on HPLC/MS analysis. As a result, these matrix-associated problems have hindered the wider acceptance of HPLC/MS for routine monitoring of labile polar organic contaminants.

Tandem mass spectrometry (MS/MS), using either quadrupole or ion-trap (IT) mass spectrometers, when coupled with high-performance liquid chromatography (HPLC/MS/MS) addresses some of these shortcomings. In MS/MS, a precursor ion of a single m/z is selected from all the ions produced and transmitted to the mass analyzer. In HPLC/MS/MS under positive electrospray ionization conditions, this precursor ion typically is a protonated molecule or adduct ion of the compound of interest. This precursor ion is fragmented, and one or more of the characteristic product fragment ions are analyzed and used for specific identification. This selective, multistage analysis provides a higher order of mass spectral information that can be used to identify and quantify unknown compounds in the presence of complex sample matrixes, a shortcoming for HPLC coupled to a single level of MS analysis. In some cases, chromatographic separation is not required, which provides the additional advantage of rapid analysis.

In this chapter, two examples of the unique capabilities of quadrupole and ion-trap tandem mass spectrometers for environmental HPLC/MS/MS analysis are described. Quadrupole HPLC/MS/MS is applied to a complex agricultural runoff sample where MS/MS was used to determine the identity of high molecular weight nonionic nonylphenol ethoxylate surfactants (NPEOs). Ion-trap HPLC/MS/MS is applied to the identification and quantitation of pharmaceuticals and antimicrobial surfactants in sediment extracts containing a complex mixture of natural and anthropogenic organic compounds.

Methods

Trace organic constituents in the agricultural water samples were isolated using C_8 solid-phase extraction cartridges eluted with 95:5 methanol:water, and the extracts concentrated. Commercial mixtures of NPEOs used to confirm identification in water samples were diluted and used as is. The sample extracts and standards were analyzed by injection of 5-µL aliquots into a stream of 30-percent acetonitrile in water, at a flow rate of 0.6 mL/min. The flow-injection stream was interfaced to a triple-quadrupole mass spectrometer by an atmospheric-pressure chemical ionization (APCI) interface. The APCI vaporizer was held at 475°C, the capillary at 225°C, and the sheath gas pressure was 551.6 kPa. The MS was operated in the positive ion mode.

Pharmaceuticals and antimicrobials in sediment were extracted by pressurized-fluid extraction of wet sediment with a 60-percent acetonitrile:40-percent water mixture, using the procedure of Ferrer and Furlong (15). The sediment extracts were analyzed without further isolation or purification, although they were diluted to the composition of the starting eluent of the HPLC gradient (15-percent acetonitrile in water). A 150 mm by 2.0 mm i.d. octadecylsilane reversed-phase (5-µm particle diameter) HPLC column was used. A linear gradient from 15- to 100-percent acetonitrile in an aqueous ammonium formate/formic acid buffer (10 mM, pH 3.7) was used to separate the pharmaceuticals. Each extract was analyzed twice. The first analysis used a quadrupole HPLC/MS system with electrospray ionization operated in the positive ion mode and with selected-ion monitoring to enhance sensitivity and decrease chemical noise. The second analysis made use of ion-trap mass spectrometry (ITMS), also with electrospray ionization operated in the positive ion mode. For the ITMS, two analyses were required, the first by full-scan MS analysis, followed by an MS/MS analysis based on full-scan MS results. The full-scan MS analysis was used to identify the protonated molecule or adduct ion for each pharmaceutical. This step was followed by an MS/MS analysis of each precursor ion to confirm the identification of each pharmaceutical. Antimicrobial surfactants were determined using only ion-trap tandem mass spectrometry, described by Ferrer and Furlong (15).

Results

Agricultural Samples

An initial full-scan MS analysis (Figure 1) of the surface-water sample, suspected to contain agricultural runoff, suggested that the sample potentially contained three homologous series of compounds. Each series was separated by a diagnostic 44 mass-to-charge units (m/z), which in turn suggested three series of polyethoxylated polymers (16). The first series, ranging from 353 to 573 m/z, was suspected to be a series of nonylphenol ethoxylate (NPEO) surfactant homologues containing 3 to 8 ethoxy units, based on the expected protonated molecules for NPEO surfactants. The mass spectrum of the protonated molecules of a mixture of NPEO surfactants was determined by full-scan MS analysis of a commercial NPEO mixture, which indicated that the suspected NPEO protonated molecules in the water samples also were present in the standard (Figure 2). However, the distributions of NPEO homologues in this and other commercial mixtures were substantially biased towards higher NPEO homologues relative to the homologues present in the environmental samples.

Tandem mass spectrometry then was used to confirm identification of NPEOs in this sample by comparison of MS and MS/MS spectra from the environmental sample to MS and MS/MS spectra of a commercial mixture of NPEOs (Figures 3 and 4). In both standard and environmental samples, the same series of putative protonated molecules were observed, each separated by 44 m/z. In the MS/MS spectra of the environmental sample and commercial mixture, the same product fragment ions are observed, although there are differences between the relative abundances of the same mass ion in each spectrum. These differences likely are a function of the absolute abundance of the protonated molecular ion used for the MS/MS analysis (m/z 485) of the environmental sample in relation to the pure standard. The pattern of NPEO homologues in the environmental water samples did not closely match any of the standards and might reflect biotic or abiotic alteration of the NPEO signature under environmental conditions.

A less abundant second series differing by m/z 44, from 370 to 529 m/z, tentatively was identified as a homologous series of polyglycol ethers (PGEs), described as minor components of commercial NPEO mixtures (17). The number of ethoxy units in the homologous series ranges from 8 to 12. A third homologous series, tentatively identified as carboxylated degradation products of NPEOs (CNPEOs), also was observed. The CNPEO homologues contain 6 to 8 ethoxy units. The CNPEOs previously have been reported as primary degradation products of the NPEOs in wastewater, surface water, and agricultural settings (18-20), although homologues with lower degrees of

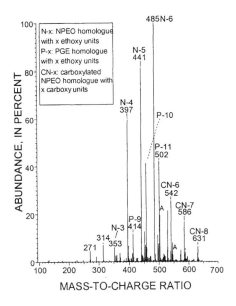

Figure 1. Positive ion atmospheric-pressure chemical ionization spectrum from a surface-water sample containing agricultural runoff.

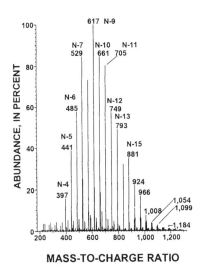

Figure 2. Positive ion atmospheric-pressure chemical ionization spectrum of a commercial nonylphenol polyethoxylate mixture, NP-10.

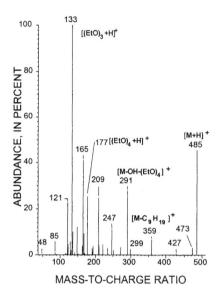

Figure 3. Positive ion APCI/MS/MS spectrum of the m/z 485 ion isolated from the MS spectrum of the surface-water sample.

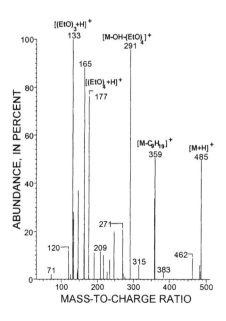

Figure 4. Positive ion APCI/MS/MS spectrum of the m/z 485 ion isolated from the MS spectrum of the commercial nonylphenol ethoxylate mixture.

ethoxylation typically have been observed (4). At the time of analysis, standards were not available, so the presence of CNPEOs or PGEs in the environmental samples could not be confirmed.

Sediment Samples

Typical sediment extracts are much more complex than environmental water extracts. Fine-grained sediments often contain organic carbon concentrations ranging between 1 and 3 percent by weight, most of which consist of naturally present organic matter. This natural organic matter consists of a complex mixture of heterogeneous polyelectrolytes, a substantial fraction of which is water- and acetonitrile-extractable (21). This material contains numerous chromophores, and as a result, HPLC chromatograms using ultraviolet detection often show a substantial chromatographic "hump." This hump interferes with ultraviolet spectral identification of compounds, reduces the accuracy of identification, and raises method-reporting levels. A single level of low-resolution mass spectrometry, as in a single quadrupole HPLC/MS system, can eliminate much of this inherent chemical noise. However, the level of chemical interferences present in complex matrixes, including sediment, can overwhelm even the capabilities of HPLC/MS. Tandem spectrometry, using triple quadrupole or ion-trap systems, can resolve this problem of chemical interferences. Note that high-resolution mass spectrometry systems, such as HPLC/time-of-flight mass spectrometry (HPLC/TOF-MS), also can identify trace contaminants in complex sediment matrixes (Thurman and Ferrer, this volume), but are expensive and not widely available for environmental analysis.

In this study, pharmaceutical concentrations in sediment samples initially were estimated by HPLC/SIM-MS. Selected-ion monitoring (SIM) analysis permitted an initial, although incomplete, assessment of the level of interferences present in the sample extract. This assessment is incomplete in that the SIM windows usually monitor between two and ten ions, all derived from one or more of the compounds of interest, while the matrix interferences typically produce an envelope of ions at every nominal mass measured. However, this initial analysis permits evaluation of the level of interferences and the degree to which they contribute to the diagnostic ions of the compounds determined under SIM analysis.

Figure 5 shows the total ion current chromatogram for a full-scan HPLC/ESI-IT-MS analysis of an extract of sediment collected from the South Platte River, near Denver, Colorado. A series of peaks are observed eluting at regularly spaced time intervals. These peaks, most identified as ethoxylate surfactants or polyethylene glycols, are underlain by a broad unresolved complex mixture of natural and anthropogenic organic constituents that interfere with the identification of the trace pharmaceuticals present in the sample. A chromatogram produced by a full-scan HPLC/quadrupole-MS analysis under similar chromatographic conditions (Figure 6) differs from the ion-trap chromatogram (Figure 5) in the total ion current signal pattern, caused by differences in the elution program and the several months that elapsed between each analysis of the extract. However, the chromatograms are similar in that they contain few resolved peaks and are dominated by an unresolved complex mixture.

Figure 7 shows a mass chromatogram of m/z 291 from the initial full-scan HPLC/IT-MS analysis of the sediment extract. The m/z 291 ion is the protonated molecules of the antibiotic trimethoprim initially identified by quadrupole HPLC/SIM-MS. Comparing the mass chromatogram to Figure 5 demonstrates that a substantial fraction of resolved nonselected compounds and the unresolvable hump are excluded. However, there are other peaks eluting in the same general region of the mass chromatogram, and until the identification is confirmed, in this case by MS/MS analysis, the identification could only be provisional. Figure 8 is the MS/MS spectrum of the peak identified in the m/z 291 mass chromatogram shown in Figure 7. The radiofrequency voltage used to induce fragmentation of the isolated ion was optimized so that m/z 291 is only a few tenths of a percent of the abundance of the fragment peaks, and is not visible in the mass spectrum at this scale. The largest ions in the mass spectrum can be rationalized to the structure of trimethoprim, as is indicated in the inset structure in Figure 8. Note that the single and double methoxy losses cannot be assigned to specific positions of the trimethoprim structure. This example shows the value of multiple MS for verifying the identity of a trace constituent initially identified by quadrupole HPLC/MS with selected-ion monitoring even though substantial chemical interferences were present.

Another advantage of HPLC/IT-MS over a SIM-HPLC/quadrupole-MS analysis of sediment is the potential to identify contaminants not initially identified by SIM-HPLC/quadrupole-MS analysis. Figure 9 shows a mass chromatogram for the ion m/z 202. One chromatographically resolvable peak was detected that did not correspond to any of the selected pharmaceuticals. Several possibilities were considered for this compound, but the even-mass protonated molecule, suggesting an odd number of nitrogens, and other information suggested thiabendazole, a compound used in agriculture for its antifungal properties and in medicine as an antihelmenitic. Multiple MS/MS analysis of this peak produced a spectrum that could be rationalized to the structure of thiabendazole (Figure 10). The MS/MS spectrum of the putative thiabendazole peak identified in Figure 10 was verified by comparison with an analysis of an authentic standard under identical instrumental conditions. To our knowledge, this is the first identification of thiabendazole in aquatic sediment. This example shows the unique capabilities of the ion trap for the application of multiple MS for identification of unknown compounds that are initially detected in a full-scan MS analysis (see Ferrer et al., this volume).

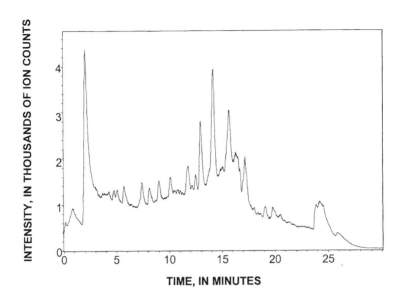

Figure 5. Positive ionization total ion current chromatogram of South Platte
River sediment extract by HPLC/ESI-IT-MS.

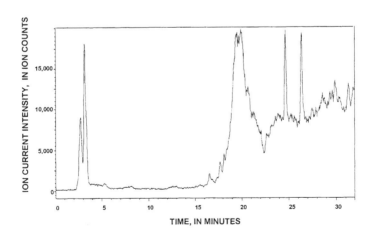

Figure 6. Positive ionization total ion current chromatogram of South Platte
River sediment extract by HPLC/ESI-Quadrupole MS.

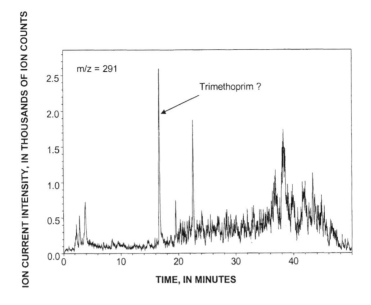

Figure 7. Mass chromatogram of ion m/z 291 in South Platte River sediment extract by HPLC/ESI-IT MS.

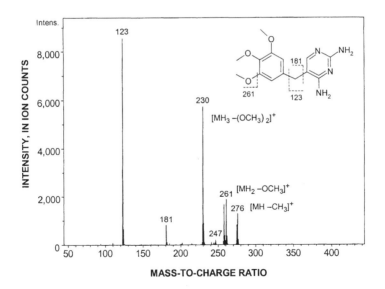

Figure 8. Tandem mass spectrum of the m/z 291 ion peak tentatively identified as trimethoprim in Figure 7.

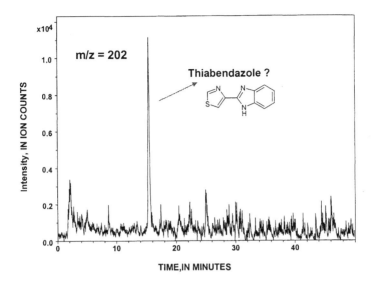

Figure 9. Mass chromatogram of the ion m/z 202 in South Platte River sediment extract by HPLC/ESI-IT/MS.

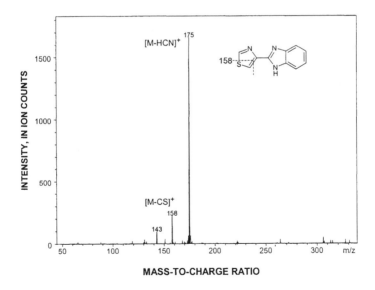

Figure 10. Tandem mass spectrum of the putative thiabendazole peak tentatively identified in Figure 9.

These examples demonstrate the multifaceted advantages of tandem mass spectrometry to identify unknown compounds in complex matrixes or to confirm identification of multiple series of unknown compounds in the presence of chemical interferences. The reliability of determining known compounds by HPLC/MS also is improved. Tandem mass spectrometry provides tools for inferring and confirming identification of unknown compounds. Widespread use of tandem mass spectrometry for the analysis of environmental samples will require development of systematic approaches that take advantage of the characteristic MS/MS fragmentations observed from the unique ionization conditions found in atmospheric-pressure ionization (Ferrer et al., this volume).

Acknowledgements

Kathryn Kuivila, U.S Geological Survey, Sacramento, California, provided the agricultural sample and ancillary information. Jeffery D. Cahill, U.S. Geological Survey, National Water Quality Laboratory, Denver, Colorado, provided initial HPLC/MS analysis of sediment extracts. Support for this research was provided in part by the Toxic Substances Hydrology Program, U.S. Geological Survey. The use of trade, product, or firm names in this chapter is for descriptive purposes only and does not imply endorsement by the U.S. Government.

References

1. Voyksner, R. D. *Environ. Sci. Technol.* **1994**, *28*, 118A-127A.
2. Ferrer, I.; Barcelo, D. *Analusis* **1998**, *26*, M118-M122.
3. Di Corcia, A. *J. Chromatogr. A.* **1998**, *794*, 165-185.
4. Richardson, S. D. *Anal. Chem.* **2000**, *72*, 4477-4496.
5. Richardson, S. D. *Chem. Rev.* (Washington, DC, U. S.) **2001**, *101*, 211-254.
6. Furlong, E. T.; Burkhardt, M. R.; Gates, P. M.; Werner, S. L.; Battaglin, W. A. *Sci. Total Environ.* **2000**, *248*, 135-146.
7. Hogenboom, A. C.; Hofman, M. P.; Jolly, D. A.; Niessen, W. M. A.; Brinkman, U. A. T. *J. Chromatogr. A* **2000**, *885*, 377-388.
8. Sterner, J. L.; Johnston, M. V.; Nicol, G. R.; Ridge, D. P. *J. Mass Spectrom.* **2000**, *35*, 385-391.
9. Cole, R. B.; Zhu, J. H. *Rapid Commun. Mass Spectrom.* **1999**, *13*, 607-611.
10. Choi, B. K.; Hercules, D. M.; Gusev, A. I. *J. Chromatogr. A* **2001**, 907, 337-342.
11. Gustavsson, S. A.; Samskog, J.; Markides, K. E.; Langstrom, B. *J. Chromatogr. A* **2001**, *937*, 41-47.
12. Petrovic, M.; Barcelo, D. *J. Mass Spectrom.* **2001**, *36*, 1173-1185.

13. Ternes, T.; Bonerz, M.; Schmidt, T. *J. Chromatogr. A* **2001**, *938*, 175-185.

14. Ito, S.; Tsukada, K. *J. Chromatogr., A* **2002**, *943*, 39-46.

15. Ferrer, I.; Furlong, E. T. *Anal. Chem.* **2002**, *74*, 1275-1280.

16. Parees, D. M.; Hanton, S. D.; Clark, P. A. C.; Willcox, D. A. *J. Am. Soc. Mass. Spectrom.* **1998**, *9*, 282-291.

17. Pattanaargsorn, S.; Sangvanich, P.; Petsom, A.; Roengsumran, S. *Analyst* **1995**, *120*, 1573-1576.

18. Field, J. A.; Reed, R. L. *Environ. Sci. Technol.* **1996**, *30*, 3544-3550.

19. Ding, W. H.; Tzing, S. H. *J. Chromatog. A* **1998**, *824*, 79-90.

20. Mihaich, E. M.; Naylor, C. G.; Staples, C. A. *Pesticide Formulations and Application Systems: A New Century for Agricultural Formulations, Twenty First Volume*; Mueninghoff, J. C., Viets, A. K., Downer, R. A., Eds.; American Society Testing and Materials: West Conshohocken, PA, 2001, pp 147-159.

21. Leenheer, J. A. *Env. Sci. Technol.* **1981**, *15*, 578-587.

Chapter 12

The Determination and Quantification of Human Pharmaceuticals in Aqueous Environmental Samples

Kimberly D. Bratton, Amy S. Lillquist, Todd D. Williams, and Craig E. Lunte[*]

Department of Chemistry, University of Kansas, 1251 Wescoe Hall Drive, Lawrence, KS 66045–7582

Analytical methods were developed to determine several basic and acidic pharmaceutical compounds including anti-hypertensives, anti-inflammatories, and anti-depressants, in various surface waters. The methods include solid phase extraction (SPE) that acts as a cleanup and concentrates the sample a thousand-fold. The samples were analyzed by LC/MS/MS using two mass spectrometry techniques, MS/MS to confirm the presence of the pharmaceuticals in the samples and MRM to quantitate. Identification of nine out of ten compounds was found in both the city of Lawrence, KS influent and effluent wastewater samples. Kansas River water showed the presence of seven out of ten compounds and in tap water five out of ten compounds were detected.

Introduction

Over the years, investigations of potential environmental pollutants have been a major concern. Many analytical methods have been developed to identify chemical pollutants in the environment. Much of the focus has been on agricultural compounds such as herbicides, pesticides, and animal antibiotics. These compounds can pose an immediate or long-term threat to humans and other organisms. However, pharmaceutical compounds appearing in surface water resulting from human uptake and excretion could also pose a threat to living organisms. Pharmacokinetics studies have shown that more than half of all compounds consumed will be excreted from the body in their original form. Often times unused or expired pharmaceutical compounds will be improperly disposed of and leach out of landfills infiltrating water supplies. Sewage treatment plants are effective at eliminating bacteria and treating sewage for reintroduction into the surface water supply. However, there is no set protocol for the degradation of pharmaceutical compounds and these substances can potentially survive sewage treatment systems because of their high stability against biological degradation (1). While the potential affects of these compounds on humans and aquatic organisms are unknown, a method of detection would be the first step in the investigation toward defining the potential risks. Trace amounts (ng/L) of the anti-inflammatory drug diclofenac have been found in tap water in Berlin (2) and surface water in Austria (3). Ketoprofen, another anti-inflammatory, was found in surface water in Germany at a concentration of 0.12 μg/L (4). Also naproxen, an anti-inflammatory, was also found in Austria (3) and Germany's surface waters with levels reaching 0.39 μg/L (4).

United States Geological Survey researchers have provided the first nationwide reconnaissance of the occurrence of a wide variety of contaminants. They have investigated the occurrence of agricultural antibiotics, hormones, and a few human pharmaceuticals in streams across the nation (5).

The compounds in this study were chosen from a list of the top two hundred most prescribed drugs in the United States (6). The selected compounds were albuterol a bronchodilator, fluoxetine an anti-depressant, atenolol, metoprolol, and propranolol, anti-hypertensives, diclofenac, indomethacin, ketoprofen, naproxen, anti-inflammatories, and sulfamethoxazole an antibiotic (see Figure 1). These compounds were selected due to their availability and their possession of an acidic or basic functional group. Sample cleanup and pre-concentration was performed by solid phase extraction using a strong anion or cation exchange cartridge. This takes advantage of the acidic and basic properties of the compounds. Analysis was performed by LC/MS/MS because it offers both the specificity and detection limits needed in this study. To qualitatively confirm the identity of the compounds in wastewater, MS/MS experiments were done to correlate fragmentation patterns from the compounds in wastewater to standards. The quantitative mass spectrometry analysis involved performing a multiple reaction monitoring (MRM) experiment to selectively detect the most abundant product fragment of each selected precursor ion.

Experimental

Chemicals and Reagents

The basic pharmaceutical standards were salbutamol (albuterol), atenolol, metoprolol, propranolol, and fluoxetine. The acidic pharmaceutical standards were diclofenac, naproxen, ketoprofen, sulfamethoxazole, and indomethacin. All standards were purchased from Sigma (St. Louis, MO, USA). The internal standard, d_5-fluoxetine, was purchased from C/D/N Isotopes Inc. (Quebec, Canada). HPLC-grade methanol was purchased from Merck (Darmstadt, Germany). HPLC-grade acetonitrile, ammonium acetate, ammonium hydroxide, and acetic acid were purchased from Fischer (Fairlawn, NJ, USA). High-purity water was prepared by a Water Pro Plus water purification system (Labconco, Kansas City, MO, USA).

Instrumentation

The LC system used was a Waters 2690 separations module (Milford, MA, USA). The triple quadrupole mass spectrometer used was a Micromass Limited, Quattro Ultima, equipped with a Z-spray source (Manchester, UK).

Water Collection

Between February and September 2002, water samples were collected from the influent and effluent sites at the Lawrence, Kansas sewage treatment plant. The treatment plant takes sewage water in from Lawrence and discharges the effluent into the Kansas River. River water samples were taken approximately one mile before the effluent discharge into the river and also 100 yards after the discharge. Tap water samples were taken from the water supply of Lawrence,

KS. Samples were collected in plastic bottles pre-rinsed with ultra-pure water. Analysis of basic compounds began by acidifying water samples to a pH of 3.5 with acetic acid. Conversely for acidic compound analysis, the water samples were brought up to a pH of 10.5 using ammonium hydroxide. All samples were then filtered with Whatman filter paper, double thick, followed by vacuum filtration with 0.22 micron filter paper and stored at 0 °C for preservation. Each sample batch was then divided into three aliquots of 300 mL each for precision studies. Solid phase extraction was performed as soon as possible to ensure no degradation would occur.

Solid Phase Extraction

The extraction focusing on the basic compounds was performed under vacuum with a vacuum manifold and strong cation exchange cartridges (10 mL, 500 mg) from Supelco (Bellefonte, PA, USA). The optimized SPE procedure began by pre-conditioning the cartridge with 3x1 mL of methanol followed by 2x1 mL of acidified ultra-pure water (pH ~ 3.5). A volume of 300 mL acidified water sample was applied to the cartridge. A wash step with 3x1 mL of 50/50 methanol/acidified water was passed through prior to elution. The cartridge was allowed to dry for about 1 minute to remove excess water. The analytes were eluted with 3x1 mL 3% ammoniated methanol. The eluate was evaporated to dryness using a Savant speedvac and refrigerated condensation trap, then reconstituted in 300 µL of ultra-pure water for injection or spiked with d_5-fluoxetine and brought to a volume of 300 µL. Recoveries as evaluated in spiked nanopure as well as in spiked real samples are found in Table I.

The extraction utilized for the sample preparation of acidic compounds was performed as follows. The filtered water samples were thawed to room temperature. Solid phase extraction (SPE) was performed using a Supelco vacuum manifold with Oasis MAX cartridges (6mL, 60mg) (Waters Inc., Milford, MA, USA). The cartridge was conditioned with 2x1 mL of methanol, followed by 2x1 mL water, pH 10.5. The flow was kept at no greater than 2 mL/min. Three hundred milliliters of the water samples, with pH adjusted to 10.5 using ammonium hydroxide, were loaded onto the cartridge. After completion of this step, the cartridge was washed with 2x1 mL of water, pH 10.95. The cartridge was allowed to dry for five minutes. Next, an elution step for any basic or neutral compounds was performed in order to reduce interference with the target acid pharmaceuticals. This was accomplished by adding 2x1 mL of methanol. The cartridge dried for another five minutes before the elution step. To elute the analytes, 2x1 mL of methanol with 2% formic acid was passed through the extraction cartridge. This fraction was evaporated under a stream of helium. The samples were then reconstituted in 300 μL of water preceding analysis. Recoveries as evaluated in spiked nanopure as well as in spiked real samples are found in Table II.

LC Analysis

For the analysis of the basic analytes a gradient was employed. Mobile phase A was a solution of 2.5% acetonitrile, 10 mM ammonium acetate (pH 7.0). Mobile phase B consisted of 70% acetonitrile, 10 mM ammonium acetate (pH 7.0). The gradient program began with a hold for three minutes at 7% B, followed by a step to 50% B. A linear ramp was then applied from 50% to 85% mobile phase B in five minutes and held at 85% B for one minute. The re-equilibration time was ten minutes. The flow rate was set to 0.3 mL/min. The column was a Phenomenex Luna, 3 micron, C8 (100 X 2.0 mm) equipped with a Phenomenex SecurityGuard guard cartridge system.

For the analysis of the acidic analytes a gradient was employed for LC analysis. Mobile phase A consisted of 10% acetonitrile, 25mM ammonium acetate (pH 7.0). Mobile phase B consisted of 60% acetonitrile, 25mM ammonium acetate (pH 7.0). A linear gradient from 25% to 50% mobile phase B was performed over 10 minutes (flow rate 0.3 mL/min) at which time the LC column was re-equilibrated for 10 minutes prior each run. The column used for analysis was a Phenomenex Prodigy, 3 micron, C18 (100 X 2.0 mm) along with a Phenomenex SecurityGuard guard cartridge (Torrence, CA, USA).

LC/MS/MS Analysis

The following parameters were used for the analysis of basic compounds. Electrospray source block: 100°C; Probe desolvation temperature: 350 °C; Cone voltage: 25 V; Argon collision gas: ca. 1.7 x10^{-3} mbar (as measured on a gauge near the collision cell); Cone nitrogen gas flow: ca. 100 L/hr; Desolvation gas: ca. 500 L/hr; Both mass analyzer resolutions: 0.8 amu FWHH. Transitions chosen for the MRM experiment are shown in Figure 1 and utilized a dwell time of 0.5 or 0.1 seconds. The collision energies determined experimentally were 18 eV for albuterol, metoprolol and propranolol, 25 eV for atenolol, and 10 eV for fluoxetine.

The following parameters were used for the analysis of acidic compounds. Electrospray source block: 100 °C; Probe desolvation temperature: 300 °C; Cone voltage: 35 V; Argon collision gas: ca. 1.7 x10^{-3} mbar, Cone nitrogen gas flow: ca. 100 L/hr; Desolvation gas: ca. 500 L/hr; Both mass analyzer resolutions: 0.8 amu FWHH. Transitions monitored of precursor-product ions for the MRM

Table I: Recovery of basic compounds after a strong cation exchange solid phase extraction

Spiked sample	Albuterol % Recovery	Atenolol % Recovery	Metoprolol % Recovery	Propranolol % Recovery	Fluoxetine % Recovery
Nanopure	97.9 ± 13.0	75.1 ± 1.9	94.8 ± 2.7	69.5 ± 1.0	25.7 ± 4.9
Influent	10.6 ± 1.3	21.5 ± 3.4	87.6 ± 5.8	31.2 ± 5.2	14.4 ± 2.1
Effluent	4.6 ± 0.4	18.5 ± 0.3	60.9 ± 2.6	38.2 ± 1.6	12.5 ± 1.4
River	7.4 ± 0.4	14.3 ± 0.2	99.3 ± 2.6	68.0 ± 0.9	29.3 ± 1.2
Tap	11.6 ± 0.2	20.2 ± 0.6	101.2 ± 6.7	30.2 ± 1.3	32.7 ± 1.7

Table II: Recovery of acidic compounds after a strong anion exchange solid phase extraction

Spiked sample	Sulfamethoxazole % Recovery	Naproxen % Recovery	Ketoprofen % Recovery	Diclofenac % Recovery	Indomethacin % Recovery
Nanopure	91.1 ± 13.0	86.5 ± 4.9	100.0 ± 7.9	61.9 ± 6.7	11.5 ± 6.1
Influent	15.0 ± 2.4	23.9 ± 3.4	11.2 ± 1.0	2.0 ± 0.3	1.0 ± 0.2
Effluent	25.0 ± 5.3	27.7 ± 1.7	30.1 ± 3.9	9.7 ± 2.9	5.9 ± 1.5
River	32.7 ± 2.9	32.5 ± 2.9	46.4 ± 7.9	12.0 ± 4.0	1.0 ± 0.2

experiment are shown in Figure 1 and had a dwell time of 0.05 seconds. The collision energies determined experimentally were 15 eV for ketoprofen, naproxen, sulfamethoxazole, and 20 eV for diclofenac and indomethacin.

Results and Discussion

Two separate SPE and LC/MS/MS methods were developed to detect both basic and acidic analytes in water supplies around Lawrence, KS. The basic and acidic compounds have very different properties, pKa's, functionality groups, and Log p's, which would make it difficult to find one solid phase extraction or LC separation method capable of treating them all. Therefore subsequent analyses were performed by dividing each compound into an acid or base category based on respective functional groups.

The solid phase extraction procedures were evaluated using standards prepared in nanopure water as well as in wastewater influent, wastewater effluent, river and drinking water. The recoveries for standards in nanopure water were all above 60 % except for fluoxetine and indomethacin (Tables I and II). For all of the compounds except metoprolol, significantly lower recoveries were determined in actual water samples, including tap water. These low recoveries introduce uncertainty into the determination of the concentration of the compounds in the actual samples, but do not impact the determination of their presence or absence. At low concentrations for the evaluated compounds in water samples, the high uncertainties in quantitation are not surprising.

Water samples were injected three times each into the LC/MS/MS to ensure adequate reproducibility. Wastewater influent samples were analyzed and showed the presence of all five basic compounds (Figure 2) at the appropriate retention times. The wastewater effluent, river water (before and after the effluent discharge), and tap water, also showed the presence of all five basic compounds at the appropriate retention times (Figure 2). For the acidic compounds four out of the five compounds studied were detected in influent and effluent samples from the sewage treatment plant, and two out of the five compounds were detected in the river water samples (see Figure 3).

To confirm the presence of the compounds in influent and effluent extracts an MS/MS experiment was performed. River water samples contained far less analyte than influent and effluent, making MS/MS analysis impossible because of interfering background ions. For the MS/MS experiment, the first quadrupole was fixed at the selected precursor m/z at the appropriate LC retention time. Fragmentation was by collision induced dissociation with argon in a collision hexapole. The final quadrupole was set to scan for fragments. The spectra collected for standard samples and wastewater samples were

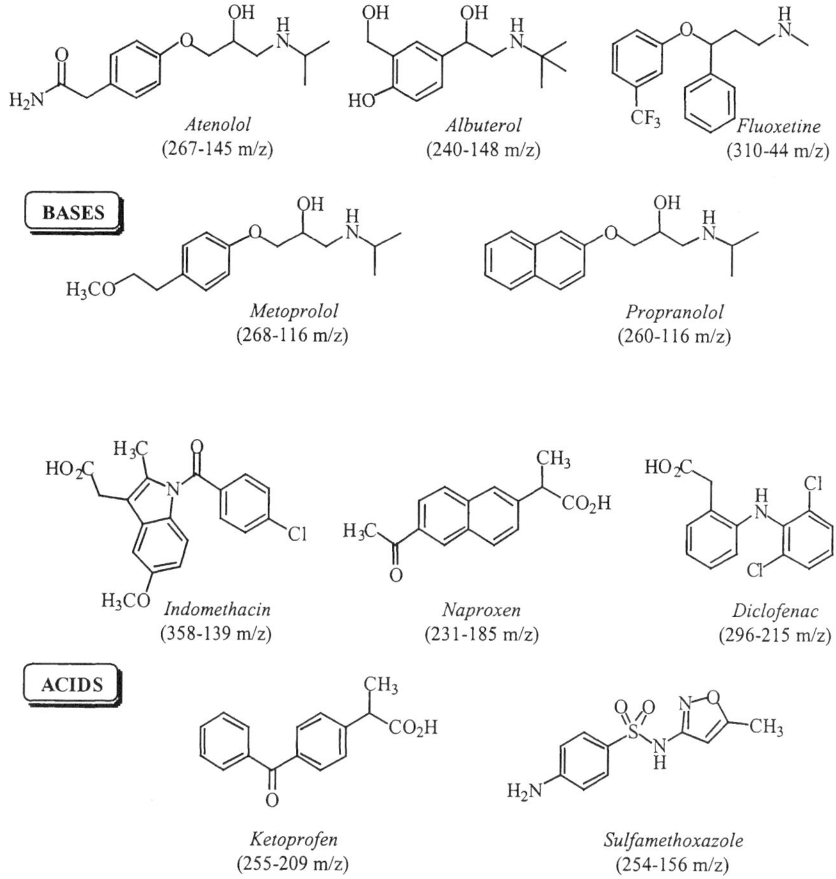

Figure 1. Structures and precursor [M+H]$^+$-product transitions of the target human pharmaceuticals.

compared for each drug. For a positive confirmation, two criteria were met: retention times of compounds had to match those of corresponding standards, and all of the major ions present in the mass spectra for the standard needed to be present in the water sample spectra. Because the limits of detection were higher with the scanning experiment, additional ions present in the wastewater product ion spectra appeared and were due to the matrix components not removed by SPE or are part of the background. Positive confirmation of all five basic compounds and four acidic compounds in wastewater samples were obtained. Sample spectra are shown in Figure 4 (bases) and Figure 5 (acids).

After confirmation of the target compounds in water samples, the next step was to determine the concentration levels in the samples. To increase the

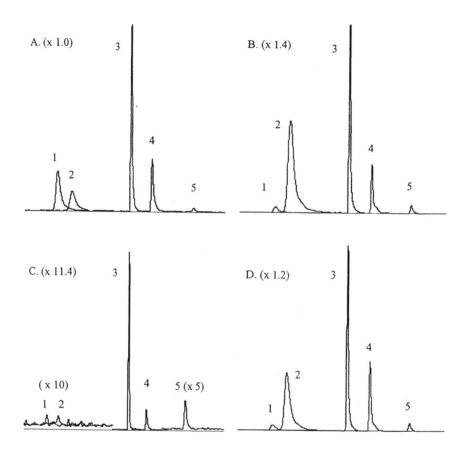

Figure 2. Overlay of typical total ion chromatograms for the basic pharmaceuticals. Shown is a 50 nM Standard (A), influent wastewater (2/13/02) (B), tap water (9/10/02) (C), effluent wastewater (2/13/02) (D), river (sampled on 9/10/02 before effluent discharge) (E), and river (sampled on 9/10/02 after effluent discharge) (F), total ion chromatogram. Peak identification is as follows: albuterol (1), atenolol (2), metoprolol (3), propranolol (4), and fluoxetine (5).

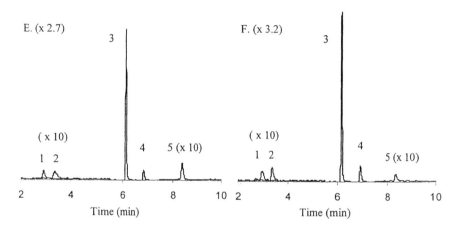

Figure 2. *Continued.*

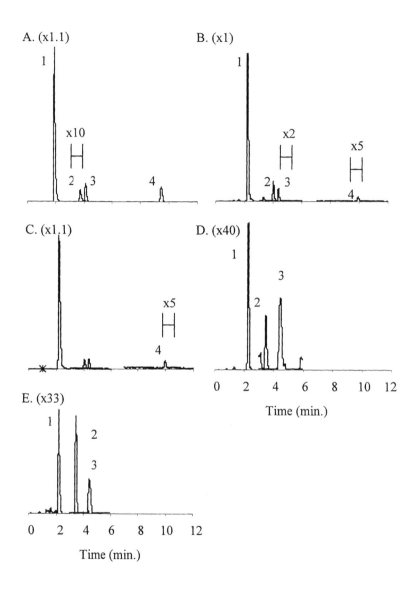

Figure 3. Overlay of typical total ion chromatograms for the acidic pharmaceuticals. Shown is a 250 nM Standard (A), influent wastewater (4/12/02) (B), effluent wastewater (4/12/02) (C), river (sampled before effluent discharge, 5/14/02) (D), and river (sampled after effluent discharge, 5/14/02) (E), total ion chromatogram. Peak identification is as follows: sulfamethoxazole (1), naproxen (2), ketoprofen (3), and diclofenac (4).

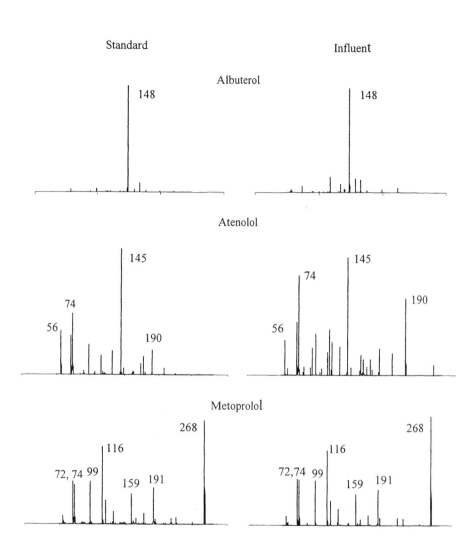

Figure 4. Mass spectra of product ions from basic compounds obtained from LC/MS/MS showing confirmation of species in real water samples compared to known standards.

Continued on next page.

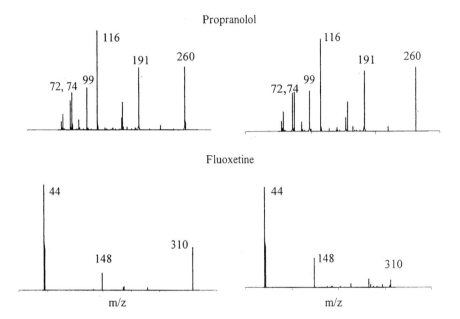

Figure 4. *Continued.*

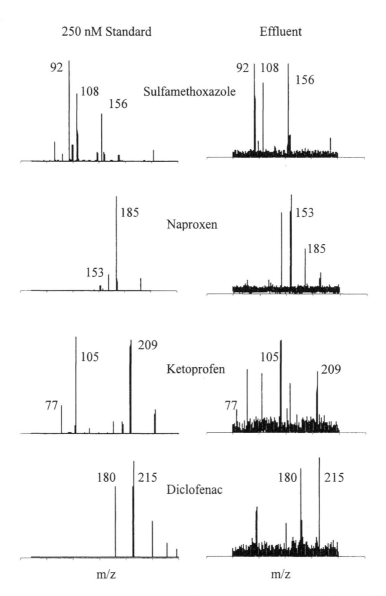

Figure 5. Mass spectra of product ions from acidic compounds obtained from LC/MS/MS showing confirmation of species in real water samples compared to known standards.

sensitivity and lower the limits of detection for quantitative analysis, an MRM experiment was performed. The first quadrupole was fixed to a selected precursor m/z during the time expected for LC separation. Fragmentation was induced in the collision hexapole with argon gas and an energy offset on the cell of 10-20 eV. The MS2 quadrupole was fixed to transmit the most abundant product ion (in this case, the [M+H]$^+$ ion). Two criteria were met for qualitative confirmation: the retention times of the compounds from water samples correlated with the standards, and only the compounds with the selected precursor-product transition made it to the detector. The precursor [M+H]$^+$-product transitions as found in the literature and confirmed experimentally were as follows: albuterol (240-148 m/z), atenolol (267-145 m/z)(7), metoprolol (268-116 m/z)(7, 8), propranolol (260-116 m/z)(7, 9), fluoxetine (310-44 m/z)(10), diclofenac (296-215 m/z), naproxen (231-185 m/z)(11), ketoprofen (255-209 m/z)(12,13), sulfamethoxazole (254-156 m/z), and indomethacin (358-139 m/z)(14).

For quantitation experiments, standards were run to determine the limits of detection of the compounds studied and to establish a concentration curve. Limits of detection (LOD) were determined at an S/N of 3 and limits of quantitation (LOQ) were determined at an S/N of 10. The concentration curve for the basic compound analysis was from 1 nM to 400 nM injected. All five basic compounds responded linearly. The LODs for all five basic compounds were found to be 0.5 nM injected, which is equivalent to 0.12 to 0.15 ng/L in water samples. This was a great improvement over the μg/L LODs obtained using UV absorbance detection or a single quadrupole mass spectrometer and the μg/L concentrations previously used for reporting levels (5). The concentration curve obtained for the acidic compound analysis was run from 5 nM to 1000 nM injected. The LODs ranged from 5 nM-100 nM injected, depending on the compound, corresponding to 1.3 to 23 ng/L in water samples (Table III). The concentrations of the pharmaceutical compounds determined in the water samples collected are listed in Table III (acids) and Table IV (bases).

While the preferred method to quantitate with LC/MS involves spiking each sample and standard with a deuterium-labeled internal standard, such standards were not readily available for each compound involved in this study. Therefore quantitation of the basic analytes was initially accomplished by a least squares analysis of externally run calibration curves. Because of reproducibility issues from the complex wastewater samples run on the LC/MS/MS, a single deuterium-labeled compound, d$_5$-fluoxetine, was selected and used as the internal standard for all of the compounds. The internal standard concentration was kept constant in each standard and sample. Concentration curves were plotted as the peak area ratio (internal standard area / standard area) vs. concentration of the standard. The peak area ratio corrects for the changing ionization efficiency or drop in signal over the course of injections. The peak area ratio (internal standard area / sample area) from the real samples was then

Table III: Concentrations of acidic compounds in water extracts

Sample (date collected)	Sulfamethoxazole (ng/L)	Naproxen (ng/L)	Ketoprofen (ng/L)	Diclofenac (ng/L)	Indomethacin (ng/L)
Detection Limit[a]	1.3	23.0	1.3	1.5	1.8
Influent (4/12/02)	100 ± 32	2800 ± 1200	14.9 ± 1.2	18.5 ± 4.9	BLD[b]
Effluent (4/12/02)	143.0 ± 13.1[c]	1100 ± 210	26.1 ± 7.3	32.2 ± 19.8	BLD
Influent (5/14/02)	194 ± 11	588 ± 106	BLD	BLD	BLD
Effluent (5/14/02)	64.5 ± 15.0	61 ± 17	BLD	BLD	BLD
River Before (5/14/02)	4.6 ± 1.2	BLD	15.5 ± 2.8	BLD	BLD
River After (5/14/02)	4.3 ± 2.1	BLD	22.7 ± 6.9	BLD	BLD
River Before (9/19/02)	120.0 ± 8.8	BLD	BLD	BLD	BLD
River After (9/19/02)	754 ± 146	BLD	BLD	BLD	BLD
Tap Water (5/16/02)	BLD	BLD	BLD	BLD	BLD
Tap Water (9/20/02)	BLD	BLD	BLD	BLD	BLD

[a]Corresponds to water sample values, not injected values, determined for S/N = 3.

[b]BLD = below limit of detection.

[c]Two extractions analyzed.

Table IV: Concentrations of basic compounds in water extracts

Sample (date collected)	Albuterol (ng/L)	Atenolol (ng/L)	Metoprolol (ng/L)	Propranolol (ng/L)	Fluoxetine (ng/L)
Influent (2/13/02)[a]	1.0 ± 0.8	73.7 ± 3.8	12.1 ± 2.3	8.6 ± 3.3	15.3 ± 4.0
Influent (5/14/02)[b]	0.6 ± 0.2	57.3 ± 5.0	33.9 ± 3.2	16.1 ± 1.5	12.9 ± 0.7
Effluent (2/13/02)[a]	1.4 ± 1.4	46.8 ± 3.6	14.2 ± 1.1	14.3 ± 5.0	16.4 ± 3.2
Effluent (5/14/02)[b]	BLQ[c]	21.9 ± 6.9	8.4 ± 3.8	7.3 ± 3.2	12.8 ± 2.0
River Before (7/20/02)[b]	BLQ	0.9 ± 0.1	BLQ	BLQ	1.5 ± 0.9
River Before (9/10/02)[b]	1.2 ± 0.3	1.0 ± 0.2	7.7 ± 4.8	1.6 ± 0.7	1.8 ± 0.5
River After (7/20/02)[b]	BLQ	1.0 ± 0.2	0.5 ± 0.2	0.6 ± 0.3	0.4 ± 0.2
River After (9/10/02)[b]	1.1 ± 0.2	1.1 ± 0.1	10.9 ± 3.8	2.2 ± 0.3	1.2 ± 0.1
Tap Water (7/20/02)[b]	BLQ	BLD[d]	BLQ	BLQ	BLQ
Tap Water (9/10/02)[b]	BLQ	0.7 ± 0.1	4.1 ± 2.7	1.3 ± 0.6	1.5 ± 0.3

[a]Concentration determined using an external standard curve.

[b]Concentration determined using a deuterated internal standard.

[c]Peak detected, but below the limit of quantitation (BLQ).

[d]BLD = Below limit of detection.

compared to the standard curve to determine the concentration of each compound in the real water samples. For each water sample, three extractions were performed in parallel and three LC/MS/MS injections were made of each extract. Table IV therefore lists the uncertainty in the extraction and injection reproducibility.

The acidic compound quantitation was performed using standard addition. An approximate determination of concentration in water samples was determined by a simple five-point external calibration curve. Then the samples were spiked with concentrations that increased the signal approximately 1.5 times the original sample. After analysis of all spiked samples was accomplished, results were plotted as analyte response vs. concentration of added analyte. The original analyte concentration was determined through extrapolation of the line created by the data points through the x-axis. Table III lists the concentrations determined for the acidic compounds.

The concentrations of the basic and acidic compounds in effluent were equivalent to or slightly lower than that in influent. While levels of these compounds exist in the wastewater effluent, it was unknown if these levels decreased through the treatment process. No conclusions could be made concerning the ability of the sewage treatment plant to remove pharmaceuticals from wastewater because the influent and effluent samples came from two different batches of water. The influent had only recently been at the plant and had not made its way through the treatment process while the effluent had been through the various stages and was ready to be introduced back into the environment.

Conclusions

LC/MS/MS methods have been developed to successfully detect and quantify human pharmaceuticals in the environment. The low limits desired for the study are achieved with the MRM technique. Confirmation of compounds was possible in three dimensions: correlating LC retention times, complementary mass spectra, and the shared characteristic precursor-product transition. Solid phase extraction provides a good method for sample cleanup and pre-concentration.

Detectable levels of several pharmaceuticals have been found in both the wastewater influent and effluent. Subsequent studies of environmental samples (river and tap water) also show the presence of these compounds, although in general these concentrations are greatly reduced. Further studies are needed to determine the ramifications of this phenomenon.

Acknowledgements

The authors wish to acknowledge Homigol Biesiada for her support with the mass spectrometer and Joel Moore for help with structures.

References

1. Richardson, M.L.; Bowron, M.J. *J. Pharm. Pharmacol.* **1985,** 37, 1-12.
2. Heberer, Th.; Schmidt-Baumler, K.; Stan, H.J. *Acta Hydrochim. Hydrobiol.* **1998,** *26,* 272-278.
3. Ahrer, W.; Scherwenk, E.; Buchberger, W. *J. Chromatogr. A.* **2001,** *910,* 69-78.
4. Daughton, C.G.; Ternes, T.A. *Environmental Health Perspectives* **1999,** *107,* 907-938.
5. Kolpin, D.W.; Furlong, E.T.; Meyer, M.T.; Thurman, E.M.; Zaugg, S.D.; Barber, L.B.; Buxton, H.T. *Environ. Sci. Technol.* **2002,** *36,* 1202-1211.
6. www.rxlist.com/top200.htm
7. Gergov, M.; Robson, J.N.; Duchoslav, E.; Ojanpera, I. *J. Mass Spectrom.* **2000,** *35,* 912-918.
8. Walhagen, A.; Edholm, L.E.; Heeremans, C.E.M.; Van Der Hoeven, R.A.M.; Niessen, W.M.A.; Tjaden, U.R.; Van Der Greef, J. *J. Chromatogr.* **1989,** *474,* 257-263.
9. Beaudry, F.; Yves Le Blanc, J.C.; Coutu, M.; Ramier, I.; Moreau, J.P.; Brown, N.K. *Biomed. Chromatogr.* **1999,** *13,* 363-369.
10. Sutherland, F.C.W.; Badenhorst, D.; de Jager, A.D.; Scanes, T.; Hundt, H.K.L.; Swart, K.J.; Hundt, A.F. *J. Chromatogr. A* **2001,** *914,* 45-51.
11. Morden, W.; Wilson, I.D. *Rapid Commun. Mass Spectrom.* **1996,** *10,* 1951-1955.
12. Hoke II, S.H.; Pinkston, J.D.; Bailey, R.E.; Tanguay, S.L.; Eichold, T.H. *Anal. Chem.* **2000,** *72,* 4235-4241.
13. de Kanel, J.; Vickery, W.E.; Diamond, F.X. *J. Am. Soc. Mass Spectrom.* **1998,** *9,* 255-257.
14. Taylor, P.J.; Jones, C.E.; Dodds, H.M.; Hogan, N.S.; Johnson, A.G. *Ther. Drug Monit.* **1998,** *20,* 691-696.

Chapter 13

Using the LC–Atmospheric Pressure Ionization–Ion Trap for the Analysis of Steroids in Water

Paul Zavitsanos

Agilent Technologies, 2850 Centerville Road, Wilmington, DE 19808

The determination of many synthetic estrogens in aquatic environments can be a difficult task. The combination of high potency, severe biological impact and poor analytical characteristics makes for a dangerous environmental situation. The innate sensitivity of modern triple-quadrupole mass spectrometers, in MRM mode, can often overcome the analytical shortcomings of such compounds, but that success is usually at the cost of the spectral data. Ion trap mass spectrometers offer the promise of high sensitivity while retaining the full spectral data, as well as the potential for simpler MS/MS spectra and the possibility of automatic MS/MS. This work uses a model set of estrogens, both synthetic and endogenous, to show that the ion trap allows the routine capture of full MS/MS spectra with less than 5 ng of compound on column. With a suitable extraction method, and in full MS/MS scan mode, a level of 50 ng/L in water is feasible. Chromatographic fidelity is very good at these levels and the performance is maintained even in the automatic MS/MS mode. The advantages of scan data in the reconstruction of compounds with complex MS/MS spectra are shown. The ionization is investigated in APCI and APPI and the impact of vaporizer temperature is studied. The reverse phase retention characteristics of the model compounds indicate that direct large volume injection from water is feasible.

Introduction

The presence of female sex hormones, both endogenous and synthetic, in the aquatic environment has raised great environmental concern over the past decade. In concentrations as low as 0.1μg/L, these compounds have been shown to induce changes in endocrine function. (1-3). One of the potential routes of exposure for aquatic fauna is through contact with contaminated surface water. Mammals excrete the largest portion of both the natural and the synthetic estrogens and progestogens via the urine.

These compounds, by way of excreted urine, can enter the surface water through discharge of industrial and domestic wastewater, sewage treatment plants and both direct agricultural drainage and runoff into streams from rainfall. These compounds have been reported to occur in the low ng/L range up to the tens of ng/L range (4). In a few cases there were reports of μg/L levels of some of these compounds (5-7). In addition, the matrices range from the fairly complex case of surface water to the extremely challenging case of sewage and sediments.

The analysis of these compounds, at these low levels present in the environment, is a difficult undertaking given the complexity of the typical matrix. It is especially difficult considering that the analysis of environmental samples has both a quantitative and a qualitative, or confirmatory, component. The impact of this situation on the practice of mass spectrometry for the analysis of environmental samples is profound. Not only does the analysis technique have to display chromatographic sensitivity and selectivity, but it must also supply sufficient qualitative information to satisfy the qualitative, or confirmatory, aspects of the analysis. Chromatographic retention time, an adequate measure of identification in less demanding analysis, is not considered sufficient as a measure of confirmation for these complex samples.

Typically quantitative analysis for these samples is carried out in selected ion monitoring (SIM) using a single stage linear quadrupole mass spectrometer or multiple reaction monitoring (MRM) using a triple linear quadrupole mass spectrometer. Qualitative aspects are most often addressed by monitoring several ions or in the case of MS/MS, several product ions. The ratios between these ions constitute an overly simplified approximation of the spectrum.

The ideal mass spectrometric technique would embody both, a high chromatographic sensitivity for the quantitative aspects and a high spectral sensitivity for the qualitative aspects; it would supply highly sensitive chromatograms and full, high sensitivity spectra of the compound throughout the analytical range. In addition MS^2 spectra would be reproducible from instrument to instrument and independent of concentration. Such an instrument would allow the conception of useful MS^2 spectral libraries and confirmations could occur on the basis of library matches and not upon simple ion ratio measurements.

Ion trap mass spectrometers have very high scan sensitivity especially in MS/MS mode and are able to give excellent spectra at levels that linear quadrupoles cannot give. While both single quadrupoles and triple quadrupoles can give excellent chromatographic sensitivity in SIM and MRM respectively, that sensitivity is rapidly degraded once the instruments are operated in their respective scan modes. The linear quadrupoles must be satisfied with ion ratios as a crude approximation to the spectrum of the compound. Ion traps can use the full MS/MS spectrum for the confirmation aspects of these analytes.

Ion trap mass spectrometers can also perform data dependent experiments, where the outcome of one scan defines the following scan experiment. An MS scan may, for instance, identify a number of precursor ions as exceeding a preset threshold. That information will then define a subsequent MS/MS experiment for each of those ions. This mode is generally referred to as "automatic MS/MS". Both experiments can be concluded in about 300 msec. allowing accurate representation of the chromatographic data, in both MS (MS^1) and MS/MS (MS^2) modes, simultaneously. Such a capability is useful if one is engaged in the search for true unknowns, where the precursor ion's mass to charge ratio is not known. In linear triple quadrupole instruments such data dependent experiments are difficult to conduct in chromatographic time frames.

Product ions can also be isolated and fragmented leading to MS^3 and higher order experiments at the cost of time. Since the chromatographic time frame is limited, there is a useful limit to the number of experiments that can be conducted on a given parent ion and still maintain chromatographic fidelity.

There are advantages to acquiring the full scan data in post-run processing as well. In MS^2 experiments in the ion trap, the entire product ion spectrum is available and multiple product ions can be used in an additive fashion to reconstruct the chromatographic signal. This characteristic can be valuable in cases where the onset of fragmentation is abrupt and the extent of fragmentation is extensive. When an MS^2 spectrum becomes very complex each individual product ion is a small portion of the total product ion signal. Such extensive fragmentation can limit sensitivity in triple quadrupole instruments. As the data in this article will illustrate, many steroids exhibit those very fragmentation characteristics.

Ion traps can also take advantage of another unique property of their principle of operation. After a precursor ion is isolated, a range of fragmentation energies can be brought to bear on the ion. Moreover, unlike instruments based on linear collision cells, once the product ions are formed they do not continue to undergo fragmentation. Simpler product ion spectra can result.

Both GC/MS and LC/MS techniques have been applied to this analysis in both single stage and tandem MS modes with a variety of analyzers (1-11). Typically the majority of these studies were quantitative analysis. Ion trap analyzers have been used in the analysis of steroids in environmental water samples (11).

The analysis of synthetic and naturally occurring steroids in water would be an excellent application to test the unique capabilities of the ion trap for environmental work. It would test the ion trap's ability to supply high chromatographic and spectral sensitivity and to perform useful data dependent experiments while generating simple yet informative spectra. In addition we wished to investigate the different APCI and APPI modes of operation for sensitivity, spectral content, and selectivity.

Target compounds (Fig.1) were selected from the natural estrogens, their metabolites, and the synthetic compounds common in contraceptive pills. Compounds with phenolic A rings are known to have good responses in ESI whereas compounds with 3-keto functionality are more challenging. Moreover these 3-keto contraceptives comprise a large portion of prescribed contraceptives.

Experimental Methods

The instrument used in this work was an Agilent G2440AA 1100 LC-DAD/MSD TRAP. The system is comprised of a binary pump with vacuum degassing unit, a well-plate auto-sampler with needle wash, a heated/cooled column compartment, a diode-array detector DAD, an ion trap mass spectrometer with the APCI source and an isocratic pump for post-column solvent addition. An APPI source was used for some parts of this study. All modules were controlled by the Agilent ChemStation data system. The binary pump had the optional mixer removed to reduce the already small delay time even further but there were no other modifications to the system.

Liquid Chromatography

The column used was a Zorbax Eclipse XDB-C18, 5cm long by 2.1mm I.D. filled with 3.5μm particles. The chromatography was performed in gradient mode with solvent A being 0.25mM Ammonium Acetate in water. Solvent B was acetonitrile. The total flow was 0.25 ml/min. Total run time was 9.8

minutes with a 6-minute equilibration time. The gradient followed the time points listed in Table I.

Reducing the length of each time segment by factor 2 and increasing the flow rate to 0.5 ml/min could halve the analysis times and still retain excellent peak shape and chromatographic resolution. The more complex MS^n experiments could then have been compromised however due to the narrow chromatographic peak-widths. The longer analysis times were selected based on spectrometer sampling considerations and not on any column flow rate limitations on the source.

The APPI experiments used methanol instead of acetonitrile as the strong solvent. The ionization potential of methanol allowed better ionization than acetonitrile under APPI conditions. The column temperature was set to 30°C. The DAD was set to collect spectra from 200 to 450 nm and was also set to acquire a number of UV signals.

Table I. Gradient table

Time, min.	%ACN
0.0	20
0.2	35
8.0	90
9.0	90
9.7	20

Mass Spectrometry: Source Conditions

APCI Positive Ion

Drying gas was set at 350°C and at a flow of 8 L/min. Nebulizer gas pressure was set at 30 psi. With the vaporizer temperature set at 475°C. No post column reagents were added. Capillary voltage was set at 3000 V with the corona current set at 5 µA. Under these conditions the corona voltage settled at approximately 2200 V.

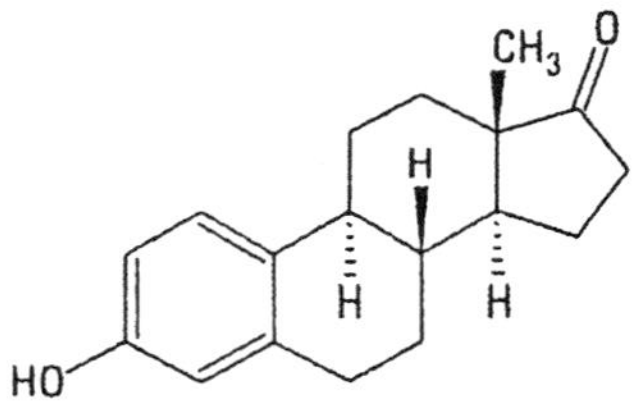

Estradiol

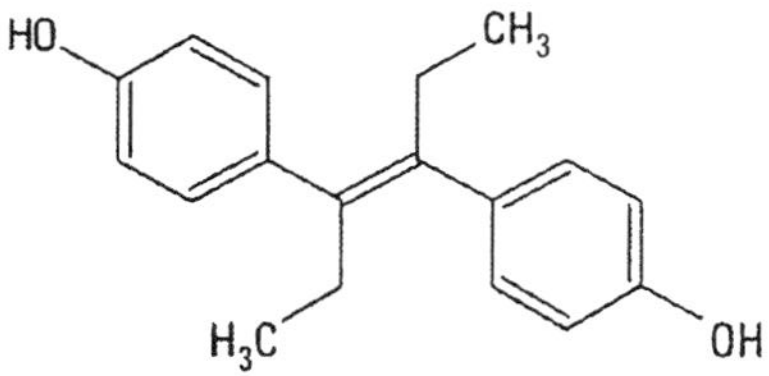

Diethylstilbestrol (DES)

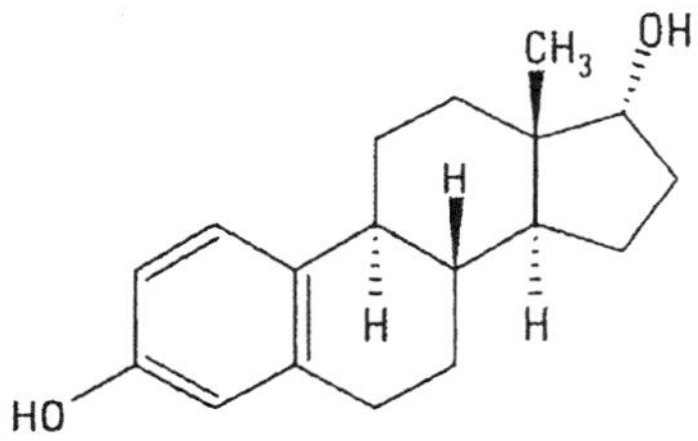

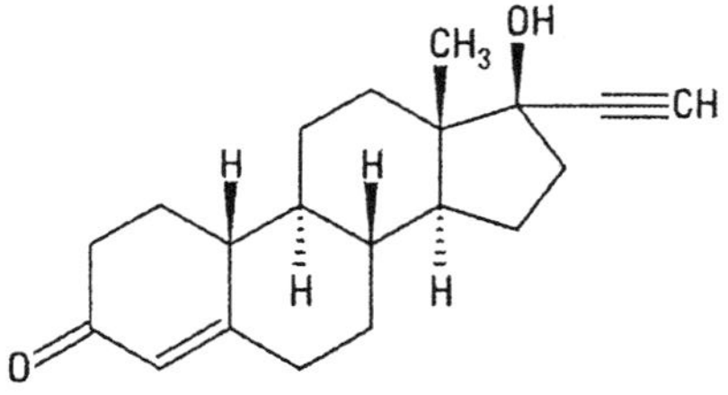

Estriol

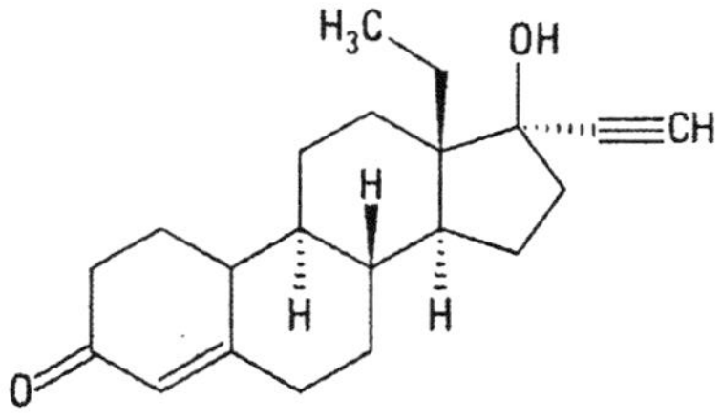

Norgestrel

Estrone

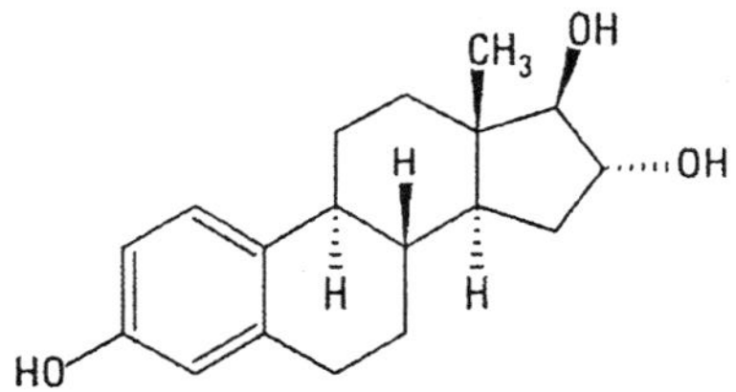

Norethindrone

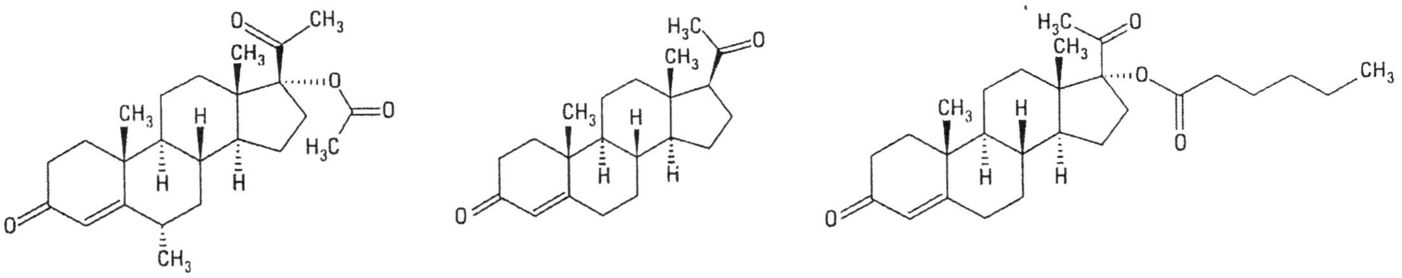

Figure 1. Structures of compounds used in this study

Medroxyprogesterone acetate

Progesterone

Hyroxyprogesterone caproate

APCI Negative Ion

Negative ion conditions were similar to the positive ion conditions with the exception of the capillary voltage and the fact that best conditions were obtained with a 1% addition post column of CH_2Cl_2. The capillary voltage was set to 1500 V and under these conditions the corona voltage was approximately 1800 V.

APPI Negative Ion

Drying gas was set at 350°C and at a flow of 8L/min. Nebulizer gas pressure was set at 30psi with the vaporizer temperature set at 475°C. Acetone, at a rate of 10% of total flow, was added as post column reagent to aid with ionization. Capillary voltage was set to 3000 V.

Mass Spectrometry: Analyzer Conditions

MS^1

Scan range was 150-500 m/z. Fragmentor voltage set at 70 volts. Trap Drive set at 48. Maximum accumulation time was set at 100 msec with spectral averaging set to 3 spectra per data point. The MS^2 modes employed in-trap CID fragmentation conditions that were defined by the amplitude voltage and the voltage ramp on the end-cap electrodes. The voltage ramp ensured that a range of fragmentation energies was applied to each precursor ion. Such a mode increased the likelihood of generating useful spectra as compared to a single voltage setting. The fragmentation voltage was set to 1 volt with the voltage ramp running from 30 to 200% of that value over the 40-msec fragmentation period.

AUTO MS^2

In Auto MS/MS mode the both FastCalc modes and SmartFrag modes of operation were used to set the cutoff value and to optimize the fragmentation voltage automatically. The number of spectral averages was set to one to maximize the number of automatic MS/MS experiments for the expected chromatographic peak width. Thresholds and exclusion masses were set based on the background. The number of allowable precursor ions was set to two.

MRM

The precursor ions for each compound were set in the MRM table and product ion scans were collected. The MRM chromatogram was re-constructed in data analysis by extracting the selected product ion.

Chemicals and Reagents

All steroid standards were purchased from the USP as reference standards and were of the highest purity. Diethylstilbestrol was purchased from Aldrich and was of technical grade. Solvents were purchased from Burdick and Jackson. Water was generated by a Milli-Q system. All stock solutions and dilutions were made up in isopropyl alcohol at approximately 0.75 mg/ml with the exception of the water solutions. The water solutions were made by volumetric dilutions of isopropyl alcohol stock solution aliquots with Milli-Q water. Nitrogen used for MS drying and nebulization gas was from an in-house source of boil-off from a large liquid nitrogen dewer.

Results and Discussion

Figure 1 shows the structures of the compounds used in this study. Four of the compounds (estrone, estriol, estradiol, and diethylstilbestrol) posses the phenolic "A" ring functionality but diethylstilbestrol does not possess the typical 4 ring backbone of steroids. These compounds ionize well in negative ion ESI and there is a fair body of work that details that success; moreover the phenolic group is readily derivatized to enhance ionization if required. There are five synthetic compounds that contain the 3-keto functionality (norethindrone, norgestrel, progesterone, medroxyprogesterone acetate, and hydroxyprogesterone caproate). The 3-keto compounds are more difficult to ionize in ESI than those that contain the phenolic functionality in the "A" ring. The high potency of these synthetics and the challenging ionization characteristics combine to yield a potentially dangerous situation in environmental analysis. This study focuses on investigating techniques that will give satisfactory performance for all the compounds especially the difficult to ionize, such as 3-keto compounds.

Two of the 3-keto based compounds, norgestrel and norethindrone, contain acetylenyl moiety in the 17 position whereas two others are based on esterified 17 hydroxy functional groups. It is of interest to see how the change of groups in the 17 position affects MS/MS fragmentation.

The MS data for all techniques is summarized in Table II. Positive ion APCI was the only technique in this study that gave uniform ionization for all compounds under a single set of experimental conditions. Norgestrel and norethindrone did not ionize in negative ion mode under these and other similar negative ion APCI conditions. diethylstilbestrol was also well ionized in negative ion APCI but the impurity in diethylstilbestrol (DES1) was more sensitive than DES itself. APPI was extremely selective, and only ionized estrone, estradiol and estriol with any effectiveness. The comparative ionization of two, very similar compounds, medroxyprogesterone acetate and hydroxyprogesterone caproate, serves as a demonstration of APPI selectivity. Hydroxyprogesterone caproate ionizes very well whereas medroxyprogesterone acetate does not ionize significantly. Positive ion APCI displays a consistent ionization across the entire compound in the study and is the mode selected for MS^2 investigations.

Figure 2 shows a full scan positive ion APCI MS chromatogram of a mixture of all the compounds in Figure 1 at the 50ng on-column level. Diethylstilbestrol (DES) shows two peaks DES1 and DES2. The first peak DES1 is a minor impurity in the standard but it has greater response factor than the major component DES2. There are a number of other minor impurities related to the diethylstilbestrol standard in the 4-5.8 minute retention time range. One of these impurities co-elutes with diethylstilbestrol and has a m/z at 267 rather than at the 269m/z corresponding to diethylstilbestrol. There is another 267m/z impurity at an earlier retention time as well, but it does not interfere with other compounds. There was the concern that the 267m/z ion may be the result of a dehydrogenation, or a neutral loss of H_2, that may be occurring in the APCI source. The relative elution behavior of the diethylstilbestrol at 269m/z and the interfering impurity at 267m/z, as evidenced by MS^1 extracted ion profiles, indicated that there are two separate compounds eluting at slightly different times.

Moreover, diode array detector UV data confirms that these are indeed two different compounds, with different UV spectra, and that they elute at slightly different retention times. In the UV, the interfering impurity is small (ca.1-5%) when compared to the diethylstilbestrol; the MS data indicates that the impurity is much more sensitive in positive ion APCI than the diethylstilbestrol and consequently appears as a significant interference.

Table II. Ions Observed Using Different Atmospheric Pressure Ionization Modes

No.	Steroid	MW	APCI positive ion			APCI negative ion, CH2Cl2			APPI negative ion, acetone		
			Expected m/z	Measured m/z	Ion	Expected m/z	Measured m/z	Ion	Expected m/z	Measured m/z	Ion
1	Estriol	288	289	271	$[M+H-H_2O]^+$	287	287	$[M-H^+]^-$	287	287	$[M-H^+]^-$
2	Estradiol	272	273	255	$[M+H-H_2O]^+$	271	271	$[M-H^+]^-$	271	271	$[M-H^+]^-$
3	Estrone	270	271	271	$[M+H]^+$	269	269	$[M-H^+]^-$	269	269	$[M-H^+]^-$
4	Norethindrone	298	299	299	$[M+H]^+$	297	Not detected		297	Not detected	
5	Norgestrel	312	313	313	$[M+H]^+$	311	Not detected		311	Not detected	
6	Progesterone	314	315	315	$[M+H]^+$	313	313	$[M-H^+]^-$	313	Not detected	
7	Medroxyprogesterone acetate	386	387	387 327	$[M+H]^+$ $[M+H-60]^+$	385	385	$[M-H^+]^-$	385	Not detected	
8	Hydroxyprogesterone caproate	428	429	429 313	$[M+H]^+$ $[M+H-116]^+$	427	427	$[M-H^+]^-$	427	427	$[M-H^+]^-$

218

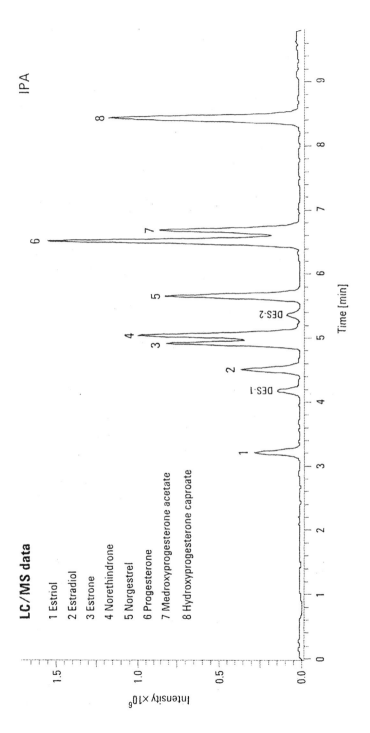

Figure 2. *Typical chromatogram in positive ion APCI MS. Amount injected was 50ng on-column*

The sensitivity, both chromatographic and spectrometric, is very good for all compounds by published standards, including the 3-keto compounds, especially considering that these are scan data. All compounds with the exception of DES gave good response with 5 ng on column. Most sensitivity numbers in the literature are comparable, but they are based on SIM or MRM experiments using linear quadrupole instruments; the spectral information is lost. The data in this work was acquired in scan mode with the spectra full stored for later inspection. In automatic MS/MS experiments, both the MS^1 and MS^2 spectra are stored. The chromatographic fidelity and peak shape is also good indicating that the sampling statistics were favorable.

Figure 3 shows the spectral quality that is afforded by the ion trap at these concentrations and with these compounds. These data are collected in automatic MS/MS mode; both MS^1 and MS^2 data are collected for each chromatographic peak. MS^2 data are collected for up to 2 independent precursor ions from the MS^1 scan. Base peak intensity is factor 100 or more above spectral noise indicating excellent spectral sensitivity. This spectral sensitivity suggests that the ion trap may allow the use of full scan data for the qualitative identification of pollutants whereas instruments based on linear quadrupoles would defer to multiple reaction monitoring from a single precursor.

Most of these compounds are characterized by a strong $[M+H]^+$ ion in positive ion APCI. Both estriol and estradiol demonstrate a facile loss of water from $[M+H]^+$ with base peak being $[M+H-H2O]^+$. This loss is difficult to stop; even by lowering the up-front CID energy to the point where ion transmission is affected.

It is interesting to note that, as can be seen in Figure 3, the 3-keto, 10 hydroxy esters (medroxyprogesterone acetate and hydroxyprogesterone caproate) both demonstrate a facile loss of the acid moiety. This provides a significant up-front CID fragment ion for the corresponding 17 OH compounds. The contribution of this fragment ion to total ionization can be significantly reduced with very low up-front CID voltages, but ion transmission may be adversely affected.

The spectrum of diethylstilbestrol clearly shows the contamination of the spectrum by a minor but intensely sensitive related compound at 267m/z. This is indeed a contamination and not a product of up-front CID. The ratio of 269/267 is not constant over the width of the chromatographic peak; moreover LC/DAD data confirm that the DES peak is not homogeneous. While it may be possible to get a neutral loss of H_2, this seems unlikely at this low an up-front CID energy figure. Figure 4 shows the MS^2 data for the $[M+H]^+$ ion from the MS^1 data detailed in Figure 3. The MS^2 spectra for the $[M+H]^+$ largely mirror the up-front CID data. Medroxyprogesterone acetate and hydroxyprogesterone caproate both undergo a neutral loss of their relative acid ester groups as the major product ion with the simple MS^2 spectra dominated by $[M+H-60]^+$ (loss of

APCI Positive ion

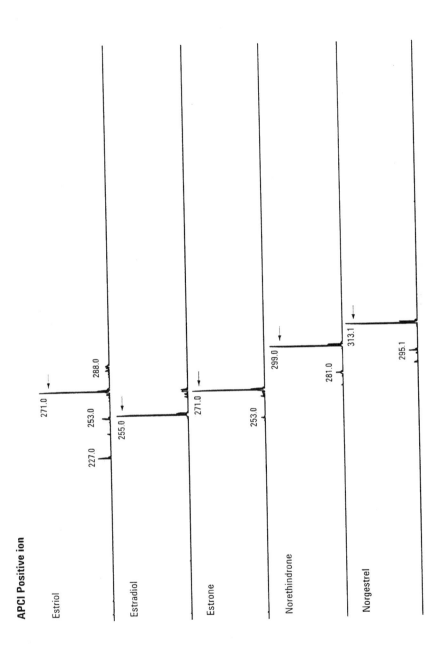

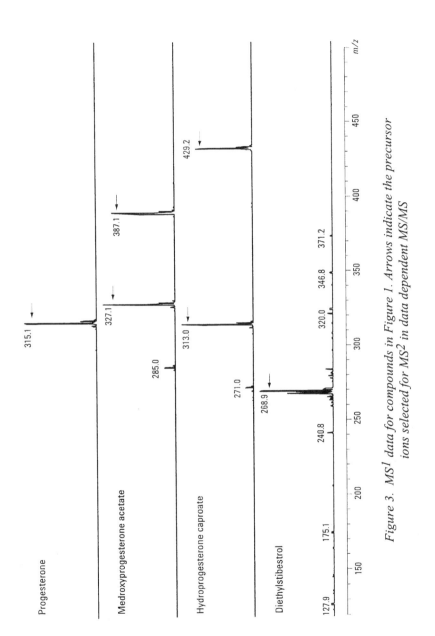

Figure 3. MS^1 data for compounds in Figure 1. Arrows indicate the precursor ions selected for MS^2 in data dependent MS/MS

222

APCI Positive ion MS/MS

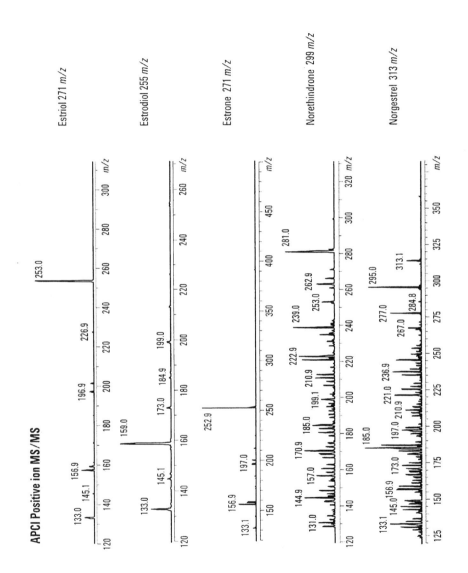

Estriol 271 *m/z*

Estrodiol 255 *m/z*

Estrone 271 *m/z*

Norethindrone 299 *m/z*

Norgestrel 313 *m/z*

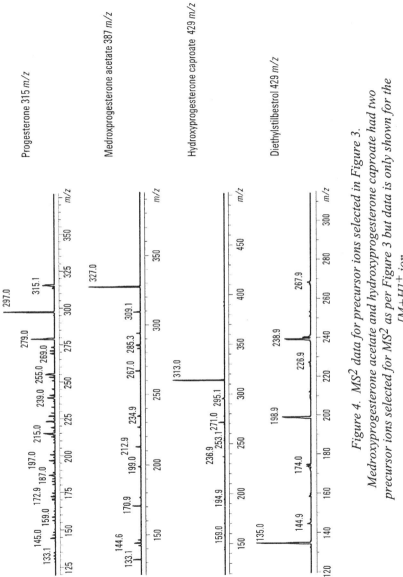

*Figure 4. MS² data for precursor ions selected in Figure 3.
Medroxyprogesterone acetate and hydroxyprogesterone caproate had two
precursor ions selected for MS² as per Figure 3 but data is only shown for the
[M+H]⁺ ion*

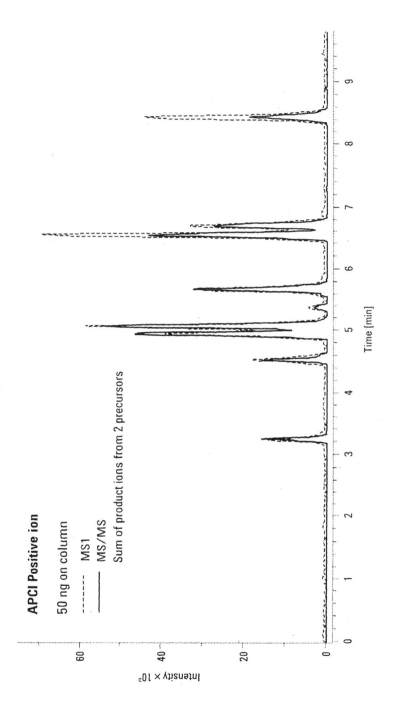

APCI Positive ion

50 ng on column

- - - - - MS1
——— MS/MS

Sum of product ions from 2 precursors

Time [min]

Intensity × 10³

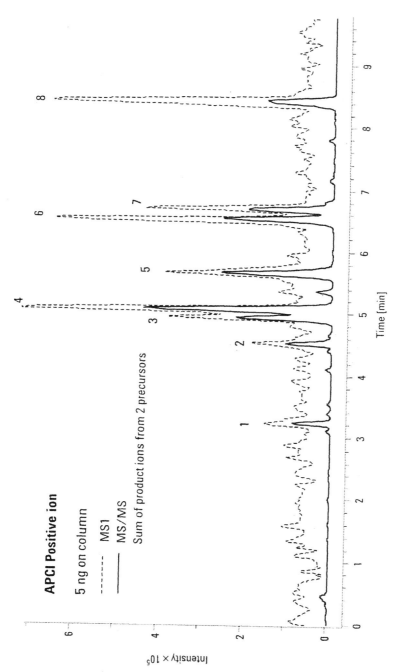

Figure 5. *Positive ion APCI chromatograms using data dependent scanning. Both MS¹ and MS² spectra are acquired*

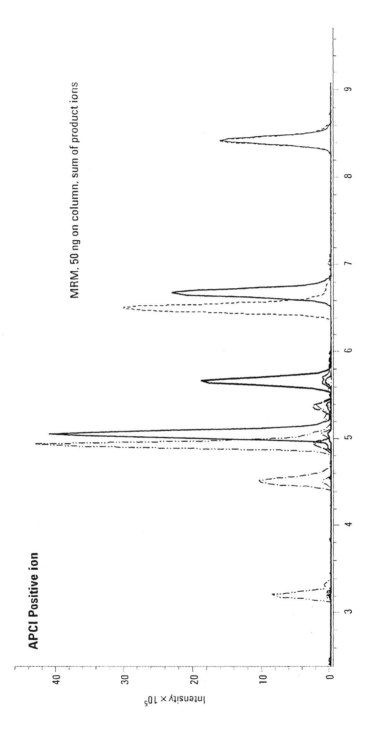

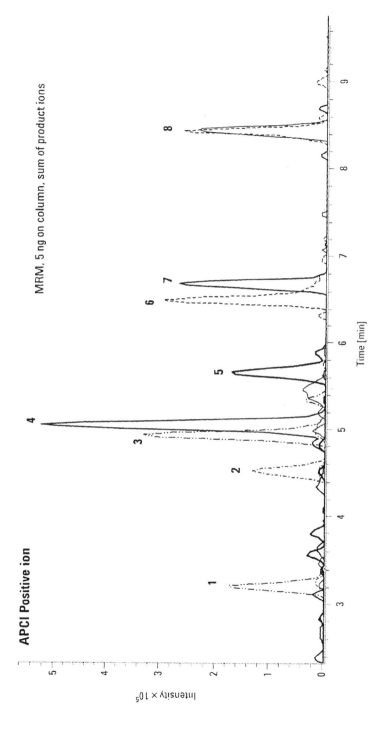

Figure 6. Chromatograms in MRM mode at the 50ng (top) and 5 ng (bottom) on-column level. The ion trap produces full scan MS² spectra in MRM mode.

228

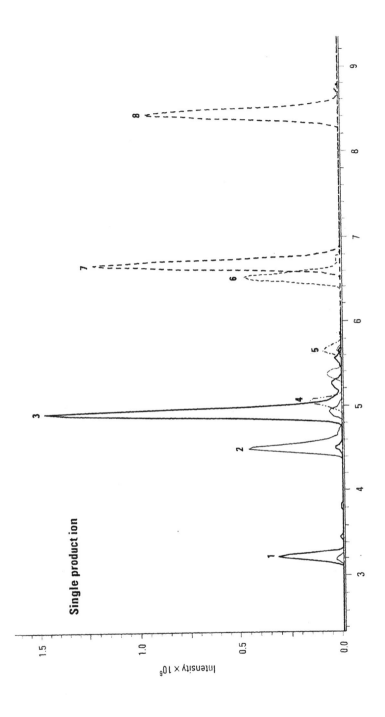

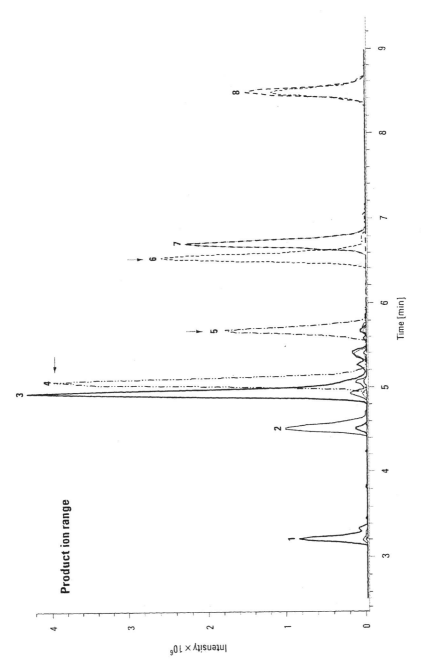

Figure 7. Effect of summing product ion responses in MS². Peaks 4, 5, 6 are norethindrone, norgestrel, and progesterone respectively.

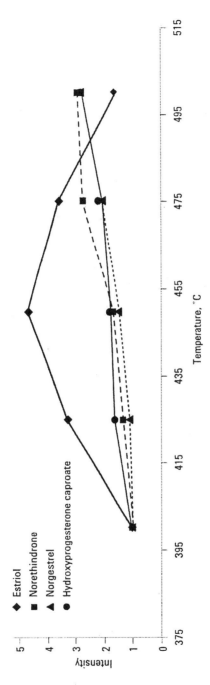

Figure 8. Ion abundance versus APCI vaporizer temperature at 60psig nebulizer pressure

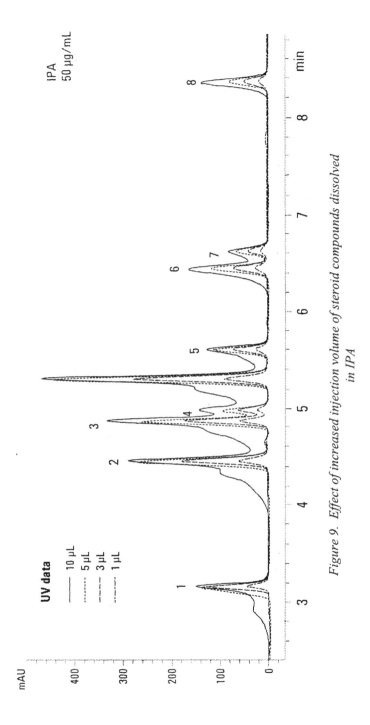

Figure 9. Effect of increased injection volume of steroid compounds dissolved in IPA

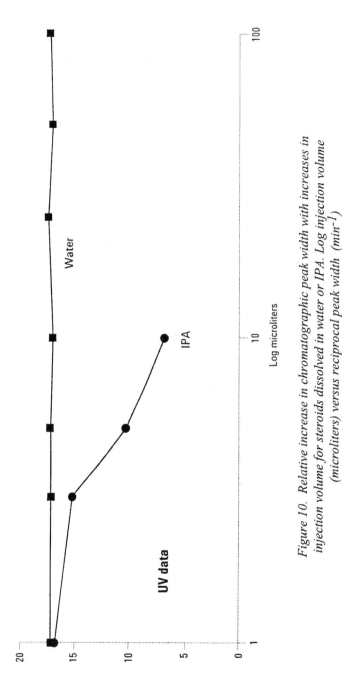

Figure 10. Relative increase in chromatographic peak width with increases in injection volume for steroids dissolved in water or IPA. Log injection volume (microliters) versus reciprocal peak width (min⁻¹)

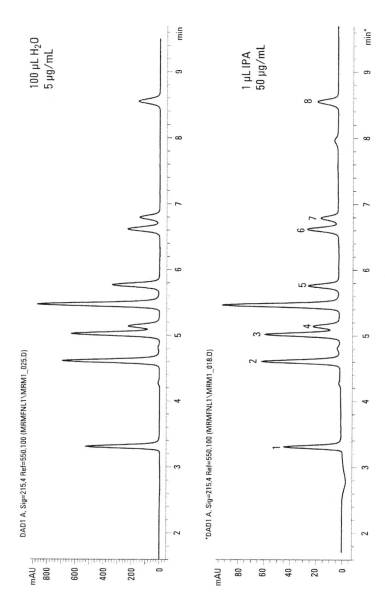

Figure 11. Chromatogram of 100ul of an aqueous solution of steroids. The 1ul injection of an IPA standard mix is included for resolution comparison

acetic acid) and $[M+H-116]^+$ (loss of caproic acid) respectively. There are a number of other ions at 5-10% each of total ionization. Neutral loss of water is the dominant fragmentation pathway for estrone, just as was indicated in MS^1 up-front CID. As indicated above, estriol and estradiol did not give abundant $[M+H]^+$ ions; the $[M+H-H_2O]^+$ ion is the dominant precursor ion in both instances. In the case of estriol, the major MS^2 product ion of the $[M+H-H_2O]^+$ precursor results from a second water loss. Estradiol however shows a major product ion that is not the result of water loss from the precursor ion. Progesterone shows a water loss as the major product ion from the $[M+H]^+$ precursor with the second most abundant ion being the $[M+H-2H_2O]^+$ ion; however the MS^2 spectrum is beginning to show significant complexity at this in-trap CID energy.

The most complex MS^2 data is generated by compounds that possess the 17 acetyleneyl moiety, norethindrone and norgestrel. Both compounds show extensive fragmentation beyond the $[M+H-H_2O]^+$ and the $[M+H-2H_2O]^+$ ions. No single product ion has more than 20% of total ionization. This kind of fragmentation is reminiscent of the EI fragmentation of steroids. MRM experiments with this type of fragmentation would not yield the chromatographic sensitivity typically associated with MRM. Diethylstilbestrol has a major product ion at 239 corresponding to $[M+H-C_2H_6]^+$.

Figure 5 shows chromatograms from the ion trap operated in a data dependent manner, or in this case automatic MS/MS, at the 50ng and 5ng on-column levels. Both MS^1 and MS^2 data are collected in this mode. When MS^1 produced an ion above the set threshold level then, on the subsequent scan, the ion is isolated, fragmented and the product ions scanned producing an MS^2 spectrum. The spectra clearly show sensitivity in scan mode more commonly associated with SIM data in published work.

MRM mode, as displayed in Figure 6, yields excellent data at 5ng on column; even at this level the full MS^2 scan data is available. The full MS^2 spectrum provides more information than two or three ion ratios, especially with the more complex compounds such as norethindrone and norgestrel. The advantage of having all the MS^2 information goes beyond the availability of the MS^2 spectrum. The more complex the MS^2 spectrum is the smaller the percentage of total ionization will be contained in any one product ion. This has the effect of reducing the sensitivity of any MRM technique that monitors a single product ion with such compounds. In the ion trap's case however, the full product ion spectrum is available and a signal based on the sum of a number of product ions can be reconstructed. The resultant signal is superior for those compounds that have complex product ion spectra. The concept is illustrated by the data in Figure 7. Norethindrone, norgestrel, and progesterone show more complex product ion spectra with the first two compounds being significantly more complex than progesterone. Figure 7 shows reconstructed MS^2

chromatograms based on single product ions (top trace), and based on multiple product ions (bottom trace). It can readily be seen in the differences in the abundance of compounds 4 and 5 (norethindrone, norgestrel) and peak 6 (progesterone) that the compounds with the most complex MS^2 spectra benefit the most from summing the product ion signals.

Both APCI and APPI share column effluent nebulization and vaporization as the first step in ionization. These steroid compounds are not highly volatile and given their functionality also offer the possibility of thermal degradation; both parameters are dependent on vaporizer temperature at constant nebulizer pressure. It is logical to suspect that the peak concentration of steroid in the vapor phase would depend on the balance of vaporization and thermal decomposition set by the vaporizer temperature at constant nebulizer pressure for each compound. Figure 8 details the results of an experiment designed to test this hypothesis. Most of the steroids tested benefit from higher vaporizer temperatures; however estriol does not. Optimum vaporizer temperature is 450°C for estriol whereas the other compounds in this chart show maximum intensity at the maximum available temperature of 500°C. A compromise temperature of 475°C was set for the balance of the experiment. Vaporizer temperature and the nebulizer pressure are two instrumental parameters that control this balance between vaporization and thermal decomposition for these compounds and should be carefully studied when working with new estrogen compounds.

Given the excellent MS/MS spectral sensitivity with this trap unit, there is the possibility that large volume injection of samples such as potable water and surface waters may suffice as a sample preparation scheme. This concept is even more viable given the ease of implementation of 2D LC apparatus. The main point to consider is how effectively these steroid compounds retain on conventional 2-mm ID and 4-mm ID columns. Moreover it is vital to achieve this trapping without band broadening, peak splitting or otherwise degrading the chromatography.

Figure 9 shows the effect of injection volume of these compounds dissolved in strong solvent (IPA) on chromatographic integrity. This is overlaid LC/DAD chromatographic data. Figure 10 graphs the relationship of chromatographic peak width versus injection volume from solutions made up in either water or IPA. Loss in chromatographic resolution is dramatic with as little as 5μl of strong solvent whereas solutions made in water can tolerate 100μl injections without affecting peak shapes. Figure 11 shows an LC/DAD comparison of a 1μl injection of IPA based stocks with a 100μl injection of a water based stock. Chromatographic resolution is identical even though the injection volumes differ by 100 times. It is suspected that even larger injections can be made, from water, with these compounds without affecting resolution. LC/DAD was used to perform these experiments because the impact of linearity over a range of 100 to 1000 times on peak shape was not an issue as it would have been in LC/MS.

Conclusions

The LC/API-Trap approach gave excellent spectral sensitivity in both MS^1 and MS^2. In MS^1 the instrument gave scan data at a level usually reserved for SIM data in the recent literature. The CID fragmentation features of the spectrometer give excellent MS^2 spectra over a range of compounds; moreover this can be done in a data dependent manner looking for compounds whose parent ion mass is unknown. The LC characteristics of the compounds allow for effective off-line and on-line concentration techniques. The combination of these performance factors may yield a quantitative technique where the qualitative aspects of the analysis may be satisfied by a spectrum rather than a set of MRM ion ratios.

References

1. Lopez de Alda, M. J.; Barcelo, D., *J. Chrom A.* **2001**, *938*, 145-153.
2. Pelissero C.; Flouriot G.; Foucher J.L.; Bennetau B.; Dunogues J.; LeGac F.; Sumpter, J. *J. Steroid Biochem. Molec. Biol.* **1993**, *44*, 263.
3. Purdom, C.E.; Hardiman, P.A.; Bye, V.J.; Eno, N.C.; Tyler, C.R.; Sumpter, J.P. *Chem. Ecol.* **1994**, *8*, 275.
4. Lopez de Alda, M. J.; Barcelo, D., *J. Chrom. A.* **2000**, *892*, 391-406.
5. Shore, L.S.; Gurevitz, M.; Shemesh, M.; *Bull. Environ. Contam. Tox.* **1993**, *51*, 361.
6. Tabak, H.H.; Bunch, R.L.; Bloomhuff, R.N. *Dev. Ind. Microbiol.* **1991**, *22*, 497.
7. Ternes, T.A.; Stumpf, M.; Mueller, J.; Haberer, K.; Wilken, R-D.; Servos, M. *Sci. Total Environ.* **1999**, *225*, 81.
8. Petrovic M.; Eljarrat E.; Lopez de Alda, M. J.; Barcelo, D., *Trends in Ana.l Chem.* **2001**, *20*, 11.
9. Lopez de Alda, M. J.; Barcelo, D., *J. Chrom. A.* **2001**, *911*, 203-210.
10. Xiao, Xiao-Yao; McCalley, D.V.; McEnvoy, J. *J. Chrom. A.* **2001**, *923*, 195-204.
11. Kelly, J. *J. Chrom. A.* **2000**, *872*, 309-314.

Emerging Contaminants

Pesticides

Chapter 14

Determination of Chloroacetanilide and Chloroacetamide Herbicides and Their Polar Degradation Products in Water by LC/MS/MS

John D. Vargo

Hygienic Laboratory, University of Iowa, 102 Oakdale Campus,
Iowa City, IA 52242–5002 (john-vargo@uiowa.edu)

Chloroacetanilide (acetochlor, alachlor, metolachlor) and chloroacetamide (dimethenamid) herbicides and their ethane sulfonic acid (ESA) and oxanilic acid (OXA) degrades are determined in ground and surface waters with a limit of detection of 0.025 µg/L using reversed phase . HPLC with tandem quadrupole MS/MS detection. The analytes were isolated from water samples using carbon SPE cartridges. The effects of mobile phase composition and column temperature on analyte sensitivity and chromatographic peak shape were investigated. The sensitivities for the ESA and OXA degrades were 9-15X greater using ACN/H2O/acetic acid as the mobile phase compared to MeOH/H2O/ammonium acetate. Elevated column temperature was the most important factor in improving the chromatographic peak shape for those analytes exhibiting split peak shapes at room temperature due to stereoisomerism. No signal enhancement or suppression effects were observed when using this procedure to analyze surface water samples.

Introduction

Herbicides are the most widely used class of crop protection chemicals in the United States, accounting for nearly one-half of the total amount used (*1,2*). Corn and soybeans are the two most widely planted crops in the Midwest agricultural region of the United States. Each requires intensive use of herbicides for protection from weeds. More crop protection chemicals, in terms of pounds of active ingredient applied, are used for the production of corn than any other crop grown in the United States (*1*).

The chloroacetanilide and chloroacetamide classes of herbicides are widely used for the control of broadleaf and grass weeds. Based on a survey conducted in 1997, metolachlor, acetochlor, and alachlor (all chloroacetanilide herbicides) ranked second, fifth, and ninth for total herbicide usage (in terms of pounds of active ingredient applied) for all crops in the United States (*1*). For corn production in the state of Iowa, metolachlor and acetochlor ranked first and second, with atrazine third, for total pounds of active ingredient applied (*1*). Alachlor use was much less (only 2% of Iowa's corn acreage was treated with this chemical in 1997), no doubt related to the conditional registration of acetochlor (*3*) which required a substantial reduction in the total poundage of select corn herbicides used, including alachlor, for this new product to remain on the market. Metolachlor and alachlor ranked first and second (in terms of pounds of active ingredient applied) for all pesticides used in sweet corn in Iowa (*1,2*). Acetochlor is not registered for use on sweet corn (*3*).

The chloroacetanilide and chloroacetamide herbicides are used just prior to, or just after, corn plant emergence which corresponds to late spring/early summer. The primary degradation mechanism for the parent herbicides is metabolism in the soil. The ethane sulfonic acid (ESA) and oxanilic acid (OXA) analytes are the most significant of the metabolites originating from environmental mechanisms. These metabolites are quite water soluble and as a result are mobile in the soil column. Rain events, common at this time of the crop season, may result in the parent active ingredient and/or metabolites entering water sources by surface run-off and/or leaching through the soil into the water table. Detectable levels of the parent herbicides will normally be observed during the summer months in areas where they are used. By fall, the parent herbicides have usually degraded to the point where they are no longer detectable. When only the parent active ingredient herbicide is monitored, as is common with most water monitoring procedures, this can lead to the mistaken impression that the herbicide residues are no longer present at significant levels in the environment.

Formation and Presence of Environmental Metabolites

The ESA metabolite of alachlor was first detected in water during routine water monitoring using immunoassay screening (4,5). False positive detects for alachlor were subsequently determined to be due to the presence of the ESA metabolite. The metabolite was determined to be present in water samples from Ohio at levels varying from 4-74 µg/L in groundwater and at levels varying from the limit of detection (0.5 µg/L) to 2 µg/L in surface water. Thurman et al. (6,7) extensively monitored ground and surface water sources across the Midwest finding the ESA metabolite of alachlor at greater frequency than the parent active ingredient and at concentrations much greater than the parent herbicide. Suspecting that other chloroacetanilide herbicides may also form the ESA metabolite, Aga et al. conducted a soil metabolism study with metolachlor and identified metolachlor ESA as a significant metabolite (8). Additional soil metabolism studies showed that alachlor ESA was formed at a faster rate and at concentrations 2-4 times higher than for metolachlor ESA (9). Feng determined that acetochlor ESA was the terminal degradation product for acetochlor in soil metabolism studies (10). Demonstration that metolachlor and acetochlor also formed the ESA metabolites in soil metabolism studies, coupled with the knowledge that the ESA metabolites are water soluble and highly mobile in soil, led scientists to speculate that it was highly likely that these ESA metabolites were present in the environment, in addition to the alachlor ESA metabolite. Ferrer et al. (11) first detected the ESA metabolites in Iowa surface and groundwater by LC/MS, finding amounts as high as 1.8 µg/L. The methodology which was used could not distinguish between acetochlor ESA and alachlor ESA so residue levels for these analytes could not be reported. They also monitored for the oxanilic acid (OXA) metabolites of acetochlor, alachlor, and metolachlor in Iowa surface and groundwater, finding levels varying from approximately 0.01-1.7 µg/L (11). Kalkhoff et al. (12) conducted an extensive survey of surface and groundwater samples from Iowa, monitoring for parent herbicide along with the ESA and OXA metabolites for acetochlor, alachlor, and metolachlor. The frequency for detection, along with the concentration of analyte found, was ESA>OXA>parent. They found that the ESA and OXA metabolites accounted for more than 80% of the total residue present.

Alachlor is the only chloroacetanilide herbicide regulated by the United States Environmental Protection Agency (USEPA) under the Safe Drinking Water Act (SDWA) with a maximum contaminant level (MCL) of 2 µg/L in finished drinking water (13). In 1998, the USEPA published its Drinking Water Candidate Contaminant List (CCL) listing chemicals which they are considering for regulation by the SDWA (14). "Alachlor ESA and other acetanilide degradation products" were listed as potential candidates to add to that list based on the frequency and concentrations that chloroacetanilide degradation products have been observed in surface and groundwater sources.

Analytical Methodologies

Based on the frequency with which the ESA and OXA degradates of the chloroacetanilide herbicides have been found in various water supplies and the increasing interest from the USEPA to routinely monitor for these chemicals, the need for an accurate and sensitive analytical method is apparent. The parent chloroacetanilide herbicides are most commonly determined by GC/MS. However, the ESA and OXA metabolites are not amenable to analysis by GC due to their polarity and acidity.

Various procedures have been described in the literature for the extraction and subsequent analysis of the parent chloroacetanilide/chloroacetamide herbicides and/or their ESA and OXA metabolites in water. Heberle et al. described a combination GC/MS and HPLC/UV method for the analysis of acetochlor, alachlor, and metolachlor along with the ESA and OXA metabolites (15). A Carbopack B solid phase extraction (SPE) column was used to isolate the analytes. The analytes were sequentially eluted as three separate fractions: parents, ESA metabolites, and OXA metabolites. The parent analytes were analyzed by GC/MS. The OXA metabolites were derivatized and analyzed by GC/MS, separate from the parent analytes. The ESA metabolites were analyzed by HPLC with diode array UV absorbance detection. Ferrer et al. developed the first LC/MS method for the analysis of ESA and OXA metabolites of acetochlor, alachlor, and metolachlor (11). C18 SPE was used to isolate the analytes from the water sample. The parent analytes were eluted as a separate fraction from the ESA and OXA metabolites. The ESA and OXA metabolites were subsequently analyzed by reversed phase HPLC coupled with a single quadrupole mass spectrometer equipped with an electrospray interface (ESI). The authors noted that they were unable to qualitatively and quantitatively discriminate between acetochlor ESA and alachlor ESA as the two analytes are very difficult to chromatographically resolve and they both form a molecular ion of the same mass. Vargo described the first LC/MS/MS procedure for the analysis of the ESA metabolites of acetochlor, alachlor, metolachlor, and dimethenamid (16). A polymeric reversed phase SPE column was used to isolate the analytes from 100 mL of groundwater. A tandem quadrupole LC/MS/MS system with an ESI interface was used for the analysis. He demonstrated that the co-eluting ESA metabolites of acetochlor and alachlor could be qualitatively and quantitatively discriminated from each other using minor differences in the fragmentation spectra of the molecular ions. Yokley et al. have also demonstrated the use of a tandem quadrupole system for analyzing the parent herbicides and ESA and OXA metabolites in water *(17)*. Thurman et al. developed a single quadrupole ESI LC/MS method that permitted the qualitative and quantitative analysis of the ESA and OXA metabolites for acetochlor, alachlor, metolachlor, dimethenamid, and flufenacet in surface water *(18)*. C18 SPE was used to isolate the analytes from water. Partial chromatographic resolution of acetochlor ESA and alachlor ESA was achieved, but at the expense of analysis time, which was approximately 40 minutes. Shoemaker developed a single quadrupole LC/MS procedure for the analysis the ESA and OXA metabolites of acetochlor, alachlor, metolachlor, dimethenamid and propachlor, in addition to the sulfinyl acetic acid (SAA) metabolites of alachlor and propachlor, in drinking water *(19)*. The analytes were isolated from drinking water using a carbon SPE column. Single quadrupole ESI LC/MS was used for detection. Near baseline chromatographic resolution of the ESA metabolites of acetochlor and alachlor was achieved using a reversed phase column at a temperature of 70°C with a mobile phase of methanol/water/ammonium acetate. Thurman et al. described the use of a time of flight (TOF) LC/MS system, equipped with an electrospray interface, for analysis of alachlor ESA and acetochlor ESA *(20)*.

Single quadrupole LC/MS has limitations when attempting to analyze both acetochlor and alachlor along with their ESA and OXA metabolites. The corresponding pairs of parent herbicides, the ESA metabolites, and the OXA metabolites all have the same molecular weight. In addition, each pair is extremely difficult to chromatographically resolve. The parent and OXA metabolites can be mass discriminated from one another in single quadrupole systems by increasing the orifice potential to induce characteristic fragmentation of the molecular ion. In the negative ion monitoring mode, which is most commonly used for detecting the ESA and OXA metabolites, alachlor ESA and acetochlor ESA do not generate unique orifice-induced fragments ions of sufficient abundance to permit discrimination between each other. Thurman et al. demonstrated that unique ion fragments for these analytes could be generated at high orifice potential in the positive ion monitoring mode *(20)*.

Tandem mass spectrometric LC/MS/MS analysis permits the chromatographically unresolved analytes to be distinguished from each other based on differences in their product ion fragmentation. Vargo *(16)* demonstrated the ability of LC/MS/MS to resolve co-eluting acetochlor ESA and alachlor ESA based on unique product ion fragment ions. While these unique product ions were of low intensity (approximately 15% of the intensity of the most abundant fragment ion), the LC/MS/MS system coupled with solid phase extraction (SPE) cleanup permitted 100-mL water samples to be quantified at levels as low as 0.05 μg/L.

Other Herbicides to Monitor

There are other acetanilide and acetamide herbicides which are registered for crop use that also metabolize to form the ESA and OXA degradates and may be of interest to monitor. Dimethenamid, a chloroacetamide, is used on corn and sweet corn. While not used as extensively as metolachlor and acetochlor, in 1997 it was used on 7% of corn acreage and on 8% of sweet corn acreage in Iowa (1). Flufenacet, a fluorinated acetamide, received a conditional registration in 1998 for use on corn and soybeans. No data was available regarding the amounts of this new product used, but it would be expected to increase with time and therefore may be of interest to monitor. Zimmerman et al. have described a single quadrupole LC/MS procedure for the analysis of the ESA and OXA metabolites of flufenacet, in addition to the ESA and OXA metabolites of dimethenamid (21). Propachlor, a chloroacetanilide, is not used on corn in Iowa and has minimal overall use in the Midwest.

Objectives for this Study

The goals of the work described in this presentation were: (1) Modify existing procedures to permit the simultaneous extraction and subsequent analysis of the parent herbicides along with their respective ESA and OXA degradates, (2) Contrast analyte peak shape, resolution, and sensitivity for different mobile phase conditions and column temperatures, (3) Examine the stability of the analytes when stored refrigerated or frozen using plastic or glass bottles, (4) Determine whether any matrix effects (either suppression or enhancement of the analyte signal) occur in surface water samples, and (5) Determine the concentration of analytes present in several surface water samples from the Midwest agricultural region.

Chemical Structures

The chemical structures and their respective chemical abstracts registry numbers for all of the parent herbicides and their ESA and OXA metabolites which are analyzed using this procedure are presented in Figure 1.

244

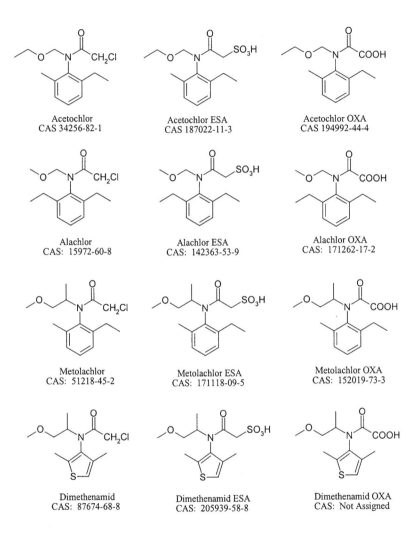

Acetochlor
CAS 34256-82-1

Acetochlor ESA
CAS 187022-11-3

Acetochlor OXA
CAS 194992-44-4

Alachlor
CAS: 15972-60-8

Alachlor ESA
CAS: 142363-53-9

Alachlor OXA
CAS: 171262-17-2

Metolachlor
CAS: 51218-45-2

Metolachlor ESA
CAS: 171118-09-5

Metolachlor OXA
CAS: 152019-73-3

Dimethenamid
CAS: 87674-68-8

Dimethenamid ESA
CAS: 205939-58-8

Dimethenamid OXA
CAS: Not Assigned

Figure 1. Chemical Structures and Chemical Abstracts Registry Numbers

Experimental

Extraction Procedure

The extraction procedure which was used was developed by Shoemaker (*19*). In this study, the volume of the SPE eluting solvent was increased to permit the parent herbicides to be eluted along with the ESA and OXA degradates. A 100-mL sample of water (no pH adjustment) was passed through a preconditioned Supelco Envi-Carb SPE column (250 mg size), rinsed with purified water, and eluted with 10 mL of methanol (10 mM in ammonium acetate). The methanol content was removed via nitrogen evaporation until only aqueous remained (approximately 0.5 mL). Acetonitrile was added (0.4 mL) and the sample diluted to a final volume of 2.0 mL with purified water. The samples were stored refrigerated until the time of analysis. For long term storage (> 1 week), it is recommended that the samples be stored at freezer temperatures.

LC/MS/MS Conditions

The samples were analyzed using a Micromass Quattro LC/MS/MS (tandem quadrupole) system with an electrospray interface. The ESA and OXA degradates were monitored as negative ions while the parent herbicides were monitored as positive ions. All separations were performed using a Zorbax SB C8 column (3.0 x 150 mm, dp = 5 μm) with a mobile phase flow rate of 0.6 mL/min. For the majority of the data generated for this presentation, a mobile phase gradient using acetonitrile (0.15% in acetic acid) and water (0.15% in acetic acid) was used with a linear gradient ramp from 20-100% ACN over 10 minutes. A column temperature of 30°C was used for most studies except those where column temperature was varied. For comparison studies, a mobile phase gradient using 20% methanol/water/20 mM ammonium acetate (A) and 90% methanol/water/20 mM ammonium acetate (B) and a column temperature of 60°C was employed. The gradient was a linear ramp from 100% A to 100% B over 15 minutes.

Unique precursor ion/product ion pairs were monitored in the multiple reaction monitoring mode (MRM) for each analyte. Argon was used as the collision gas. The precursor ions were the protonated molecules (positive ion monitoring) or the deprotonated molecules (negative ion monitoring). Data acquisition conditions are presented in Table I.

Table I. Data Acquisition Conditions for Micromass Quattro LC/MS/MS

Funct.	Analyte	Precursor Ion	Product Ion	Dwell (sec)	Cone (V)	Collision (V)
1	Acetochlor OXA	264.00	145.90	0.1	20	12
1	Alachlor OXA	264.00	159.90	0.1	20	12
1	Dimethenamid OXA	270.00	197.90	0.1	20	12
1	Metolachlor OXA	278.00	205.95	0.1	20	12
1	Acetochlor ESA	314.10	161.90	0.1	40	23
1	Alachlor ESA	314.10	159.90	0.1	40	23
1	Dimethenamid ESA	320.00	120.90	0.1	40	23
1	Metolachlor ESA	328.10	120.90	0.1	40	23
2	Acetochlor	270.10	223.90	0.15	20	10
2	Alachlor	270.10	238.00	0.15	20	15
2	Dimethenamid	276.10	243.95	0.15	20	15
2	Metolachlor	284.15	252.00	0.15	20	16

ESA/OXA metabolites monitored as negative ions.
Parent herbicides monitored as positive ions.

Function 1: Start time: 2.0 min, End time: 7.0 min

Function 2: Start time: 7.0 min, End time: 11.0 min

Results and Discussion

Method Validation Experiments

The method was validated by fortification of HPLC-grade water with all analytes at levels of 0.05 µg/L, 0.5 µg/L, and 20 µg/L and subsequent extraction and analysis of the samples. Quantitation was performed using external standards with linear regression with a calibration range from 1.25-50 pg/µL (50 µL injected). The limit of detection (based on a 100 mL water sample, final sample volume of 2 mL, and low calibration standard of 1.25 pg/µL) was 0.025 µg/L. Excellent average recoveries (ranging from 82-107%) and percent relative standard deviations (<15%) were observed for all analytes at all fortification levels. The correlation coefficients for the linear regression calibration curves typically exceeded 0.999 for all analytes. An LC/MS/MS reconstructed MRM chromatogram for the lowest standard used in the calibration curve (1.25 pg/µL) is presented in Figure 2.

Analyte Chromatographic Peak Shape and Sensitivity

The chloroacetanilide and chloroacetamide herbicides and their ESA and OXA degradates exist as a mixture of rotational isomers. Hindered rotation about the phenyl-nitrogen bond coupled with the presence of a chiral carbon in the aliphatic chain bonded to nitrogen, leads to a mixture of enantiomeric and diastereomeric isomers (22). This results in severe peak broadening and splitting for some analytes, especially the OXA degradates (see Figure 2).

At room temperature, and under the sample processing conditions used in this method, no interconversion has been seen among the stereoisomers. This is supported by no visual difference in the chromatographic peak shape for the analytical standards compared to processed water samples. At elevated column temperature, however, interconversion between the stereoisomers takes place resulting in compressed and relatively gaussian-shaped chromatographic peak shapes. Shoemaker (19) described a procedure using reversed phase HPLC with a methanol/water/ammonium acetate gradient at 70°C which not only eliminated the peak splitting observed with acetonitrile/water/acetic acid at room temperature, but also provided partial resolution of the ESA degradates of acetochlor and alachlor, thus permitting single quadrupole LC/MS to be used for detection.

The author investigated the effects of mobile phase composition and column temperature on chromatographic peak shape and analyte sensitivity. The methanol/water/ammonium acetate gradient produced superior peak shape to the acetonitrile/water/acetic acid gradient, but at a column temperature of 30°C, the OXA metabolite peaks were still quite broad. By increasing the column temperature to 60°C, the peaks became narrow and gaussian. Increasing the column temperature also greatly improved the peak shape when using the acetonitrile/water/acetic acid gradient, but not to the extent that was observed with the methanol mobile phase. Figure 3 shows the effects of column temperature and mobile phase composition on chromatographic peak shape for the OXA metabolites.

The amount of acetic acid present in the acetonitrile/water/acetic acid gradient was varied to determine the effect on analyte peak shape and sensitivity. The amount of acid present was varied from 0.05-0.25% (v/v). Analyte peak shape was not affected by the variation. Analyte sensitivity, as measured by integrated peak area, was greatest for a composition containing 0.15% acetic acid in both acetonitrile and water.

Analyte sensitivity, as a function of mobile phase composition and column temperature, was determined by injecting a 10 pg/µL calibration standard onto the analytical column and contrasting the resulting peak area observed for each analyte. The results are presented in Table II (below). The results have been normalized based on the conditions exhibiting the least sensitivity for each analyte. Column temperature did not have a significant impact on analyte sensitivity when the same mobile phase composition was used. The results

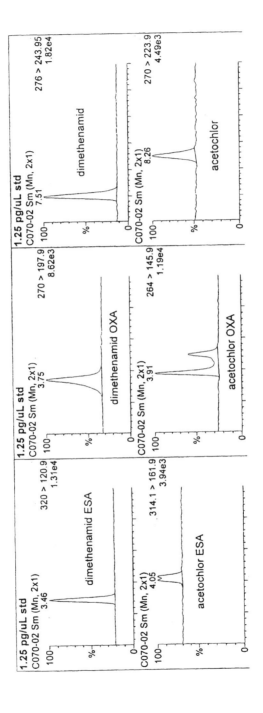

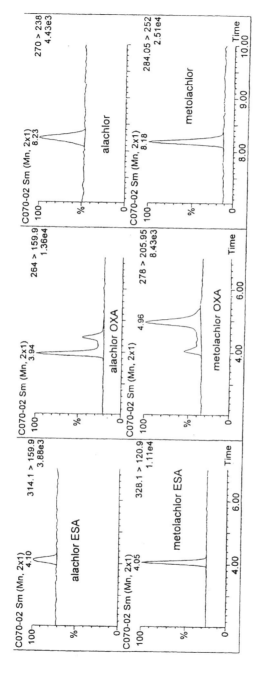

Figure 2. Lowest Calibration Standard, 1.25 pg/µL. The horizontal line across each peak indicates how the peak is integrated

250

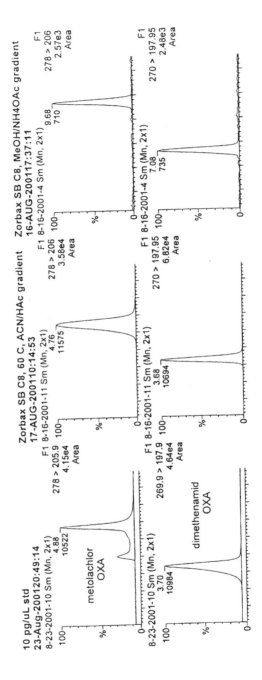

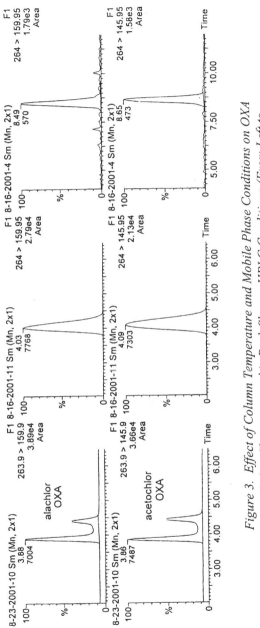

Figure 3. Effect of Column Temperature and Mobile Phase Conditions on OXA Metabolite Chromatographic Peak Shape. HPLC Conditions (From Left to Right): (Left) Acetonitrile/Water/Acetic Acid (0.15%) Gradient, 30 °C, (Center) Acetonitrile/Water/Acetic Acid (0.15%) Gradient, 60 °C, (Right) Methanol/Water/20 mM Ammonium Acetate, 60 °C

indicate that a sensitivity enhancement of 14-23X is observed for the ESA and OXA metabolites when acetonitrile/water/acetic acid is used as the mobile phase in comparison to methanol/water/ammonium acetate. Analyte sensitivity was comparable for the parent herbicides.

Table II. Effects of Column Temperature and Mobile Phase Composition on Analyte Sensitivity (as Measured by Peak Area)

Analyte	Relative Response vs. HPLC Conditions		
	ACN/H$_2$O, 0.15% HAc 30°C	ACN/H$_2$O, 0.15% HAc 55°C	MeOH/H$_2$O, 20 mM NH$_4$OAc 60°C
Dimethenamid ESA	22.3	22.7	1.0
Dimethenamid OXA	16.9	17.1	1.0
Acetochlor ESA	18.6	18.1	1.0
Acetochlor OXA	18.6	18.2	1.0
Alachlor ESA	18.2	18.4	1.0
Alachlor OXA	15.3	15.3	1.0
Metolachlor ESA	13.9	14.5	1.0
Metolachlor OXA	14.9	15.1	1.0
Dimethenamid	1.5	1.3	1.0
Acetochlor	1.1	1.0	1.0
Alachlor	1.0	0.7	1.0
Metolachlor	1.2	1.0	1.0

Matrix Effects

Matrix enhancement or suppression of the analyte signal may occur when using LC/MS techniques. Electrospray ionization tends to be more prone to matrix effects than atmospheric pressure chemical ionization (APCI). It is the author's experience that matrix effects are more likely, and more pronounced, when operating in the positive ion monitoring mode.

No matrix effects have been observed using the standard conditions outlined in the experimental section. This was evidenced by obtaining excellent recoveries for all analytes spiked into a variety of surface water samples over a time period exceeding six months. In addition, it was desired to analyze several surface water samples using the methanol/water/ammonium acetate HPLC conditions described by Shoemaker (19) and contrasting those results to results obtained using the author's standard HPLC conditions. No significant differences in the detected residue levels were observed although the limit of detection was significantly lower for the ESA and OXA metabolites using the HPLC conditions described in this procedure.

Storage Stability

Storage stability studies were conducted to determine proper storage conditions and length of storage. The analytes were fortified into groundwater at a concentration level of 5 µg/L and stored under the following conditions: refrigerated in an amber plastic HDPE bottle, frozen in an amber plastic HDPE bottle, and refrigerated in an amber borosilicate glass bottle. The samples were analyzed by direct injection into the LC/MS/MS system to eliminate any variability introduced by the sample extraction procedure.

Samples have been analyzed with storage to 161 days. No significant losses (>10%) have been observed under any of the storage conditions for the ESA and OXA metabolites. However, significant losses of the four parent herbicides were observed in the plastic bottles (stored refrigerated or frozen) in less than thirty days of storage. No significant losses of the parent herbicides were observed when the samples were stored refrigerated in an amber borosilicate glass bottle.

Iowa Surface Water Samples

Several surface water samples were collected in Iowa in mid-September, 2001, and analyzed using the standard procedure outlined in the experimental section. The results from these analyses are presented in Table III.

Table III. Residues Found in Iowa Surface Water Samples

Analyte	Residue Found (µg/L)	
	Skunk River	Coralville Reservoir
Acetochlor ESA	0.21	0.18
Acetochlor OXA	0.12	0.11
Acetochlor	<0.025	<0.025
Alachlor ESA	0.17	0.54
Alachlor OXA	0.036	0.060
Alachlor	<0.025	<0.025
Metolachlor ESA	1.5	1.6
Metolachlor OXA	0.35	0.28
Metolachlor	0.066	0.055
Dimethenamid ESA	<0.025	0.036
Dimethenamid OXA	<0.025	<0.025
Dimethenamid	<0.025	<0.025

Summary

Hundreds of Iowa surface water samples coming from over sixty sampling locations across the state have been successfully analyzed using the analytical procedure outlined in this presentation. The analytical procedure has a limit of detection of 0.025 µg/L. The method exhibits excellent specificity. Even though they are not chromatographically resolved and have the same molecular weight, acetochlor ESA and alachlor ESA are easily distinguished from one another by monitoring unique precursor ion/product ion pairs. No matrix enhancement/suppression of the analyte signal has been observed when using this procedure. The method exhibits excellent accuracy as evidenced by obtaining excellent recoveries for all analytes when spiked into surface water samples.

References

1. Gianessi, L. P.; Marcelli, M. B. "Pesticide Use in U.S. Crop Production: 1997," Washington, D.C., National Center for Food and Agricultural Policy, unnumbered pages. (Available on the internet at: www.ncfap.org)
2. Aspelin, A. L.; Grube, A. H., "Pesticide Industry Sales and Usage: 1996 and 1997 Market Estimates," USEPA, OPP, November 1999. (Available on the internet at: www.epa.gov/oppbead1/pesticides/)
3. Acetochlor Registration Agreement and Addendums, USEPA, OPP, March 1994.
4. Baker, D. B.; Bushway, R. J.; Adams, S. A.; and Macomber, C. *Environ. Sci. Technol.* **1993,** 27, pp 562-564.
5. Macomber, C.; B *J. Agric. Food Chem.* **1992,** 40, pp 1450-1452.
6. Thurman, E. M.; Goolsby, D. A.; Aga, D. S.; Pomes, M. L.; and Meyer, M. T. *Environ. Sci. Technol.* **1996,** 30, pp 569-574.
7. Kolpin, D. W.; Thurman, E. M.; and Goolsby, D. A. *Environ. Sci. Technol.* **1996,** 30, pp 335-340.
8. Aga, D. S.; Thurman, E. M.; Yockel, M. E.; Zimmerman, L. R.; Williams, T. D. *Environ. Sci. Technol.* **1996,** 30, pp 592-597.
9. Aga, D. S.; Thurman, E. M. *Environ. Sci. Technol.* **2001,** 35, pp 2455-2460.
10. Feng, P. C. C. *Pestic. Biochem. Physiol.* **1991,** 40, pp 136-142.
11. Ferrer, I.; Thurman, E. M.; and Barceló, D. *Anal. Chem.* **1997,** 69, pp 4547-4553.
12. Kalkhoff, S. J.; Kolpin, D. W.; Thurman, E. M.; Ferrer, I.; Barceló, D. *Env. Sci. Technol.* **1998,** 32, pp 1738-1740.

13. Maximum Contaminant levels for Organic Contaminants. (1999) Code of Federal Regulations, Part 141.61, Title 40; (1991) Fed. Regist. 56, p 3593.

14. Announcement of the Drinking Water Contaminant Candidate List, Fed. Regist., **1998,** 63, pp 10274-10287.

15. Heberle, S. A.; Aga, D. S.; Hany, R.; Müller, S. R. *Anal. Chem.* **2000,** 72, pp 840-845.

16. Vargo, J. D. *Anal. Chem.* **1998,** 70, pp 2699-2703.

17. Yokley, R. A.; Mayer, L.C.; Huang, B.; Vargo, J. *Anal. Chem.* **2002,** 74, pp 3754-3759.

18. Thurman, E. M.; Lee, E. A.; Kish, J. L.; Zimmerman, L. R. Methods of Analysis by the U.S. Geological Survey Organic Geochemistry Research Group-Update and Additions to the Determination of Chloroacetanilide Herbicide Degradation Compounds in Water Using High-Performance Liquid Chromatography/Mass Spectrometry: U.S. Geological Survey Open-File Report 01-10, **2001,** 17 p.

19. Shoemaker, J. "Analytical Method Development for Alachlor ESA and Other Acetanilide Herbicide Degradation Products," 49th ASMS Conference on Mass Spectrometry and Allied Topics, Chicago, IL, May 27-31, 2001.

20. Thurman, E. M.; Ferrer, I.; Parry, R. *J Chromatogr.* **2002,** 957, pp 3-9.

21. Zimmerman, L. R.; Schneider, R. J.; Thurman, E. M. *J. Agric. Food Chem.* **2002,** 50, pp 1045-1052.

22. Müller, M. D.; Poigner, T.; Buser, H-R. *J. Agric. Food Chem.* **2001,** 49, pp 42-49.

Chapter 15

Determination of Acetanilide Degradates in Ground and Surface Waters by Direct Aqueous Injection LC/MS/MS

John D. Fuhrman and J. Mark Allan

Environmental Sciences Technology Center, Monsanto Company, 800 North Lindbergh Boulevard, St. Louis, MO 63167

Acetanilide herbicides are widely used throughout the United States to control annual grasses in corn, soybeans and other crops. The herbicides studied, include the acetanilides; acetochlor, alachlor, metolachlor, propachlor and the acetamide; dimethenamid. The biological degradation of these herbicides produces a myriad of polar metabolites the most prominent of which are the ethanesulfonic acids (ESA) and oxanilic acids (OX) moieties. These degradates have been found in surface waters from agricultural runoff and in shallow groundwater due to leaching through vulnerable soils. Several less significant degradates have recently been identified and incorporated into a single multi-residue methodology. A high throughput method has been developed for the analysis of fourteen soil degradates of the specified herbicides. The method uses direct aqueous injection (DAI) liquid chromatography / mass spectrometry / mass spectrometry (LC/MS/MS) to analyze these materials without any sample pretreatment or concentration. MS/MS provides a very specific and highly sensitive mode of detection using TurboIonSpray in the negative ion mode. Thirteen of the fourteen degradates were validated at the 0.05 µg/L level in three different water matrices, raw surface water, finished surface water (drinking water) and groundwater.

Introduction

Acetochlor, alachlor, and propachlor are soil-applied herbicides manufactured by Monsanto for pre-emergence and early post-emergence control of annual grasses and broadleaf weeds in crops. These chloroacetanilides degrade readily and extensively in soil, mainly through displacement of chlorine followed by further metabolism to numerous degradation products (1-3). In aerobic soil studies conducted with alachlor, acetochlor, and propachlor, the most abundant metabolites have been identified as the water-soluble oxanilic, sulfonic, and sulfinylacetic acids. Public environmental concerns and government regulatory requirements continue to prompt the need for reliable methods to determine residues of these herbicide metabolites.

The objective of this study was to develop and validate a multi-residue confirmatory method for the major soil degradates of acetochlor, alachlor, metolachlor, propachlor, and dimethenamid in water from ground and surface water sources. This multi-residue method includes the major chloroacetanilide and chloroacetamide soil degradates and thereby ensures accurate mass spectral resolution, identification and quantification of these degradates. The development of methods for chloroacetanilide soil degradates is challenging due to the separation and detection of numerous analytes of similar chemical structure. Additionally, degradates exist as rotational isomers due to restricted rotation at the amide bond or the bond to the aromatic ring, when the ring is asymmetrically substituted (4). These rotamers, due to their restricted rotation about the amide bond, generally interconvert rapidly, but in some cases separate into two distinct peaks during HPLC analysis.

Several methods have been developed for the analysis of the most common chloroacetanilide degradates, the ethanesulfonic and oxanilic acids of acetochlor, alachlor and metolachlor, in environmental waters (5-8). This methodology has expanded the number of analytes to fourteen and includes other known degradates such as the sulfinylacetic acids, alachlor secondary oxanilic acid and degradates from the herbicides dimethenamid and propachlor. The method presented here was designed for rapid analysis of water samples by direct aqueous injection reversed-phase liquid chromatography tandem mass spectrometry (LC/MS/MS). No pretreatment or concentration of the sample is necessary prior to analysis. The method may be used to analyze for a single degradate or any combination of the fourteen degradates. The acetanilide degradate structures are presented in Figure 1 and the acetamide degradate structures are in Figure 2.

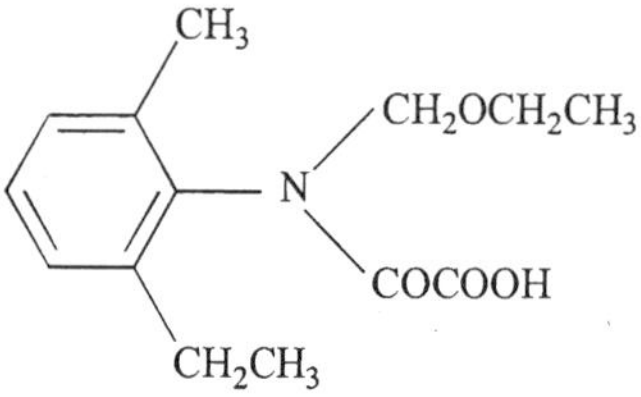

Acetochlor - OX

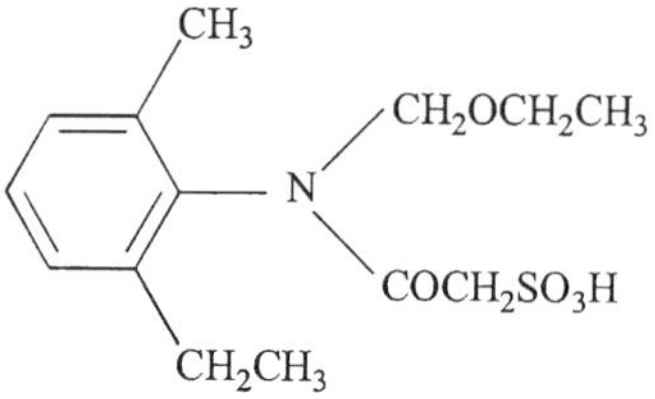

Acetochlor - ESA

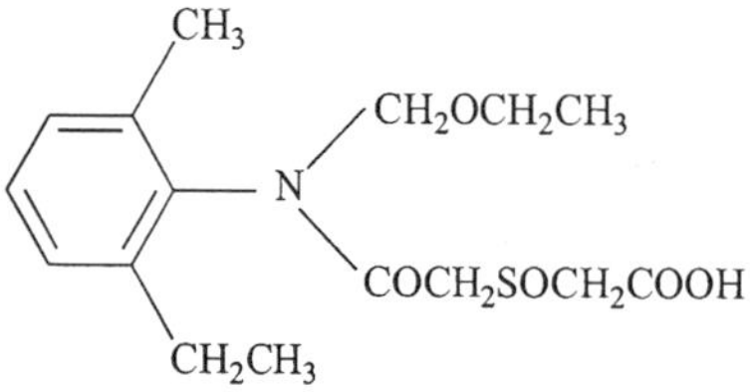

Acetochlor - SAA

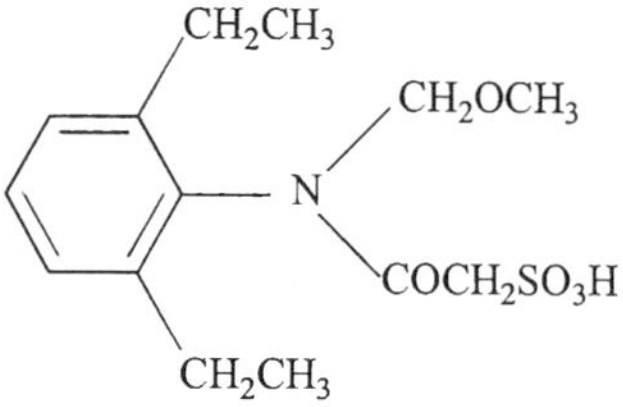

Alachlor - ESA

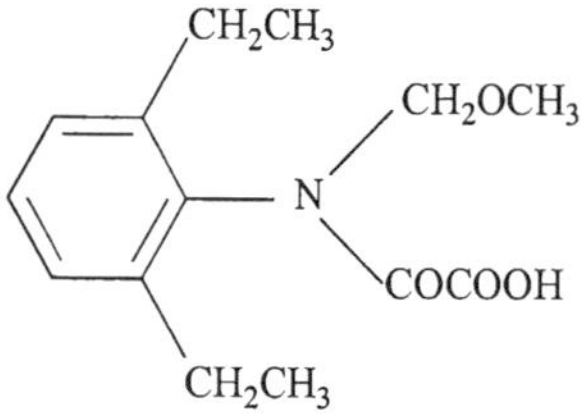

Alachlor - OX

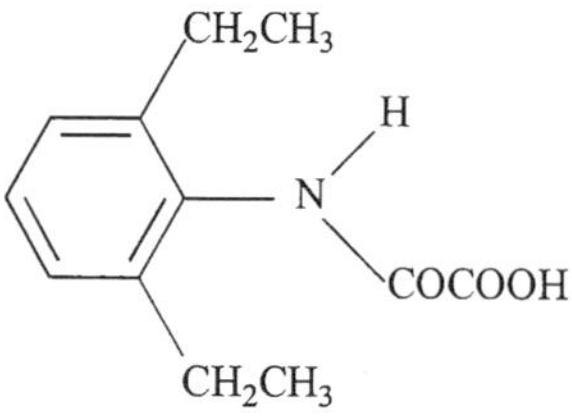

Alachlor - sOX

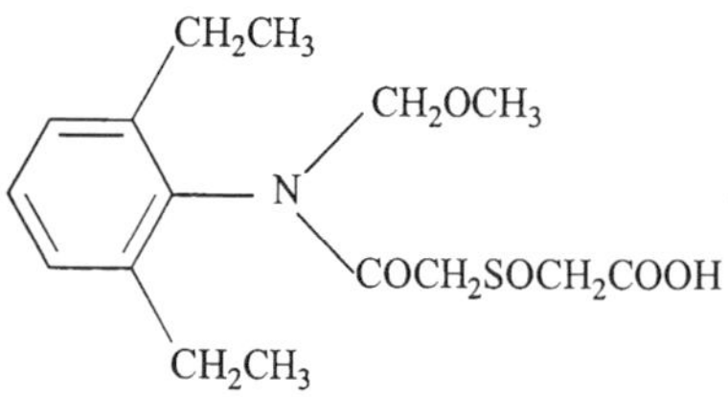

Alachlor - SAA

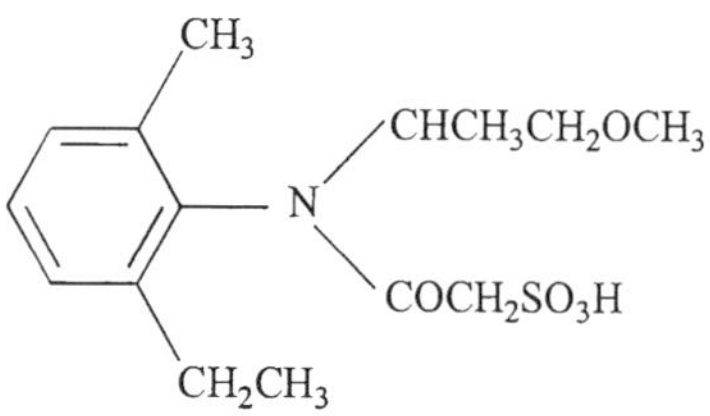

Metolachlor - ESA

Figure 1: Acetanilide Degradate Structures

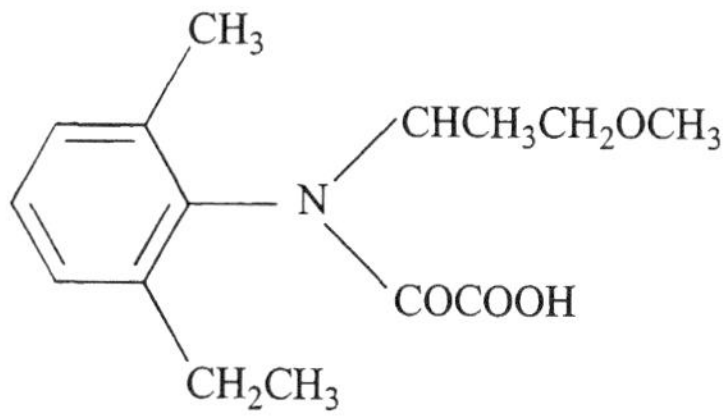

Metolachlor - OX

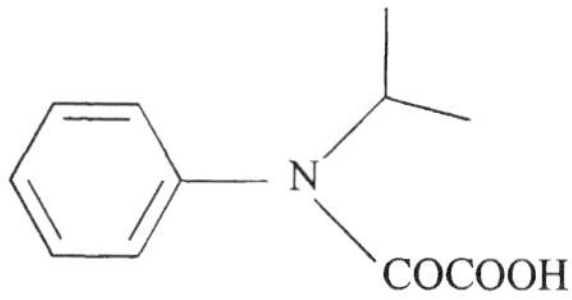

Propachlor - OX

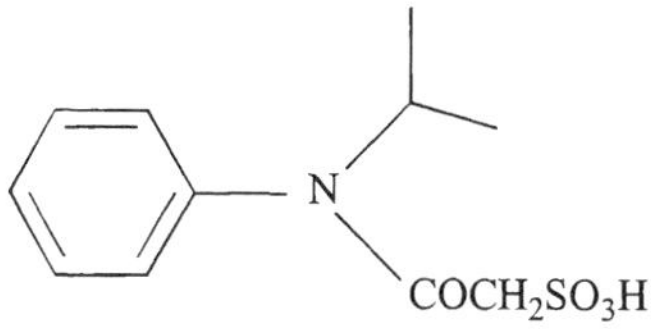

Propachlor - ESA

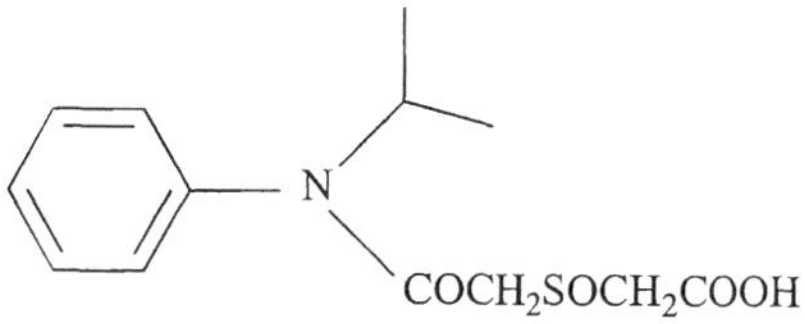

Propachlor - SAA

Figure 1: Acetanilide Degradate Structures (cont.)

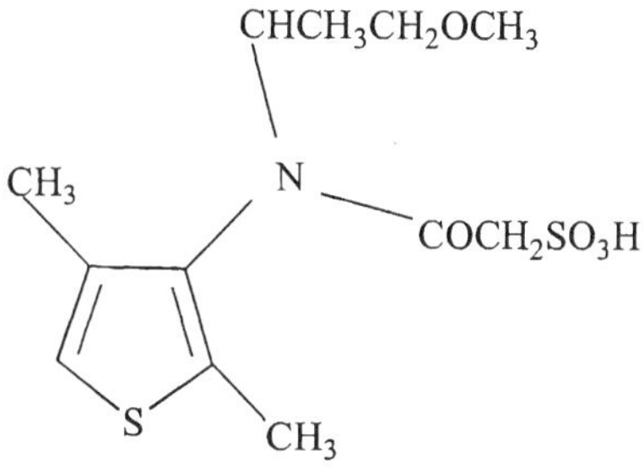

Dimethenamid - ESA

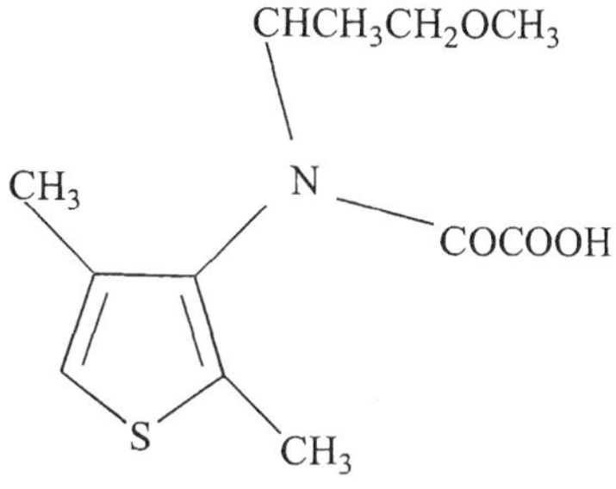

Dimethenamid - OX

Figure 2. Acetamide Degradate Structures

Methods

Reagents

Acetonitrile, methanol and reagent water were Optima® grade from Fisher Scientific. Absolute ethanol was obtained from Aaper Alcohol and Chemical Company. Glacial acetic acid was obtained from J.T. Baker. Water for mobile phase preparation was from a Milli-Q water purification system.

Reference Standards

All fourteen acetanilide and acetamide degradates used in this method were obtained from the Standards Reference Officer at Monsanto Company (Saint Louis, Missouri, USA). All compounds were either synthesized internally or custom synthesized by external contractors. All degradates were used as received without further purification and were certified and issued as their respective sodium salts. Purity of the reference compounds should generally be greater than 95%, but in no instance less than 90%. All degradate solution concentrations were prepared purity corrected as the free acid. A list of the reference compounds used in this study is in Table I.

Stock Solutions

Stock solutions of the individual degradates were prepared in absolute ethanol at nominal concentrations of 1000 µg/mL (weight adjusted for purity of the free acid). Mixed stock solutions for the fourteen degradates were prepared to facilitate calibration and fortification. A mixed degradate stock solution at 1.0 µg/mL was prepared from the individual stock solutions. From this mixed stock solution serial dilutions at 100.0 µg/L and 10.0 µg/L were prepared in absolute ethanol.

Working solutions were prepared in reagent water in order to fortify control matrices to determine analytical accuracy and to calibrate the response of the analyte in the mass spectrometer. All standard solutions (stock, fortified, and calibration) were stored refrigerated (2-10°C) in clean amber glass bottles with Teflon-lined screw caps. The stock solutions are adequate to prepare the fortification and calibration standards in the range of 0.010-20.0 µg/L of each analyte.

Table I: Reference Compounds

Degradate Name – IUPAC and Common	Acronym	CAS Registry No.
[(ethoxymethyl)(2-ethyl-6-methylphenyl) amino]-2-oxoacetic acid, (acetochlor oxanilic acid)	AcOX	194992-44-4
2-[(ethoxymethyl)(2-ethyl-6-methylphenyl)amino]-2-oxoethanesulfonic acid, (acetochlor sulfonic acid)	AcESA	187022-11-3
{2-[(ethoxymethyl)(2-ethyl-6-methylphenyl)amino]-2-oxoethylsulfinylacetic acid, (acetochlor sulfinyl-acetic acid)	AcSAA	NA
[(2,6-diethylphenyl)(methoxymethyl)amino]oxo-acetic acid, (alachlor oxanilic acid)	AlOX	140939-14-6
2-[(2,6-diethylphenyl)(methoxymethyl)amino]-2-oxoethanesulfonic acid, (alachlor sulfonic acid)	AlESA	140939-15-7
(2,6-diethylphenyl)amino-2-oxoacetic acid, (alachlor sec-oxanilic acid)	Al-sOX	NA
{2-[(methoxymethyl)(2,6-diethylphenyl)amino]-2-oxoethyl}sulfinylacetic acid, (alachlor sulfinylacetic acid)	AlSAA	NA
[(2,4-dimethyl-3-thienyl)(2-methoxy-1-methylethyl)amino]oxoacetate, (dimethenamid oxanilic acid)	DimOX	NA
2-[(2,4-dimethyl-3-thienyl)(2-methoxy-1-methylethyl)amino]-2-oxoethanesulfonate, (dimethenamid sulfonic acid)	DimESA	205939-58-8
2-[(2-ethyl-6-methylphenyl)-(2-methoxy-1-methylethyl)amino]-2-oxoacetic acid, (metolachlor oxanilic acid)	MeOX	152019-73-3
2-[(2-ethyl-6-methylphenyl)(2-methoxy-1-methylethyl)amino]-2-oxoethanesulfonic acid, (metolachlor sulfonic acid)	MeESA	171118-09-5
[(1-methylethyl)phenylamino]oxoacetic acid, (propachlor oxanilic acid)	ProOX	NA
2-[(1-methylethyl)phenylamino]-2-oxoethane sulfonic acid, (propachlor sulfonic acid)	ProESA	NA
[[(methylethyl) phenylamino] acetyl] sulfinylacetic acid, (propachlor sulfinylacetic acid)	ProSAA	NA

(NA = not available)

Fortification Solutions

In order to estimate the analytical accuracy of the method within a given set of water samples, it was necessary to fortify control water samples with known amounts of each degradate. Fortification solutions were prepared by serial dilution from the mixed stock solutions into the appropriate control water. The fortification samples were prepared at 0.025, 0.05, 0.10, 0.25, 0.50, 1.0 and 2.0 µg/L by serial dilution from the mixed stock solutions.

Calibration Standards

The calibration standards were prepared at 0.010, 0.025, 0.05, 0.10, 0.25, 0.50, 1.0, 2.0 and 5.0 µg/L by serial dilution from the mixed stock solutions. Fisher Optima® grade reagent water was used as the matrix for all calibration standards.

Instrumentation

An LC/MS/MS was used for separation and quantitation of degradates. Using multiple reaction monitoring (MRM), in the negative ion electrospray ionization mode (-ESI) the LC/MS/MS gives superior specificity and sensitivity when compared to conventional LC/MS techniques. The improved specificity eliminates interferences typically found in LC/MS or LC/UV (ultraviolet detection) analyses. The Sciex Analyst software, v1.1β, provided complete control of the mass spectrometer as well as data acquisition and processing.

Two ions were monitored for each degradate, one transition (deprotonated molecule) ion and one quantitation (fragment) ion. The transition and quantitation ions for the degradates are listed in Table II.

HPLC Description and Conditions

The degradates were chromatographed on a Zorbax StableBond C_8 column, 50 mm x 4.6 mm x 3.5 µ, in combination with a Zorbax StableBond C_8 guard column, 12.5 mm x 4.6 mm x 5 µ. The liquid chromatograph was a Hewlett Packard 1100 system, including a binary pump, degasser, column heater and autosampler. The column was maintained at 70°C to minimize or eliminate chromatographic separation of the rotational isomers. A solvent gradient was

Table II. Multiple Reaction Monitoring Ion Selection

Degradate	MRM Transition Ion (Daltons)	Quantitation Ion (Daltons)
acetochlor oxanilic acid	264	146
acetochlor ethanesulfonic acid	314	162
acetochlor sulfinylacetic acid	340	146
alachlor oxanilic acid	264	160
alachlor s-oxanilic acid	220	148
alachlor ethanesulfonic acid	314	176
alachlor sulfinylacetic acid	340	160
dimethenamid oxanilic acid	270	198
dimethenamid ethanesulfonic acid	320	121
metolachlor oxanilic acid	278	206
metolachlor ethanesulfonic acid	328	121
propachlor oxanilic acid	206	134
propachlor ethanesulfonic acid	256	121
propachlor sulfinylacetic acid	282	134

used comprising a mixture of mobile phase A: 95:5 water: methanol (with 0.2% acetic acid) and
mobile phase B: 50:50 acetonitrile: methanol (with 0.2% acetic acid). Initial conditions were 95:5 A:B, to 50:50 A:B at 3 minutes, to 30:70 A:B at 6.5 minutes and hold to 7.5 minutes. Re-equilibration to 95:5 required an additional 2.5 minutes. The flow rate was 700 µL / minute and the column effluent was split approximately 14:1 at the ion source (~50 µL / minute of flow to the ion source). The injection volume was 100 µL. To minimize source contamination the column flow was diverted to waste for 3.5 minutes following injection.

Mass Spectrometer Description and Conditions

An Applied BioSystems PE Sciex API-3000 tandem mass spectrometer using Analyst software, v1.1β was used in this study. The API-3000 was coupled to the HP1100 LC system through a TurboIonSpray ion source. A Valco, Model EHMA, electrically actuated 6 port switching valve was used to divert the column flow from the ion source prior to and following elution of the analytes of interest. The MS was operated in the negative ion mode with a

TurboIonSpray voltage of 4.2 kilovolts. The TurboIonSpray gas flow was 6 L/minute (nitrogen) at a temperature of 350°C. Nitrogen was used for both the curtain and nebulizer gases. From injection to injection the total run time was 10 minutes.

The MRM experiments did not require chromatographic separation of the degradates. Compounds were identified based on their molecular weight and specific quantitation ion. Therefore, other LC conditions, columns, gradient, and injection volumes may be used provided there is adequate sensitivity, specificity and the chromatographic quality is not compromised.

Sample Preparation

Sample preparation was not necessary for direct aqueous injection. The samples were transferred directly to 2 mL autosampler vials for analysis. No preconcentration, sample cleanup or (in most cases) filtration was required prior to analysis.

Detector Calibration

A calibration curve was generated for every set of samples. The standards were placed among the analytical samples for each set. The first and last sample in each analytical sample set was a standard. The calibration curve was generated by plotting the response of each analyte in a calibration standard against its concentration. Least squares estimates of the data points was used to define the calibration curve. Linear, exponential, or quadratic calibration curves may be used. An example calibration curve for propachlor oxanilic acid, a $1/x$ weighted linear fit of response versus concentration, is in Figure 3.

Quantitation Criteria

Each individual standard, control and environmental sample was analyzed one time only. Analyte calibration was performed for each chromatographic set using a multi-point calibration curve. The complete chromatographic set containing calibration standards, control, fortified control, and fortified field samples was arranged such that the set began and ended with a calibration standard (i.e., control, fortified, and fortified field samples are bracketed by calibration standards) with the standards evenly distributed throughout the set. The minimum correlation coefficient for the linearity of the analyte calibration curve is 0.99. The average analytical recovery of fortified samples for each analyte in each set should range between 70-110% of the amount fortified. Each chromatographic set shall contain at least one laboratory control sample and one fortified laboratory control sample.

ProOx : "Linear" Regression ("1 / x" weighting): y = 1.74e+005 x + -288 (r = 0.9999)

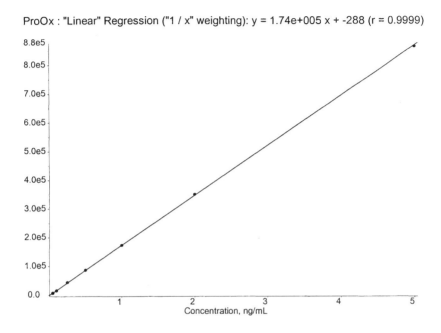

Figure 3. Propachlor Oxanilic Acid Calibration Curve

Results and Discussion

During this study, control finished surface water, raw surface water, and ground water samples were fortified with known concentrations of the fourteen degradates, at 0.05 µg/L, 0.1 µg/L, 0.25 µg/L, 0.50 µg/L, and 1.0 µg/L, and carried through the analytical procedure. Results confirm that reliable quantitation in these water matrices can be achieved by direct aqueous injection LC/MS/MS at these low part per billion levels.

Five replicates each, in ground water, raw surface water and finished surface water, were analyzed by the method described herein. No significant differences in recovery or precision were apparent across the three different matrices. Therefore, the independent matrix data were pooled to produce an average recovery and RSD value for all water types.

The LOQ (limit of quantitation) for each analyte is defined as the lowest concentration tested at which an acceptable mean recovery (70%-110%) is obtained with a relative standard deviation (RSD) of ≤20%. The LOQ criteria was met at 0.05 µg/L for all degradates with one exception, AlOX. Following in Table III are the mean analyte recoveries and RSD for the fourteen degradates at the three lowest concentration levels evaluated. These recoveries are not corrected for background, with the exceptions noted below.

Recovery for AlOX was unacceptable at the 0.05 µg/L level at 52.1% although the RSD at this level was very good at 6.7%. Even at the 0.10 µg/L level AlOX recovery was marginally acceptable at 69.5%. This was surprising in that AlOX was one of the most sensitive components in terms of raw response and that all analytes at this concentration were spiked from a single solution. With sample handling ruled out as the cause for this low recovery we suspect the MS parameters were not optimal for this analyte, in effect inducing a threshold effect on the response.

MeESA and MeOX were the only analytes requiring background subtraction due to trace level contamination in the surface water matrices at 0.027 and 0.009 µg/L, respectively. Recoveries and relative standard deviations for these analytes meet the guideline criteria at 0.05 µg/L when recoveries are background corrected. Based on a screen of several potential matrix control sources, we determined these detections of MeESA and MeOX to be random interferences and not representative of ambient concentration. It would be very difficult to obtain a matrix surface water source completely devoid of all 14 degradate analytes. Background concentrations of the other 12 analytes were generally less than 0.005 ppb in surface water and less than 0.001 ppb in ground water, when detected at all, for the matrix sources used in this study.

The water matrices used in this study were finished surface water collected in Montpelier, VT (sourced from the Berlin Pond Reservoir), raw surface water collected from the Berlin Pond Reservoir in Montpelier, VT, and ground water collected from a 400 foot deep well in St. Charles County, MO.

The tertiary alachlor and acetochlor degradates are isomeric. While the AlOX and AcOX compounds produce unique fragment ions, the primary fragment ion for the two ESA compounds is m/z 121. Generally, differentiation of these isomers would require chromatographic separation for precise quantitation. Using MS/MS these analytes can be differentiated using their

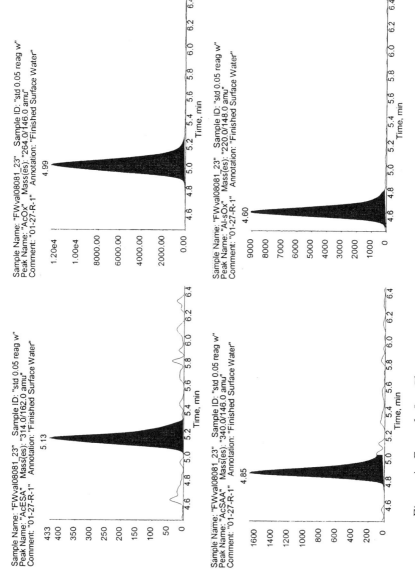

Figure 4: Example Ion Chromatograms – 0.05 µg/L Standard   Continued on next page.

268

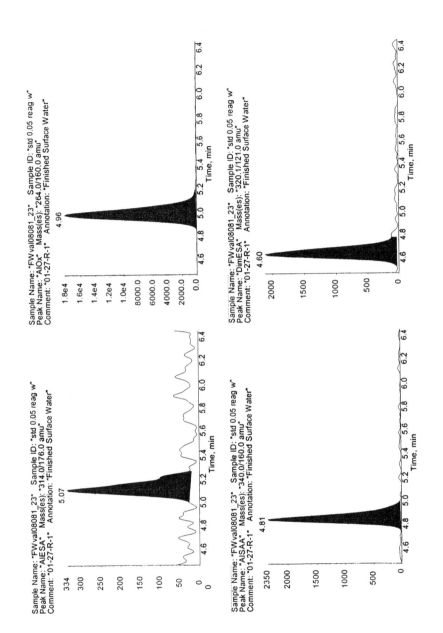

Sample Name: "FWval08081_23"   Sample ID: "std 0.05 reag w"
Peak Name: "AlESA"   Mass(es): "314.0/176.0 amu"
Comment: "01-27-R-1"   Annotation: "Finished Surface Water"

Sample Name: "FWval08081_23"   Sample ID: "std 0.05 reag w"
Peak Name: "AlOx"   Mass(es): "264.0/160.0 amu"
Comment: "01-27-R-1"   Annotation: "Finished Surface Water"

Sample Name: "FWval08081_23"   Sample ID: "std 0.05 reag w"
Peak Name: "AlSAA"   Mass(es): "340.0/160.0 amu"
Comment: "01-27-R-1"   Annotation: "Finished Surface Water"

Sample Name: "FWval08081_23"   Sample ID: "std 0.05 reag w"
Peak Name: "DimESA"   Mass(es): "320.1/121.0 amu"
Comment: "01-27-R-1"   Annotation: "Finished Surface Water"

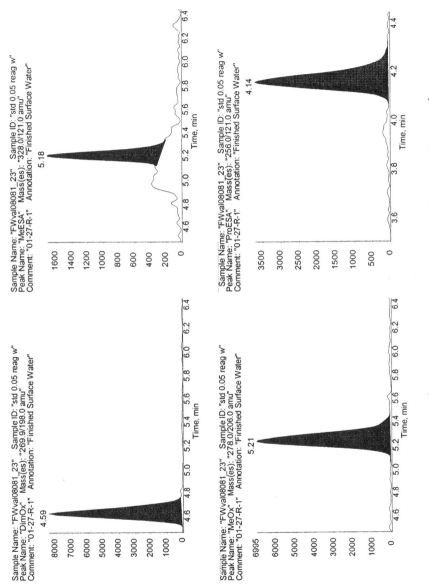

Figure 4: Example Ion Chromatograms – 0.05 µg/L Standard (cont.) Continued on next page.

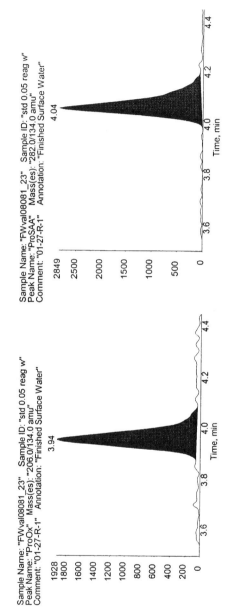

Sample Name: "FWval08081_23"   Sample ID: "std 0.05 reag w"
Peak Name: "ProOx"   Mass(es): "206.0/134.0 amu"
Comment: "01-27-R-1"   Annotation: "Finished Surface Water"

Sample Name: "FWval08081_23"   Sample ID: "std 0.05 reag w"
Peak Name: "ProSAA"   Mass(es): "282.0/134.0 amu"
Comment: "01-27-R-1"   Annotation: "Finished Surface Water"

Figure 4. *Continued.*

Table III. Percent Recovery ± RSD: All Matrices

Degradate	0.05 μg/L	0.10 μg/L	0.25 μg/L
AcESA	97.1% ± 8.6%	93.4% ± 9.5%	105.3% ± 6.6%
AcOX	94.6% ± 4.2%	87.6% ± 3.2%	99.5% ± 2.6%
AcSAA	101.7% ± 9.9%	92.0% ± 11.0%	104.5% ± 6.9%
AIESA	87.0% ± 5.3%	99.2% ± 9.3%	100.5% ± 9.1%
AIOX	52.1% ± 6.7%	69.5% ± 5.5%	95.0% ± 4.2%
AISAA	99.1% ± 7.6%	97.5% ± 5.6%	106.2% ± 4.7%
AI-sOX	98.9% ± 4.9%	93.0% ± 4.6%	101.1% ± 3.0%
ProESA	100.1% ± 3.8%	94.3% ± 5.6%	103.3% ± 4.1%
ProOX	102.9% ± 8.1%	95.4% ± 4.1%	103.4% ± 3.6%
ProSAA	96.0% ± 3.9%	94.0% ± 4.7%	103.5% ± 2.5%
MeESA	97.9% ± 9.9%	92.6% ± 8.2%	104.8% ± 6.7%
MeOX	108.0% ± 4.8%	104.4% ± 5.0%	106.9% ± 4.3%
DimESA	94.2% ± 6.2%	96.3% ± 10.4%	106.0% ± 4.7%
DimOX	101.4% ± 3.3%	96.1% ± 3.9%	104.2% ± 4.0%

secondary fragment ions m/z 176 and 162, respectively. While this provides confirmatory identification of the degradates, the secondary fragment ions are much less intense than the m/z 121 ion resulting in approximately ten fold poorer sensitivity for these compounds. Reconstructed ion chromatograms of all 14 degradate standards, at the 0.05 μg/L level (LOQ) are presented in Figure 4.

Ion suppression and ion enhancement effects are often found in MS/MS methods where there is limited analyte resolution and co-elution of both known and unknown components. To test the method for these matrix effects, a series of standard addition experiments were conducted. Real environmental samples covering all three water matrices from approximately 50 different sites were analyzed as received. Aliquots of the samples were also spiked with known and variable amounts (0.10 to 2.0 μg/L) of mixed degradate fortification solution and reanalyzed. The individual degradate results from both samples were compared and the percent spike recovery calculated. At all sites and for all degradates, the spike recoveries were well within the 70%-110% range. This was further evidence there were no matrix induced suppression or enhancement effects in the data. The standard addition summary included only results at or above the LOQ of 0.05μg/L, the level at which recovery and variability met or exceeded the quantitation criteria. The environmental samples were analyzed in a previous monitoring study and were selected because they generally had known residues of several of the common degradates.

Advantages of DAI LC/MS/MS

The primary advantage of direct aqueous injection LC/MS/MS is the absence of any type of sample preparation. The high ultimate sensitivity of the MS obviates the need for laborious, costly and time consuming sample concentration and/or cleanup. This is in contrast to previous methodologies requiring concentration from 100-200 (or larger) mL sample volumes. The only samples requiring filtration are those with visible particulates, most often ground water. Without these sample handling constraints, a simple and rapid high throughput analysis was attained. The multiple reaction monitoring technique provides a high level of selectivity and low incidence of interference due to the significant reduction of chemical background and noise. The result is a simple, straightforward analytical methodology providing both confirmatory identification and a precise quantitative assessment of chloroacetanilide degradates in aqueous environmental samples.

References

1. Roberts, T. R.; Hutson, D. H.; Lee, P. W.; Nicholls, P. H.; Plimmer, J. R., Metabolic Pathways of Agrochemicals; Part One, Herbicides and Plant Growth Regulators, Royal Society of Chemistry, 1998, pp 179-218
2. Ferrer, I; Thurman, E.M.; Barcelo, D. Anal. Chem. 1997, 69, 4547-4553
3. Field, J. A., Thurman, E. M. Environ. Sci. Technol. 1996, 30, 1413-1418
4. Heberle, S.; Aga, D.S.; Hany, R.; Muller, S.R., Rentsch, D., Environ. Sci. Technol. 1999, 33, 3462-3468
5. Heberle, S.A.; Aga, D.S.; Hany, R.; Muller, S.R., Anal. Chem. 2000, 72, 840-845
6. Lee, E.A.; Kish, J.L.; Zimmermen, L.R.; Thurman, E.M. United States Geological Survey Open-File Report 01-10, 23 p.
7. Vargo, J. D.; Anal. Chem., 1998, 70, 2699-2703
8. Shoemaker, J., "Analytical Method Development for Alachlor ESA and Other Acetanilide Herbicide Degradation Products", 49th ASMS Conference on Mass Spectrometry and Allied Topics, Chicago, IL, May 27-31, 2001

Chapter 16

Interlaboratory Comparison and Validation of Methods for Chloroacetanilide and Chloroacetamide Soil Degradates in Environmental Waters

John D. Vargo[1], Edward A. Lee[2], and John D. Fuhrman[3,*]

[1]Hygienic Laboratory, University of Iowa, Iowa City, IA 52242
[2]U.S. Geological Survey, 4821 Quail Crest Place, Lawrence, KS 66049
[3]Monsanto Company, 800 North Lindbergh Boulevard, St. Louis, MO 63167

This paper describes a corroborative evaluation of three different analytical methods for the analysis of soil degradates of chloroacetanilide and chloroacetamide herbicides in environmental waters. Independent methodologies were developed at the University of Iowa Hygienic Laboratory (IHL), U. S. Geological Survey Laboratory (USGS) (Lawrence, KS.) and at Monsanto Company (MON) with the goal of quantitating the common soil degradates at concentrations below 0.1 µg/L. Liquid chromatography / mass spectrometry (LC/MS) is the fundamental basis of these methods which use differing approaches at concentration and MS detection. All methods quantified the ethanesulfonic acids (ESA) and oxanilic acids (OXA) of acetochlor (Ac), alachlor (Al), dimethenamid (Di) and metolachlor (Me). Correlation of results between laboratories was excellent. At residue levels >0.05 µg/L and using relative standard deviation (RSD) as an indicator of variability, the average RSD across labs for all surface water samples was 9.3% for the eight primary degradates. Overall, the average recovery from the laboratory fortified samples was nearly identical between laboratories.

Introduction

Chloroacetanilides are soil-applied herbicides used for pre- and early post-emergence control of grasses and broadleaf weeds in a variety of crops. The major chloroacetanilides, acetochlor, alachlor and metolachlor are used extensively worldwide. The major soil metabolites of these materials are occasionally found in surface water and ground water but usually at very low concentrations. There now exist a variety of highly sensitive analytical methods to determine residues of these compounds in environmental waters.

Three laboratories participated in a study designed to evaluate potential sources of interlaboratory variation in the analysis of chloracetanilide herbicide soil metabolites. The participating laboratories included, Iowa Hygienic Laboratory, Iowa City, Iowa, United States Geological Survey, Lawrence, Kansas and Monsanto Company, Saint Louis, Missouri. Each laboratory had developed in-house analytical methodology based on LC/MS. While the LC/MS fundamentals are similar between laboratories, each lab took a unique and different approach to quantitation. The three participating laboratories used different extraction procedures and all used different LC/MS instrumentation. The Iowa Hygienic Lab used Envi-CARB SPE to concentrate the water samples with quantitation using a Micromass Quattro tandem quadrupole system. The USGS used C_{18} SPE to concentrate water samples and a Hewlett-Packard single quadrupole system for quantitation. Monsanto directly injected the water samples into an HPLC system with quantitation using a PE Sciex API-3000 tandem quadrupole system. All three LC/MS systems used pneumatically assisted electrospray interfaces. Each method targeted a unique spectrum of soil degrades and / or parent herbicide for detection and quantitation.

Actual field samples were collected for use in this study. Some samples were laboratory fortified to ensure measurable residues were present for analysis. All methods quantified the ethanesulfonic acids (ESA) and oxanilic acids (OXA) of acetochlor (Ac), alachlor (Al), dimethenamid (Di) and metolachlor (Me) which will be the primary focus of this paper and be referred to as "common degradates". Both USGS and MON quantified additional degradates, some common between both methods. The common analytes included; propachlor (Pr) ESA and OXA, alachlor sulfinylacetic acid (SAA) and acetochlor SAA. Analytes unique to USGS included flufenacet ESA and OXA. Analytes unique to the Monsanto method included propachlor SAA and alachlor s-oxanilic acid (s-OXA). The IHL included parent herbicides in their methodology. A number of papers have been published in recent years regarding the analysis of chloroacetanilide herbicide degradates in environmental waters (1-5). Structures are shown generically in Figure 1 and detailed in Table I.

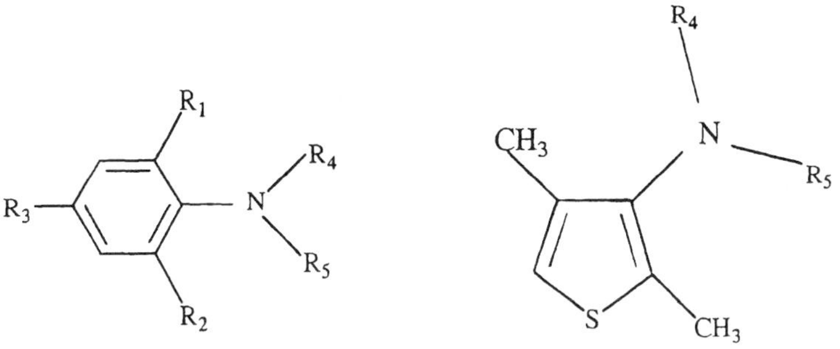

Acetanilide   Acetamide

Figure 1: Generic Degradate Structures

Table I. Reference Compounds

Abbrev-iation	CAS Registry No.	R_1	R_2	R_3	R_4	R_5
(acetanilide)						
AcOX	194992-44-4	Me	Et	H	$CH_2OCH_2CH_3$	$COCO_2H$
AcESA	187022-11-3	Me	Et	H	$CH_2OCH_2CH_3$	$COCH_2SO_3H$
AcSAA	NA	Me	Et	H	$CH_2OCH_2CH_3$	$COCH_2SOCH_2CO_2H$
AlOX	140939-14-6	Et	Et	H	CH_2OCH_3	$COCO_2H$
AlESA	140939-15-7	Et	Et	H	CH_2OCH_3	$COCH_2SO_3H$
AlsOX	140939-17-9	Et	Et	H	H	$COCO_2H$
AlSAA	140939-16-8	Et	Et	H	CH_2OCH_3	$COCH_2SOCH_2CO_2H$
FlOX	NA	H	H	F	$CH(CH_3)_2$	$COCO_2H$
FlESA	NA	H	H	F	$CH(CH_3)_2$	$COCH_2SO_3H$
MeOX	152019-73-3	Me	Et	H	$CH(CH_3)CH_2OCH_3$	$COCO_2H$
MeESA	171118-09-5	Me	Et	H	$CH(CH_3)CH_2OCH_3$	$COCH_2SO_3H$
PrOX	70628-36-3	H	H	H	$CH(CH_3)_2$	$COCO_2H$
PrESA	123732-85-4	H	H	H	$CH(CH_3)_2$	$COCH_2SO_3H$
PrSAA	12373286-2	H	H	H	$CH(CH_3)_2$	$COCH_2SOCH_2CO_2H$
(acetamide)						
DiOX	NA	-	-	H	$CH(CH_3)CH_2OCH_3$	$COCO_2H$
DiESA	205939-58-8	-	-	H	$CH(CH_3)CH_2OCH_3$	$COCH_2SO_3H$

Methods

Samples: Selection / Preparation / Storage

Each laboratory provided samples for the round robin analysis. These were intended to be representative surface and ground water samples from the corn-growing region of the Midwest.

IHL provided two surface water samples, one from the English River and the other from Coralville Lake, both in Iowa. Samples were provided to all the laboratories as field controls and laboratory fortified field controls. The English River sample was fortified at 3.33 µg/L for all IHL analytes. The Coralville Lake sample was fortified at 1.0 µg/L for all IHL analytes except AcESA, which was spiked at 12.72 µg/L.

The USGS lab provided surface water samples from Clinton Lake, in Kansas. Three samples were distributed among the participating labs, a field control and laboratory fortified field controls spiked at two concentration levels. The fortified samples were spiked at 0.50 and 2.50 µg/L, respectively, for all USGS analytes.

MON provided samples collected from surface water sources in Illinois. Included were a finished (drinking) water sample from the Kankakee River in Kankakee and a raw surface water sample from the Silver Lake reservoir in Highland. These samples had low level detections of most of the common degrades as determined by prior analysis in a separate monitoring study. These samples were provided, as field controls only, no fortifications were made as many of the analytes were already known to be present.

Samples were shipped to the participating labs within one day of preparation. Samples were shipped under chilled conditions (ice packs) and stored at <10°C between receipt and analysis.

Two samples of ground water from Minnesota were included in the round robin study. These samples were collected at an agricultural dealer site with known point source contamination and degradates at very high levels (50-1000 µg/L). These samples were included to evaluate method performance at these extreme concentration levels. The Minnesota samples were provided by the Minnesota Department of Agriculture.

Limit of Quantitation (LOQ)

The LOQs were compound dependent. IHL reported a common LOQ of 0.025 µg/L. Both USGS and MON reported a common LOQ of 0.05µg/L. All methods were validated with acceptable (70-120%) recoveries for all analytes over the range of 0.05 to 20 µg/L.

Reference Standards

All reference standards were synthesized by applicable registrants and obtained directly or via EPA National Pesticide Standard Repository in Fort Meade, MD.

Experimental – Iowa Hygienic Laboratory

The extraction procedure used was developed by Shoemaker (5). In this study, the volume of the SPE eluting solvent was increased to permit the parent herbicides to be eluted along with the ESA and OXA degradates. A 100-mL sample of water (no pH adjustment) was passed through a preconditioned Supelco Envi-Carb SPE column (250 mg), rinsed with purified water, and eluted with 10 mL of methanol (10 mM in ammonium acetate). The methanol content was removed via nitrogen evaporation until only aqueous remained (approximately 0.5 mL). Acetonitrile was added (0.4 mL) and the sample diluted to a final volume of 2.0 mL with purified water. The samples were stored refrigerated until the time of analysis. For long-term storage (> 1 week), it is recommended that the samples be stored at freezer temperatures.

The samples were analyzed using a Micromass Quattro LC/MS/MS (tandem quadrupole) system with an electrospray interface. The ESA and OXA degradates were monitored as negative ions while the parent herbicides were monitored as positive ions. All separations were performed using a Zorbax SB C_8 column (3.0 x 150 mm, dp = 5 µm) with a mobile phase flow rate of 0.6 mL/min. A mobile phase gradient using acetonitrile (0.15% in acetic acid) and water (0.15% in acetic acid) was used with a linear gradient ramp from 20-100% ACN over 10 minutes. A column temperature of 30°C was used. The injection volume was 50 µL.

Unique precursor ion/product ion pairs were monitored in the multiple reaction monitoring mode (MRM) for each analyte. The precursor ions were the protonated molecular ions (positive ion monitoring) or the deprotonated molecular ions (negative ion monitoring). Data acquisition conditions are presented in Table II.

Table II. Data Acquisition Conditions for Micromass Quattro LC/MS/MS

Scan Func.	Analyte	Precursor Ion	Product Ion	Dwell (sec)	Cone (V)	CE (V)
1	Acetochlor OXA	264.00	145.90	0.10	20	12
1	Alachlor OXA	264.00	159.90	0.10	20	12
1	Dimethenamid OXA	270.00	197.90	0.10	20	12
1	Metolachlor OXA	278.00	205.95	0.10	20	12
1	Acetochlor ESA	314.10	161.90	0.10	40	23
1	Alachlor ESA	314.10	159.90	0.10	40	23
1	Dimethenamid ESA	320.00	120.90	0.10	40	23
1	Metolachlor ESA	328.10	120.90	0.10	40	23
2	Acetochlor	270.10	223.90	0.15	20	10
2	Alachlor	270.10	238.00	0.15	20	15
2	Dimethenamid	276.10	243.95	0.15	20	15
2	Metolachlor	284.15	252.00	0.15	20	16

ESA/OXA metabolites monitored as negative ions.
Parent herbicides monitored as positive ions.

CE = collision energy

Function 1: Start time: 2.0 min, End time: 7.0 min

Function 2: Start time: 7.0 min, End time: 11.0 min

Experimental - Monsanto

The Monsanto method used direct aqueous injection of the water sample. The samples were transferred directly to 2 mL autosampler vials for analysis. No preconcentration, sample cleanup or filtration was necessary prior to analysis.

The samples were analyzed using a PE Sciex API-3000 LC/MS/MS (tandem quadrupole) system with a TurboIonSpray interface. All degradate analytes were monitored as negative ions. The degradates were chromatographed on a Zorbax StableBond C_8 column, 50 mm x 4.6 mm x 3.5 μ, in combination with a Zorbax StableBond C_8 guard column, 12.5 mm x 4.6 mm x 5 μ. The liquid chromatograph was a Hewlett Packard 1100 system, including a binary pump, degasser, column heater and autosampler. The column was maintained at 70°C to minimize chromatographic separation of the rotational isomers. A solvent gradient was used comprising a mixture of mobile phase A: 95:5 water: methanol (with 0.2% acetic acid) and mobile phase B: 50:50 acetonitrile: methanol (with 0.2% acetic acid). Initial conditions were 95:5 A:B, to 50:50 A:B at 3 minutes, to 30:70 A:B at 6.5 minutes and hold to 7.5 minutes. Re-equilibration to 95:5

required an additional 2.5 minutes. The flow rate was 700 µL / minute and the column effluent was split approximately 14:1 at the ion source (~50 µL / minute of flow to the ion source). The API-3000 was coupled to the HP1100 LC system through a TurboIonSpray source. A Valco, Model EHMA, electrically actuated

Table III. Data Acquisition Conditions for Sciex API-3000 LC/MS/MS

Scan Func.	Analyte	Precurser Ion	Product Ion	Dwell (sec)	DP (V)	CE (V)
1	Propachlor OXA	206.0	134.0	0.10	-25	-30
1	Propachlor SAA	282.0	134.0	0.10	-30	-30
1	Propachlor ESA	256.0	121.0	0.10	-20	-26
2	Alachlor s-OXA	220.0	148.0	0.05	-16	-16
2	Dimethenamid ESA	320.0	121.0	0.05	-32	-31
2	Dimethenamid OXA	270.0	198.0	0.05	-25	-16
2	Alachlor SAA	340.0	160.0	0.05	-11	-30
2	Acetochlor SAA	340.0	146.0	0.05	-11	-30
2	Alachlor OXA	264.0	160.0	0.05	-21	-16
2	Acetochlor OXA	264.0	146.0	0.05	-21	-16
2	Alachlor ESA	314.0	176.0	0.10	-36	-34
2	Acetochlor ESA	314.0	162.0	0.10	-36	-34
2	Metolachlor ESA	328.0	121.0	0.05	-36	-30
2	Metolachlor OXA	278.0	206.0	0.05	-21	-16
DP = declustering potential						
CE = collision energy						
All metabolites monitored as negative ions.						
Function 1: Start time: 3.5 min, End time: 4.5 min						
Function 2: Start time: 4.5 min, End time: 6.5 min						

6 port switching valve was used to divert the column flow from the ion source prior to and following elution of the analytes of interest. The injection volume was 100 µL. The total run time was 10 minutes. Specific precursor ion/product ion pairs were monitored in the MRM mode for each analyte. Data acquisition conditions are presented in Table III.

Experimental – USGS

A Waters C_{18} Sep-Pak (500 mg) is conditioned by sequentially passing 3 mL methanol, 3 mL ethyl acetate, 3 mL methanol, and 3 mL distilled water through each column at a flow rate of 20 mL/min by positive pressure. The filtered water sample, 123 mL, is then passed through the preconditioned Sep-Pak. The Sep-Pak is rinsed with 3.2 mL of ethyl acetate to remove interfering compounds. The adsorbed chloroacetanilide degradates are eluted from the Sep-Pak with 3.5 mL of methanol. The solution is spiked with an internal standard (2,4-dichlorophenoxyacetic acid), evaporated under nitrogen at 50°C, and reconstituted with 125μL of 50:50 mobile phase. The sample is transferred to an autosampler vial and stored at <0°C until analyzed by LC/MS.

Table IV. Data Acquisition Conditions for Hewlett-Packard LC/MSD

Scan Func.	Analyte	Molecular Ion (M-H)	Fragment Ion 1	Fragment Ion 2	Dwell Time (secs)
1	Dimethenamid ESA	320	-	-	0.057
1	Dimethenamid OXA	270	198	-	0.057
1	Alachlor SAA	-	234	160	0.057
1	Acetochlor SAA	-	234	146	0.057
1	Alachlor OXA	264	160	-	0.057
1	Acetochlor OXA	264	146	-	0.057
1	Alachlor ESA	314	-	-	0.057
1	Acetochlor ESA	314	-	-	0.057
1	Metolachlor ESA	328	-	-	0.057
1	Metolachlor OXA	278	206	-	0.057
1	Flufenacet OXA	224	152	-	0.057
1	Flufenacet ESA	274	-	-	0.057
1	Propachlor OXA	206	134	-	0.057
1	Propachlor ESA	256	-	-	0.057
Fragmentor Voltage = 70 V					
Capillary Voltage = 3100 V					
All metabolites monitored as negative ions.					

The samples were analyzed using a Hewlett-Packard, 1100 HPLC with mass selective detector (LC/MSD) and electrospray interface. The LC/MSD was operated in the negative-ion mode for all degradates. Fragment ions were generated by source-induced dissociation of the pseudo-molecular ions, as tabulated above. All separations were performed using a Phenomenex C_{18} column, 250- x 3-mm x 5 μm with a mobile phase flow rate of 0.5 mL per minute using a 50:50 isocratic mobile phase of; A) 0.3 percent acetic acid in reagent water and B) 0.3 percent acetic acid in 1:2 methanol: acetonitrile. The column was thermostated at 65°C. The injection volume was 10 μL. Data acquisition conditions are presented in Table IV.

Results and Discussion

The concentrations found varied significantly across samples and analytes, from non-detect (ND) to levels as high as 15 μg/L. Figures 2 – 9 give a graphical comparison (log scale) of surface water results, by analyte and sample for the common degrades from the three performing laboratories. Correlation of results between labs was excellent. Despite significant differences in response between degrades the average precision, as the RSD, was comparable and varied from 5.7% to 11.0% across the eight degrades and the three labs.

Correlation of degrade recoveries from the laboratory fortifications was very good between labs. The data, as presented in Table V, show the sample concentrations and background corrected recovery values for the performing laboratories. The average recovery and RSD across analytes and samples was calculated by lab and determined to be, 100.4% ± 9.4% for IHL, 103.5% ± 11.6% for USGS and 101.8% ± 9.1% for MON for the eight common degrades.

As AcESA and AlESA co-elute chromatographically (IHL and MON) and are isomeric there was concern the compounds may not be accurately detected and reported. The primary product ion for both degrades is m/z 121. Therefore, less intense product ions are used to differentiate and quantitate the isomers. USGS uses partial chromatographic resolution to quantify AcESA and AlESA. USGS quantifies the molecular ions for these degrades, as the single quadrupole system cannot generate the unique MS/MS product ions using source induced fragmentation. In the USGS method, source induced fragmentation is used to differentiate the isomeric and co-eluting AcOXA and AlOXA degrades using unique product ions for quantitation.

To test the methods abilities to accurately quantify AlESA in the presence of a large excess of AcESA, the Coralville Lake sample was spiked with 12.72 μg/L of AcESA. AlESA was spiked at 1.0 μg/L. The results are tabulated in Table V. While both IHL and MON had slightly high recoveries for AlESA (109 and 120%), the USGS recovery was slightly low (82%). In contrast, the AcESA recoveries were again slightly high for IHL and MON (110 and 106%), and also for USGS (117%). Despite the differences, the RSD across all labs for this sample was 19.0%, still within acceptable bounds.

A significantly greater variability was found with the Minnesota (MN) ground water (GW) samples. MON diluted the GW samples, 100:1, prior to analysis. IHL processed the MN samples normally and diluted the extracts with solvent until the analytes fit within the calibration range. USGS processed samples at dilutions of 1:10, 1:100 and 1:1000 and reported analyte results that fit within the calibration curve. The Minnesota results are shown in Table VI.

Table V. Recoveries from Laboratory Fortifications

Analyte	Sample	Spike	Concentration (µg/L)			% Spike Recovery				
			IHL	USGS	MON	IHL	USGS	MON	Average	RSD
AcESA	English River, IA		0.20	0.23	0.24					
	English River, IA	3.33	3.62	3.91	3.80	102.7	110.5	106.8	106.7	3.7
	Coralville Lake, IA		0.09	0.06	0.08					
	Coralville Lake, IA	12.72	14.10	15.00	13.60	110.2	117.5	106.3	111.3	5.1
	Clinton Lake, KS		0.05	0.07	0.06					
	Clinton Lake, KS	0.50	0.50	0.71	0.56	90.6	128.0	98.6	105.7	18.6
	Clinton Lake, KS	2.50	2.47	2.69	2.54	96.8	104.8	99.0	100.2	0.4
AcOXA	English River, IA		0.13	0.16	0.16					
	English River, IA	3.33	3.66	3.51	3.56	105.9	100.6	102.1	102.9	2.7
	Coralville Lake, IA		<0.05	0.05	<0.05					
	Coralville Lake, IA	1.00	1.06	0.97	1.08	103.5	92.0	105.5	100.3	7.3
	Clinton Lake, KS		0.09	0.12	0.11					
	Clinton Lake, KS	0.50	0.58	0.66	0.67	98.2	108.0	111.2	105.8	6.4
	Clinton Lake, KS	2.50	2.66	2.84	2.79	102.8	108.8	107.1	106.2	2.4
AlESA	English River, IA		0.19	0.21	0.21					
	English River, IA	3.33	3.59	3.87	3.45	102.3	109.9	97.2	103.1	6.2
	Coralville Lake, IA		0.36	0.39	0.35					

Table V. Recoveries from Laboratory Fortifications (cont.)

Analyte	Sample	Spike	Concentration (µg/L)			% Spike Recovery			Average	RSD
			IHL	USGS	MON	IHL	USGS	MON		
AIlESA	Coralville Lake, IA	1.00	1.45	1.21	1.55	109.2	82.0	120.4	103.9	19.0
	Clinton Lake, KS		0.06	0.09	0.06					
	Clinton Lake, KS	0.50	0.48	0.63	0.53	84.6	108.0	93.8	95.5	12.3
	Clinton Lake, KS	2.50	2.30	2.69	2.27	89.6	104.0	88.4	94.0	8.9
AlOXA	English River, IA		0.03	0.00	0.03					
	English River, IA	3.33	3.55	3.11	3.47	105.9	93.4	103.5	100.9	6.6
	Coralville Lake, IA		<0.05	0.05	<0.05					
	Coralville Lake, IA	1.00	1.11	0.92	1.13	107.4	87.0	109.2	101.2	12.2
	Clinton Lake, KS		0.07	0.09	0.09					
	Clinton Lake, KS	0.50	0.61	0.65	0.69	108.2	112.0	121.6	113.9	6.1
	Clinton Lake, KS	2.50	2.89	2.77	2.98	112.8	107.2	115.8	111.9	4.2
MeESA	English River, IA		0.62	0.77	0.70					
	English River, IA	3.33	4.21	4.30	3.82	107.7	106.0	93.8	102.5	7.4
	Coralville Lake, IA		1.16	1.19	1.06					
	Coralville Lake, IA	1.00	2.12	2.20	2.02	96.0	101.0	96.0	97.7	3.0
	Clinton Lake, KS		0.17	0.23	0.19					
	Clinton Lake, KS	0.50	0.65	0.78	0.68	95.0	110.0	98.0	101.0	7.9
	Clinton Lake, KS	2.50	2.66	2.77	2.52	99.5	101.6	93.1	98.1	4.9

Continued on next page.

Table V. Recoveries from Laboratory Fortifications (cont.)

Analyte	Sample	Spike	Concentration (µg/L)			% Spike Recovery			Average	RSD
			IHL	USGS	MON	IHL	USGS	MON		
MeOXA	English River, IA		0.15	0.14	0.17					
	English River, IA	3.33	3.65	2.89	3.12	105.1	82.6	88.7	92.1	12.6
	Coralville Lake, IA		0.15	0.11	0.14					
	Coralville Lake, IA	1.00	1.19	1.00	1.03	104.2	89.0	88.8	94.0	9.4
	Clinton Lake, KS		0.21	0.24	0.24					
	Clinton Lake, KS	0.50	0.73	0.76	0.77	103.2	104.0	105.0	104.1	0.9
	Clinton Lake, KS	2.50	2.93	2.70	2.76	108.6	98.4	100.7	102.6	6.7
DiESA	English River, IA		<0.05	ND	<0.05					
	English River, IA	3.33	3.43	4.12	3.64	102.3	123.7	108.6	111.5	9.9
	Coralville Lake, IA		<0.05	ND	<0.05					
	Coralville Lake, IA	1.00	0.95	1.18	1.10	92.6	118.0	107.5	106.0	12.0
	Clinton Lake, KS		<0.05	ND	ND					
	Clinton Lake, KS	0.50	0.38	0.55	0.48	70.8	110.0	95.4	92.1	21.5
	Clinton Lake, KS	2.50	2.08	2.59	2.39	82.2	103.6	95.6	93.8	9.1
DiOXA	English River, IA		<0.05	ND	<0.05					
	English River, IA	3.33	3.46	2.85	3.01	103.2	85.6	89.6	92.8	9.9
	Coralville Lake, IA		<0.05	ND	<0.05					
	Coralville Lake, IA	1.00	0.96	0.85	0.93	93.1	85.0	90.1	89.4	4.6
	Clinton Lake, KS		<0.05	ND	ND					
	Clinton Lake, KS	0.50	0.56	0.56	0.56	105.8	112.0	111.4	109.7	3.1
	Clinton Lake, KS	2.50	2.87	2.70	2.69	113.7	108.0	107.6	109.8	4.7

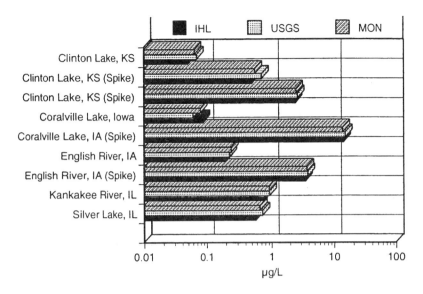

Figure 2. Acetochlor Ethanesulfonic Acid

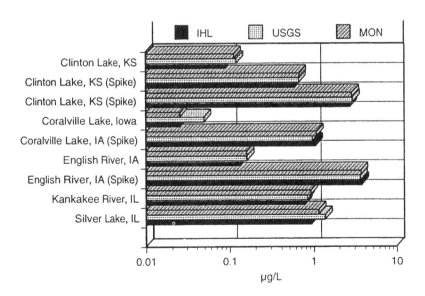

Figure 3. Acetochlor Oxanilic Acid

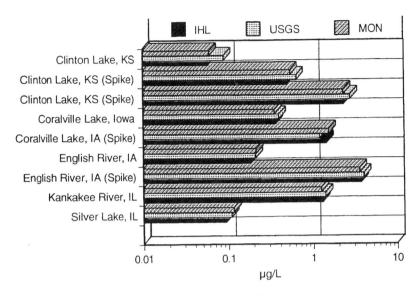

Figure 4. Alachlor Ethanesulfonic Acid

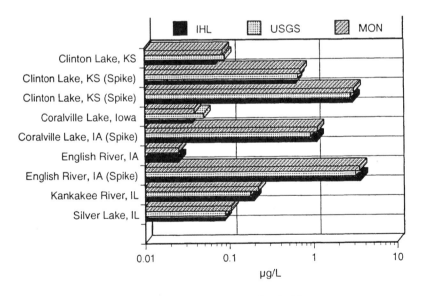

Figure 5. Alachlor Oxanilic Acid

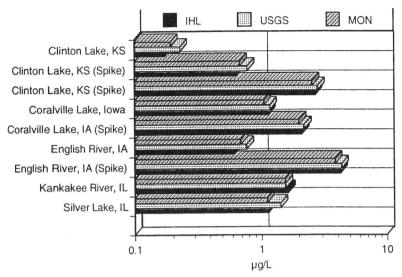

Figure 6. Metolachlor Ethanesulfonic Acid

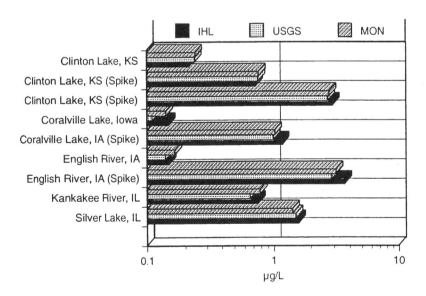

Figure 7. Metolachlor Oxanilic Acid

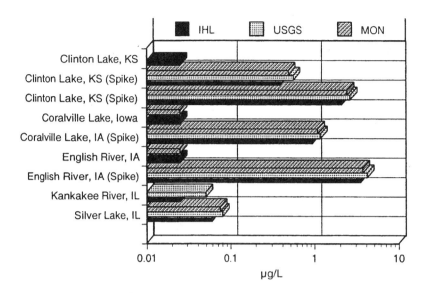

Figure 8. Dimethenamid Ethanesulfonic Acid

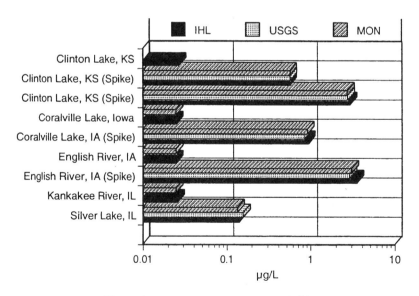

Figure 9. Dimethenamid Oxanilic Acid

Table VI. Minnesota Ground Water Results

Analyte	Sample	IHL	USGS	MON	Average	RSD
		Residue Found ($\mu g/L$)				
AcESA	Minnesota 1	17.0	13.0	17.4	15.8	15.4
AcESA	Minnesota 2	13.1	12.0	13.7	12.9	6.7
AcOXA	Minnesota 1	43.0	24.0	47.6	38.2	32.8
AcOXA	Minnesota 2	36.0	25.0	38.9	33.3	22.0
AlESA	Minnesota 1	33.8	27.0	30.9	30.6	11.2
AlESA	Minnesota 2	67.1	72.0	62.9	67.3	6.8
AlOXA	Minnesota 1	553.0	369.0	592.0	504.7	23.6
AlOXA	Minnesota 2	835.0	624.0	840.0	766.3	16.1
MeESA	Minnesota 1	8.2	12.0	9.0	9.7	20.8
MeESA	Minnesota 2	58.7	59.0	60.0	59.2	1.1
MeOXA	Minnesota 1	16.5	22.0	16.0	18.2	18.3
MeOXA	Minnesota 2	550.0	428.0	501.0	493.0	12.5
DiESA	Minnesota 1	0.2	0.4	0.3	0.3	26.3
DiESA	Minnesota 2	3.3	4.3	3.8	3.8	13.6
DiOXA	Minnesota 1	0.5	0.9	0.8	0.7	30.0
DiOXA	Minnesota 2	5.17	3.66	4.68	4.50	17.1

Acknowledgements

The authors would like to acknowledge: Lynne Sullivan for her assistance in analyzing samples at the Iowa Hygienic Laboratory, the Minnesota Department of Agriculture for donating two highly contaminated ground water samples for use in the study and Mark Allan for his assistance in analyzing samples at Monsanto. Also, David Gustafson of Monsanto and Mike Thurman and Betty Scribner of the U. S. Geological Survey for their participation in designing and implementing the study.

References

1. Ferrer, I; Thurman, E.M.; Barcelo, D. Anal. Chem. 1997, 69, 4547-4553

2. Vargo, J.D. Anal. Chem., 1998, 70, 2699-2703

3. Heberle, S.A.; Aga, D.S.; Hany, R.; Muller, S.R. Anal. Chem. 2000, 72, 840-845

4. Lee, E.A.; Kish, J.L.; Zimmermen, L.R.; Thurman, E.M. United States Geological Survey Open-File Report 01-10, 23 p.

5. Shoemaker, J. "Analytical Method Development for Alachlor ESA and Other Acetanilide Herbicide Degradation Products," 49th ASMS Conference on Mass Spectrometry and Allied Topics, Chicago, IL, May 27-31, 2001

Chapter 17

Identification of Trifluralin Metabolites in Soil Using Ion-Trap LC/MS/MS

Robert N. Lerch[1], Imma Ferrer[2,4], E. Michael Thurman[2], and Robert M. Zablotowicz[3]

[1]Agricultural Research Service, U.S. Department of Agriculture, 1406 E. Rollins Road, Room 265, Columbia, MO 65211
[2]National Water Quality Laboratory, U.S. Geological Survey, Denver Federal Center, Lakewood, CO 80225
[3]Agricultural Research Service, U.S. Department of Agriculture, P.O. Box 350, Stoneville, MS 38776
[4]Current address: immaferrer@menta.net

Trifluralin degradation in soils is complex, potentially resulting in the formation of 28 metabolites. The objective of this research was to develop an approach for the identification of trifluralin metabolites in soils using ion-trap liquid chromatography/mass spectrometry (LC/MS/MS). Authentic standards of the parent and six metabolites were used to establish appropriate instrument conditions and precursor ion (PI) fragmentation patterns to confirm their identity, as well as to facilitate confirmation of metabolites for which authentic standards were not available. Two soils from herbicide spill sites known to be contaminated with trifluralin were selected for study because of their high potential for metabolite detection. Soils were extracted with 70% aqueous acetonitrile, filtered, and directly injected into the ion-trap LC/MS/MS. Trifluralin metabolites were then identified as follows: 1) screen for PI masses; 2) compare retention time of the PI peak to standards; and 3) obtain PI spectra and compare fragmentation with standards. The validity of this approach was confirmed for the identification of metabolite, TR-20. Preliminary results identified the presence of up to eight metabolites in the soils, and the array of metabolites present were indicative of aerobic trifluralin degradation.

Introduction

Trifluralin is a dinitroaniline herbicide primarily used in soybean and cotton production, but it is also registered for use with numerous crops, vegetables, turf, and ornamentals. Trifluralin was introduced for agricultural use in the early 1960's, and it still ranks as one of the five top-selling herbicides in the U.S. It is currently registered in more than 50 countries for use on over 80 crops (*1*). Although its use has been declining since the late 1980's, 2.85 X 10^6 kg of trifluralin were used on soybeans and cotton in 2001 (*2*). It is currently the 3rd and 4th most commonly used herbicide on cotton and soybeans, respectively.

Extensive reviews of the environmental fate and chemistry of trifluralin have been reported over the last 35 years (*1,3,4*). Trifluralin is a very hydrophobic compound that strongly sorbs to soils, and therefore, its transport to surface or ground waters in the dissolved-phase is very limited. Off-site transport mainly occurs by soil erosion and subsequent sediment deposition in streams and lakes or by volatilization losses following field application.

Trifluralin degradation in soils is complex (Fig. 1), with up to 28 metabolites extracted and identified from a three year field degradation study (*5*). Predominant degradation pathways include dealkylation, reduction, oxidation, and cyclization reactions (*4-6*). Additional reactions include hydrolysis, hydroxylation, and condensation (*5*). Separate aerobic and anaerobic soil degradation pathways have been proposed for trifluralin (*3,4,7*). Aerobic degradation proceeds via the following reactions: dealkylation and partial reduction (TR-2 to TR-6); hydrolysis (TR-20); cyclization (TR-11 to TR-19); and dimeric condensation via formation of the partially reduced TR-39 to form azoxy (TR-28, TR-29, TR-31) and azo (TR-32) metabolites (*4,5*). Anaerobic degradation occurs via the following reactions: reduction and dealkylation (TR-4 to TR-9); cyclization (TR-13 and TR-14); and dimeric condensation (TR-28). Degradation rates of trifluralin have been shown to be faster under anaerobic than aerobic conditions (*4,7,8*).

To correct apparent oversights and to incorporate the findings of other studies, the degradation pathways presented in Figure 1 differ slightly from that published by Golab et al. (*5*). The commonly reported metabolites formed via dealkylation and reduction (TR-2 to TR-8) are shown as having interacting pathways. For instance, TR-4 can degrade to TR-5 via dealkylation, and TR-5 can degrade to TR-8 via reduction. Also, TR-15 could be dealkylated to form TR-18.

In addition to forming multiple metabolites, trifluralin also shows a strong propensity to form bound residues in soil. Studies reviewed by Grover et al. (*1*), reported that 10 to 72% of the applied trifluralin formed non-extractable bound

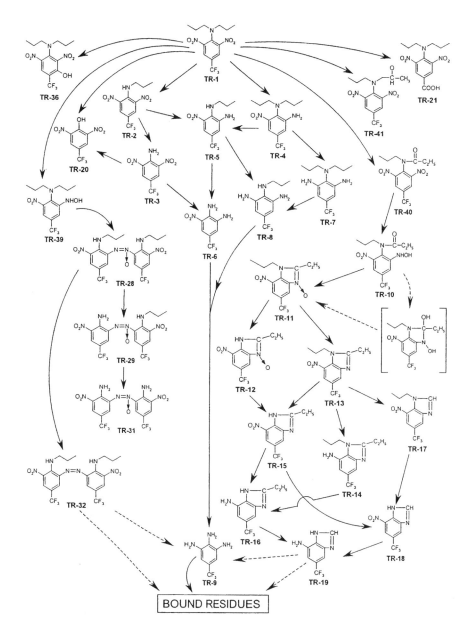

Figure 1. Proposed soil degradation pathways of trifluralin (adapted from Golab et al., 1979).

residues. Major portions of the bound residues were shown to be associated with humin, humic acid, and fulvic acid fractions of soil organic matter (1,5,9). Soil microbes also play an important role in bound residue formation (9,10). For instance, model aniline compounds have been shown to form oligomers with soil organic matter constituents when incubated in the presence of extracellular microbial enzymes (10). In an effort to better understand the nature of bound trifluralin residues, Golab et al. (5) performed adsorption-desorption studies with trifluralin and 15 of the metabolites identified in Figure 1. After 3 days, various extractants recovered at least 38% of the parent and metabolites, except TR-9 which could not be recovered by any of the extractants used. This result, combined with the trace amounts of extractable TR-9 from their field study, implicated TR-9 as the key metabolite leading to formation of trifluralin bound residues in soil.

Although a great deal of work has been conducted on the soil degradation of trifluralin, most of this work was conducted 20 to 35 years ago using radiolabeled compounds and thin-layer chromatography for separation and identification of metabolites. The study by Golab et al. (5) is still considered as the most authoritative work on the subject, yet only seven of the metabolites presented in Figure 1 were extracted in sufficient quantity to facilitate confirmation by GC/MS or direct probe MS. Thus, there is a compelling need to apply modern mass spectrometry to the identification and confirmation of the major trifluralin metabolites. Ion-trap liquid chromatography/ mass spectrometry (LC/MS/MS) offers great potential for this application because of its ability to facilitate separation, identification, and confirmation of compounds with a wide range of polarities. Ion-trap LC/MS/MS improves the identification and confirmation of unknowns compared to quadrupole LC/MS/MS systems by providing more structural information via MS^n capability.

The work by Golab et al. (5) defines the scope of possible soil degradation pathways for trifluralin, but there is no clear indication which pathways predominate. Determination of the major degradation pathways of trifluralin under field conditions needs to be achieved in order to discern the existence of stable metabolites that may contaminate soils, sediments, and waters. Given the heavy usage of trifluralin over the last 40 years, such contamination seems likely. While a comprehensive understanding of the fate of triazine and acetanilide metabolites has largely been achieved, the fate of trifluralin metabolites in the environment remains elusive. Thus, the objective of this research was to develop an approach for the identification of trifluralin metabolites in soils using ion-trap LC/MS/MS.

Materials and Methods

Trifluralin and Metabolite Standards

The analytical grade standard of trifluralin (α,α,α-trifluoro-2,6-dinitro-*N*,*N*-dipropyl-*p*-toluidine, TR-1), with a purity of 99.7%, was obtained from Axact Standards, Inc. (Commack, NY). Metabolite standards of α,α,α -trifluoro-2,6-dinitro-*N*-propyl-*p*-toluidine (TR-2), α,α,α -trifluoro-2,6-dinitro-*p*-toluidine (TR-3), α,α,α -trifluoro-5-nitro-*N*⁴-propyl-toluene-3,4-diamine (TR-5), 2-ethyl-7-nitro-1-propyl-5-(trifluoromethyl)-benzimidazole (TR-13), and 2-ethyl-7-nitro-5-(trifluoromethyl)-benzimidazole (TR-15) were originally synthesized by Lilly Research Laboratories (currently Dow-Elanco, Indianapolis, IN) or the USDA-ARS Pesticide Degradation Laboratory (Beltsville, MD). The metabolite standards were approximately 20 years old, and their original purity was at least 95% (*11*). For this study, each metabolite standard was obtained as an ~2000 mg/L solution in ethyl acetate from the USDA-ARS Southern Weed Science Laboratory (Stoneville, MS). The exact concentration of each standard was unknown because of solvent evaporation that occurred during long-term storage. However, analysis by ion-trap LC/MS indicated that the standards had remained intact. Mass spectra obtained by ion-trap LC/MS/MS and LC/MS³ also indicated that standards were acceptable for qualitative confirmation. Synthesis of α,α,α -trifluoro-2,6-dinitro-*p*-cresol (TR-20) was accomplished by mixing 10 mL of a 1 mg/mL solution of trifluralin in 50% aqueous methanol with 1 mL of 1 N NaOH. The mixture was heated to 80°C for 30 minutes in a sealed tube.

Ion-Trap LC/MS/MS Conditions

Ion-trap LC/MS/MS was used in positive and negative ion modes of operation to identify trifluralin metabolites. The analytes were separated using an HPLC (series 1100, Agilent Technologies, Palo Alto, CA) equipped with a C_{18} analytical column (Phenomenex RP18, Torrance, CA) having dimensions of 250 mm by 3 mm and 5-μm particle diameter. Column temperature was maintained at 25°C. The HPLC mobile phase consisted of acetonitrile and 10 mM ammonium formate buffer (pH 3.7), at a flow rate of 0.6 mL/min. A gradient elution was performed as follows: from 15% A (acetonitrile) and 85% B (10 mM ammonium formate) to 100 percent A and 0 percent B in 40 minutes; then 100% A was held isocratically for 5 minutes; and back to initial conditions in 10 minutes. The HPLC system was connected to an ion trap mass spectrometer (Esquire LC, Bruker Daltonics, Bellerica, MA) system equipped with an electrospray ionization (ESI) probe. The operating conditions of the MS system were optimized in full-scan mode (m/z scan range: 50–400) by flow injection analysis (i.e., HPLC column excluded) of each standard at ~20 µg/mL concentration. The source temperature was held at 350°C. Desolvation was enhanced by a countercurrent flow of nitrogen gas at 12 L/min. The capillary exit voltage was of 70 V. For each scan, 50,000 ions were accumulated, and every 5 scans were summed. Ions were accumulated in the trap for up to 0.2 s. Helium was introduced into the ion trap at 6 x 10^{-6} mbar for the purpose of ion cooling within the trap and to facilitate fragmentation of ions during MS/MS or MS^3 experiments. MS/MS and MS^3 experiments were carried out by isolating each target ion of interest and then fragmenting it. The width of the m/z window for the isolated ion was set at two. Terminology for the various mass spectral experiments are defined as follows: precursor ion (PI) - the ion fragmented for MS/MS experiments, in this work all PIs were either $[M + H]^+$ or $[M - H]^-$; PI spectra - the LC/MS/MS spectra generated from fragmentation of the PI; product ion - an ion generated by fragmentation of the PI via LC/MS/MS; and product/product ion - an ion generated by fragmentation of a product ion via LC/MS^3.

Soils

Soils collected from two herbicide spill sites in Illinois were used because of their high potential for detection of trifluralin metabolites without concentration of the soil extracts (see details below). The intent was to use samples that would provide a straightforward opportunity to test the efficacy of the ion-trap LC/MS/MS system. One sample was collected from a spill site in Piatt County, IL (Piatt soil) in 1989 (*12*). The sample was passed through a 3-mm sieve and stored at 2-4°C. This sample was contaminated with alachlor, metolachlor, atrazine, and trifluralin. The other sample was collected from a herbicide spill site near Lexington, IL (Lexington soil) in 1990 (*13*). It was processed and stored in the same manner as the Piatt soil, and it was also contaminated with the same four herbicides. Prior to collection, the Lexington site was the scene of a fire, and the soils were briefly flooded in the process of extinguishing the blaze. Although the trifluralin concentration of these soils were unusually high, trifluralin degradation occurred under natural environmental conditions.

Soil Extraction of Trifluralin Metabolites

An Accelerated Solvent Extraction system (ASE 200, Dionex Co., Sunnyville, CA) was used for the soil extractions. Ten grams of soil were packed into an 11-mL stainless steel vessel. The packed vessels were sealed at both ends with circular cellulose filters of 2.1 cm diameter (Whatman, Springfield Mill, Maidstone, Kent, UK). Extraction conditions were as follows: 20 mL of acetonitrile: water (7:3) as extraction solvent; temperature of 120°C; pressure of 1500 psi; heating time of 6 minutes; and three cycles of 5 minute static extraction. At the end of each extraction, nitrogen gas was used to expel the extract into glass collection vials (60 second purge). The total volume of extract was ~18 mL. A 1-mL volume of the extract was passed through a syringe filter for cleanup, and 50 μL were injected into the ion-trap LC/MS/MS for analysis.

Results and Discussion

Mass Spectra of Trifluralin and Metabolites

The choice of negative or positive ion mode for a particular compound was based on the intensity of the PI formed, and the signal to noise ratio of the baseline. Compounds with a completely substituted aniline group (TR-1 and TR-13) produced high intensity PIs in positive ion mode, but they were unresponsive in negative ion mode. Compounds with partially substituted or un-substituted aniline groups, or the hydrolysis product TR-20, produced PIs of high intensity in negative ion mode (Table I). Some compounds, particularly TR-15, produced high intensity PIs in both modes. While the positive ion mode resulted in higher intensity PIs for TR-15, the baseline signal to noise ratio in negative ion mode was much lower, making it the preferred mode (Table I).

PI spectra obtained by ion-trap LC/MS/MS, equipped with an electrospray interface, provided ample fragmentation of trifluralin and its metabolites (Table I). From one to four diagnostic ions were produced, providing the needed fragmentation for confirmatory analyses. The term diagnostic ions is used to indicate product ions whose tentative identification was indicative of the PI, and they were of sufficient relative abundance (>25%) in the spectra. Only TR-13 contained the PI (m/z 302) in the PI spectra, indicating the considerable stability

Table I. Tentative identification of the diagnostic product ions in the mass spectra of trifluralin and selected trifluralin metabolites.

Compound[a]	Molecular Weight[a]	Ion Mode[b]	t_R[c] Min.	LC/MS Precursor m/z; identification	LC/MS/MS Product r Ions m/z; identification (RA)[d]	LC/MS[3] Product/product Ions m/z; identification (RA)[d]
TR-1	335	+	34.6	336 [M + H]+	294* [M - C$_3$H$_6$ + H]+ (100)	252 [294 - C$_3$H$_6$]+ (42)
					276 [M - C$_3$H$_6$ - H$_2$O + H]+ (34)	248 [294 - NO$_2$]+ (34)
					252 [M - 2(C$_3$H$_6$) + H]+ (36)	236 [294 - C$_3$H$_7$NH]+ (100)
					248 [M - C$_3$H$_6$ - NO - H$_2$O + 3H]+ (47)	
					236 [M - (C$_3$H$_7$)$_2$N + H]+ (95)	
TR-2	293	-	29.9	292 [M - H]-	204 [M - C$_3$H$_6$ - NO$_2$ - H]- (100)	ND
TR-3	251	-	23.5	250 [M - H]-	220 [M - NO - H]- (100)	ND
					203 [M - HNO$_2$ - H]- (27)	
TR-5	263	-	28.3	262 [M - H]-	226 Not identified (100)	ND
					215 [M - HNO$_2$ - H]- (52)	
					204 [M - C$_3$H$_7$NH - H]- (73)	
					200 [M - NO$_2$ - NH$_2$ - H]- (83)	

TR-13	301	+	28.1	302 [M + H]$^+$	302 [M + H]$^+$ (53)	
					260* [M - C$_3$H$_6$ + H]$^+$ (100)	214 [260 - NO$_2$]$^+$ (100)
TR-15	259	-	20.1	258 [M - H]$^-$	228 [M - NO - H]$^-$ (100)	ND
TR-20	252	-	17.1	251 [M - H]$^-$	221 [M - NO - H]$^-$ (100)	ND

^{a}See Figure 1 for compound codes. ^{b}+ = positive mode; - = negative mode. ^{c}t$_R$ = Retention time. ^{d}Tentative identification; RA = relative abundance in percent. * Indicates product ion fragmented for LC/MS3. ND = not determined.

of its molecular structure. Ion-trap LC/MS³ provided an additional diagnostic ion for TR-13 by fragmentation of m/z 260 to form product/product ion, m/z 214 (Table I).

Fragmentation patterns in the PI spectra were typically characterized by losses of propyl, NO, NO_2, or propylamine groups (Table I). For those compounds with one or more propyl groups (TR-1, TR-2, TR-5, and TR-13), loss of the propyl group was key to formation of the base peaks, except TR-5. Base peak formation in the spectrum of TR-5 has not yet been resolved. For compounds with no propyl groups (TR-3, TR-15, and TR-20), base peak formation typically occurred by loss of an NO group.

Identification and Confirmation of Trifluralin Metabolites

Trifluralin metabolites were identified and confirmed by the following procedure: 1) perform selective ion monitoring to screen for PI masses; 2) compare HPLC retention time of the PI peak to standards; and 3) obtain PI spectra and compare fragmentation with standards. Hence, the procedure provided three criteria for confirmation. Tentative metabolite identifications presented in Table II were based on the first two confirmation criteria. Because this work is preliminary, PI spectra have not yet been obtained on all metabolites. The presence of a metabolite was considered confirmed if all three confirmation criteria were met. To ascertain the presence of metabolites for which no standard was available, retention times and PI fragmentation were compared to the standard representing the closest structural analogue. Overall, TR-1 and the 28 metabolites identified by Golab et al. (*5*) were screened for their presence in the soil extracts.

In the Lexington soil, trifluralin and eight metabolites were identified, and in the Piatt soil, trifluralin and three metabolites were identified (Table II). Confirmation analyses are pending for TR-6, TR-9, and TR-41 in the Lexington soil, and for TR-13, TR-15, and TR-20 in the Piatt soil. In the Lexington soil, relative intensities of the PIs for all metabolites, except TR-2 and TR-5, ranged from 10^4 to 10^6 ion counts and were comparable to the relative intensity of trifluralin in this soil. Intensity of the TR-5 PI (~900 ion counts) was insufficient to facilitate detectable product ion formation (Table II). The PI intensity of 5300 ion counts for TR-2 was, however, sufficient for detectable product ion formation. In the Piatt soil, PI intensities of TR-13, TR-15, and TR-20 were

Table II. Identification and Confirmation of Trifluralin Metabolites

Soil	Metabolite	Tentative Identification		Confirmation[a]	
		t_R[b]	LC/MS[c]	LC/MS/MS	LC/MS^3
		min.	m/z	m/z	m/z
Lexington	TR-2	30.0	292 [M - H]⁻	204[d]	ND
	TR-5	28.3	262 [M - H]⁻	*	
	TR-6	20.3	222 [M + H]⁺	ND	
	TR-9	15.3	192 [M + H]⁺	ND	
	TR-13	28.1	302 [M + H]⁺	260[d]	214[d]
	TR-15	20.1	258 [M - H]⁻	228[d]	
	TR-20	17.2	251 [M - H]⁻	221[d]	
	TR-41	36.9	352 [M + H]⁺	ND	
Piatt	TR-13	28.1	302 [M + H]⁺	ND	
	TR-15	20.1	258 [M - H]⁻	ND	
	TR-20	17.1	251 [M - H]⁻	ND	

[a]Observed diagnostic product ions (see Table I). [b] t_R = Retention time. [c]Observed precursor ion. [d]Base peak. ND = not determined. * = insufficient precursor ion intensity to perform MS/MS.

about an order of magnitude lower than the Lexington soil. This was likely an indication of its lower initial trifluralin levels compared to the Lexington soil.

Four of the eight metabolites identified in the Lexington soil were confirmed by LC/MS/MS or LC/MS^3. For all confirmed metabolites, diagnostic product ions were observed (Table II). Example precursor and product ion spectra for TR-13 and TR-15 from the Lexington soil are given in Figure 2. For TR-13 confirmation, the PI (m/z 302) was isolated in the spectrum from the soil extract and the diagnostic product ion (m/z 260) was observed. Furthermore, the product ion spectra of m/z 260 yielded the diagnostic m/z 214 ion, providing definitive confirmation of this metabolite.

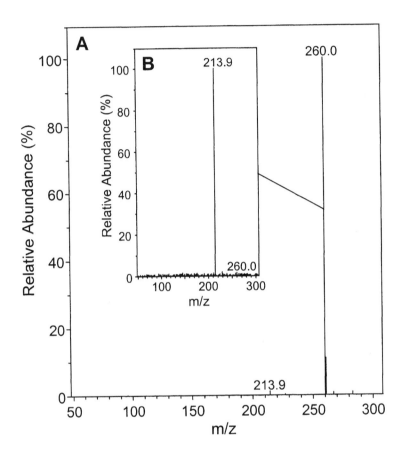

Figure 2. Confirmation of TR-13 and TR-15 in Lexington soil extract. A. Ion-trap LC/MS/MS of TR-13 precursor ion (m/z 302) yields diagnostic m/z 260 product ion. B. Ion-trap LC/MS³ of TR-13 product ion (m/z 260) yields diagnostic m/z 214 product/product ion. C. Ion-trap LC/MS/MS of TR-15 precursor ion (m/z 258) yields diagnostic m/z 228 product ion.

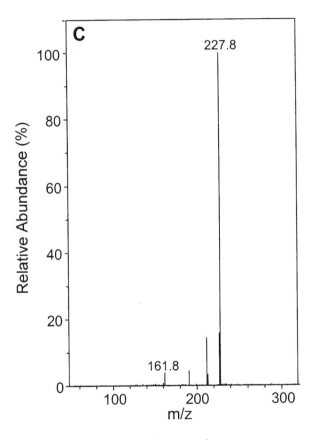

Figure 2. *Continued.*

Validation of the Procedure - Confirmation of TR-20

Upon initiation of these studies, authentic standards were not available for four of the metabolites (TR-6, TR-9, TR-20, and TR-41) tentatively identified in the soils (Table II). Of these four metabolites, TR-20 was detected in both soils with high PI intensities, and it was the most facile to synthesize because of the susceptibility of trifluralin to alkaline hydrolysis at high temperatures (14). Thus, the tentative identification of TR-20 in both soils provided an ideal opportunity to test the validity of the ion-trap LC/MS/MS procedure to identify metabolites for which standards were (initially) unavailable.

Flow injection MS of the trifluralin hydrolysis products showed that TR-20 (PI m/z 251) was the major product form, with much smaller amounts of one other product (m/z 287) (Figure 3). Subsequent PI spectra of m/z 251 showed the product ion base peak of m/z 221, formed by the loss an NO group (Figure 3; Table I). The next step was to confirm the tentative identification of TR-20 in the soil extracts against the synthesized TR-20 standard, using the same conditions as employed for the initial metabolite identification. Ion-trap LC/MS of the PI (m/z 251) showed an exact retention time match between the soil extracts and the TR-20 standard (Figure 4). Furthermore, PI spectra confirmed the presence of TR-20 by producing the diagnostic product ion, m/z 221 (Figure 4). These experiments validated the developed procedure for identifying trifluralin metabolites in soils, even for cases in which authentic standards were not available. Furthermore, the efficacy of ion-trap LC/MS/MS as a tool for identifying pesticide metabolites in environmental samples was convincingly demonstrated.

Trifluralin Degradation Pathways

The presence and absence of the 28 metabolites provided insight to the predominant trifluralin degradation pathways in these soils (Figure 1). Because of the high levels of trifluralin and other herbicides present, these soils certainly do not represent typical agronomic settings. Nonetheless, the study by Wheeler et al. (6) showed that extremely high trifluralin levels (20,000 mg/kg) did not affect degradation rates or pathways compared to levels that were one- or three-orders of magnitude less. Therefore, the metabolite information obtained from these soils may still contribute relevant information to our understanding of trifluralin degradation.

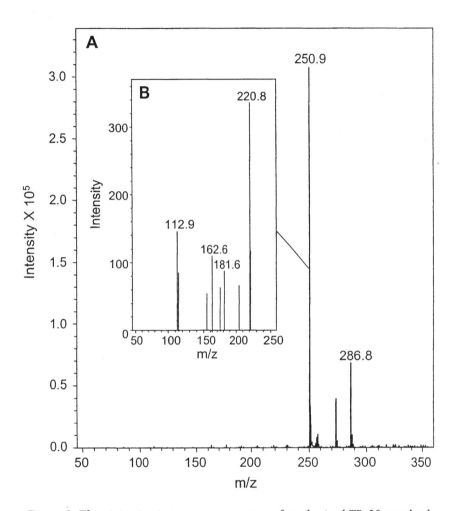

*Figure 3. Flow injection ion-trap mass spectra of synthesized TR-20 standard.
A. Negative ion mode MS of trifluralin hydrolysis products shows the TR-20
precursor ion (m/z 251). B. Precursor ion (m/z 251) spectra (MS/MS) yields
diagnostic product ion, m/z 221.*

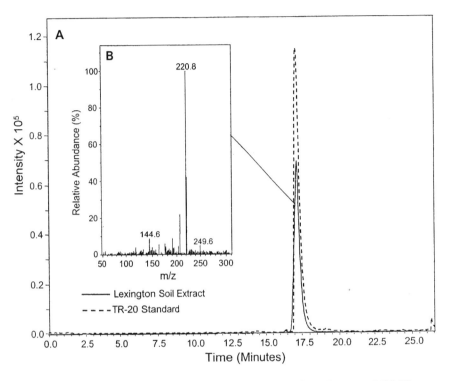

Figure 4. Confirmation of TR-20 in Lexington soil. A. Ion-trap LC/MS chromatograph of the precursor ion (m/z 251) from Lexington soil extract and TR-20 standard. B. Ion-trap LC/MS/MS of precursor ion from Lexington soil extract yields diagnostic m/z 221 product ion.

Dealkylation, Reduction, and Hydrolysis

The presence of TR-2, TR-5, TR-6 and TR-9 demonstrated the combined importance of dealkylation and reduction reactions to trifluralin degradation in the Lexington soil. The presence of these metabolites implied that aerobic degradation conditions prevailed (4). This was further supported by the absence of the completely reduced metabolites (TR-7 and TR-8). Furthermore, TR-4 was also absent, although its formation has been noted to occur under aerobic conditions (5). It is possible that TR-4 was an unstable intermediate in the pathway leading to TR-9 formation. However, Wheeler et al. (6) detected TR-2, but not TR-4, when trifluralin was degraded under aerobic conditions, a finding similar to the results of this study. Therefore, a more likely scenario was initial dealkylation of trifluralin to TR-2 followed by a combination of dealkylation and reduction reactions leading to formation of TR-9. The intensity of the TR-6 PI in the Lexington soil was about an order of magnitude greater than that of TR-9. This finding suggested that TR-6 was present in greater concentration than TR-9. Reduction of the first nitro group of a polynitro-aromatic (e.g., TR-2 or TR-3) can occur rapidly, but reduction of the second nitro group is a much slower reaction (15). Therefore, the reaction kinetics of polynitro-aromatic reduction favor the formation and stability of TR-6 over that of TR-9, especially when aerobic conditions prevail. The absence of TR-3, combined with the presence of TR-2, TR-6, and TR-20, suggested that it is either an unstable intermediate or that its formation is excluded in favor of reduction or direct hydrolysis of trifluralin. The presence of TR-20 in both soils showed the importance of its direct formation via trifluralin hydrolysis or its indirect formation via dealkylation and subsequent hydrolysis of TR-3.

Benzimidazole Formation, Condensation, and Other Reactions

The benzimidazole metabolites, TR-13 and TR-15, were two of the most prevalent metabolites detected in both soils, yet neither their antecedent metabolites (TR-40, TR-10 to TR-12) nor their subsequent degradation products (TR-14, TR-16 to TR-19) were detected. This implied that the antecedent metabolites were unstable intermediates and that TR-13 and TR-15 represent potentially stable end-points in the soil environment. This conclusion is supported by the findings of Wheeler et al. (6), and by the pathways of aerobic degradation proposed by Probst et al. (4). Oxidation of a propyl group to form TR-41 was also observed in the Lexington soil, further supporting the importance of aerobic degradation in this soil. The absence of TR-39, TR-32, and the azoxy metabolites (TR-28, TR-29, and TR-31) in either soil, eliminated condensation reactions as an important pathway for trifluralin degradation in these soils. The absence of metabolites formed by hydroxylation (TR-36) or trifluoromethyl oxidation (TR-21) indicated that neither of these pathways were significant to trifluralin degradation.

The array of metabolites present in these soils was consistent with aerobic degradation of trifluralin (*4,6*) and suggested that the predominant degradation pathways were as follows: 1) benzimidazole formation, TR-1→TR-40→TR-10→TR-11→TR-13→TR-15; 2) dealkylation and reduction, TR-1→TR-2→TR-5→TR-6→TR-9; 3) direct TR-20 formation via trifluralin hydrolysis, TR-1→TR-20; and 4) indirect TR-20 formation via dealkylation and hydrolysis, TR-1→TR-2→TR-3→TR-20.

Implications for Environmental Contamination

Stable pesticide metabolites represent potential environmental contaminants. Based on this study and other studies, the most stable metabolites of trifluralin are TR-6, TR-9, TR-13, TR-15, and TR-20. Previous studies of trifluralin degradation have emphasized that none of the identified trifluralin metabolites represented more than 3 to 4% of the applied trifluralin (*4,5*). Similarly low percentages of the atrazine metabolites, deethylatrazine and deisopropylatrazine, have been shown to form as a result of atrazine degradation in soils (*16,17*). However, these two metabolites are among the most significant contaminants of surface and ground waters in the Midwestern U.S. (*18-20*). Thus, seemingly insignificant formation of metabolites based on soil degradation studies, alone, can lead to the erroneous conclusion that the metabolites are of minimal environmental concern. With regards to off-site hydrologic transport, soil degradation studies do not take into account the importance of metabolite chemical properties, nor the impact of repeated herbicide usage on metabolite accumulation in soils.

Of the trifluralin metabolites identified in this study, TR-20 is the only metabolite that represents a potentially important dissolved-phase contaminant in surface and ground waters due to the likelihood of a low acid dissociation constant. Acid dissociation constants of structurally similar dinitrophenols are about 10^{-4}. Therefore, TR-20 should be present in the environment as an anion under the typical pH range of soils. The other stable trifluralin metabolites, TR-6, TR-9, TR-13, and TR-15, are likely to contaminate soils and sediments in watersheds with high trifluralin usage. These metabolites are likely to have high sorption intensity as a result of their hydrophobicity (TR13 and TR-15) or tendency to form bound residues (TR-6 and TR-9). All four metabolites would also have a strong propensity to form H-bonds with soil colloids. Since these metabolites will primarily be transported from fields sorbed to soil particles, their impact on the aquatic environment will likely occur via slow desorption from stream or lake sediments.

Summary

Ion-trap LC/MS/MS was successfully applied to the identification and confirmation of eight trifluralin metabolites in soils. The developed analytical procedure provided three points of confirmation based on PI formation, HPLC retention time, and diagnostic product ion formation. The procedure was validated for metabolite identification, even for cases in which standards were unavailable, by confirming the presence of a tentatively identified metabolite (TR-20). To date, four of the eight metabolites tentatively identified have been confirmed. Metabolites present in the soils represented formation via benzimidazole formation (TR-13 and TR-15), hydrolysis (TR-20), dealkylation (TR-2, TR-5, TR-6, TR-9), reduction (TR-5, TR-6, TR-9), and oxidation (TR-41). This array of metabolites was consistent with aerobic degradation of trifluralin and indicated that the predominant degradation pathways occurred via the following reactions: 1) benzimidazole formation; 2) dealkylation and reduction; 3) direct TR-20 formation via trifluralin hydrolysis; and 4) indirect TR-20 formation via dealkylation and hydrolysis. The most stable trifluralin metabolites appear to be TR-6, TR-9, TR-13, TR-15, and TR-20. These stable metabolites represent potential environmental contaminants. TR-20 represents a potentially important dissolved-phase contaminant in surface and ground waters. The other stable trifluralin metabolites are likely to contaminate soils and sediments in watersheds with current or historically high trifluralin usage. Future work will focus on completing the confirmation analyses of trifluralin metabolites in these soils, as well as soils collected from typical agronomic settings. In addition, the presence of these metabolites in soils, sediments, and waters obtained from watersheds with high trifluralin usage will be pursued.

Disclaimer— Mention of specific companies, products, or trade names is made only to provide information to the reader and does not constitute endorsement by the USDA-Agricultural Research Service or the USDI-U.S. Geological Survey.

References

1. Grover, R.; Wolt, D. W.; Cessna, A. J.; Schiefer, H. B. Rev. Environ. Contam. Toxicol. **1997,** 153, 1-64.

310

2. USDA-National Agricultural Statistics Service. *Agricultural Chemical Usage, 2001 Field Crop Summary.* 2002.
3. Probst, G. W.; Tepe, J. B. In *Degradation of Herbicides*; Kearney, P. C.; Kaufman, D. D., Eds.; Marcel Dekker: New York, 1969; pp. 255-282.
4. Probst, G. W.; Golab, T.; Wright, W. L. In *Herbicides: Chemistry, Degradation, and Mode of Action;* Kearney, P. C.; Kaufman, D. D., Eds.; Marcel Dekker: New York, 1975; pp. 453-500.
5. Golab, T.; Althaus, W. A.; Wooten, H. L. *J. Agric. Food Chem.* **1979,** 27, 163-179.
6. Wheeler, W. B.; Stratton, G. D.; Twilley, R. R.; Ou, L.-T.; Carlson, D. A.; Davidson, J. M. *J. Agric. Food Chem.* **1979,** 27, 702-706.
7. Willis, G. H.; Wander, R. C.; Southwick, L. M. *J. Environ. Qual.* **1974,** 3, 262-265.
8. Parr, J. F.; Smith, S. *Soil Sci.* **1973,** 115, 55.
9. Nelson, J. E.; Meggitt, W. F.; Penner, D. *Weed Sci.* **1983,** 31, 68-75.
10. Bollag, J-M.; Myers, C. J.; Minard, R. D. *Sci. Total Environ.***1992,** 123/124, 205-217.
11. Koskinen, W. C.; Oliver, J. E.; Kearney, P. C.; McWhorter, C. G. *J. Agric. Food Chem.* **1984,** 32, 1246-1248.
12. Felsot, A. S.; Dzantor, E. K. In *Enhanced Biodegradation of Pesticides in the Environment*; Racke, K. D.; Coats, J. R., Eds.; ACS Symposium Series 426, American Chemical Society: Washington, DC, 1990; pp. 249-268.
13. Felsot, A. S.; Miller, J. K.; Bicki, T. J.; Franks, J. E. In *Pesticide Waste Management: Technology and Regulation;* Bourke, J. B.; Felsot, A. S.; Gilding, T. J.; Jensen, J. K.; Seiber, J. N., Eds.; ACS Symposium Series 510, American Chemical Society: Washington, DC, 1992; pp. 244-262.
14. Ramesh, A.; Balasubramanian, M. *J. Agric. Food Chem.* **1999,** 47, 3367-3371.
15. Larson, R. A.; Weber, E. J. *Reaction Mechanisms in Environmental Organic Chemistry;* Lewis Publishers: Boca Raton, FL, 1994; pp. 182-185.
16. Winkelmann, D. A.; Klaine, S. J. *Environ. Toxicol. Chem.* **1991,** 10, 335-345.
17. Kruger, E. L.; Rice, P. J.; Anhalt, J. C.; Anderson, T. A.; Coats, J. R. *J. Environ. Qual.* **1997,** 26, 95-101.
18. Thurman, E. M.; Goolsby, D. A.; Meyer, M. T.; Mills, M. S.; Pomes, M. L.; Kolpin, D. W. *Environ. Sci. Technol.* **1992,** 26, 2440-2447.
19. Kolpin, D. W.; Thurman, E. M.; Goolsby, D. A. *Environ. Sci. Technol.* **1996,** 30, 335-340.
20. Lerch, R. N.; Blanchard, P. E.; Thurman, E. M. *Environ. Sci. Technol.* **1998,** 32, 40-48.

Surfactants and Natural Products

Chapter 18

Charge Characteristics and Fragmentation of Polycarboxylic Acids by Electrospray Ionization–Multistage Tandem Mass Spectrometry

Jerry A. Leenheer[1], Imma Ferrer[1,2], Edward T. Furlong[1], and Colleen E. Rostad[1]

[1]U.S. Geological Survey, Building 95, MS 408, Federal Center, Denver, CO 80225
[2]Current address: immaferrer@menta.net

The majority of dissolved organic matter (DOM), both natural and anthropogenic in water is dominated by polycarboxylic acid structures. When these polycarboxylic acids are subjected to either positive or negative ESI/MS, a mixture of molecular aggregates, multiply-charged species, and cationic adducts (Na^+, K^+, NH_4^+) result. These ions obscure the molecular-weight distributions of DOM. ESI/MS/MS studies of polyacrylic acid found that multiple charges formed every 8-10 carbon atoms in this linear polymer, and that singly charged ions could be distinguished from multiply charged ions in the MS/MS spectra by ion loss differences caused by water loss (18/n) and carbon dioxide loss (44/n) where n is the number of charges. ESI/MS/MS studies of glycyrrhizic acid and Suwannee River fulvic acid found that multiply-charged ions fragmented into singly charged ions that could be structurally elucidated by progressive fragmentation by ESI/MS/MS. Methylation minimized both multiple charges and molecular aggregate ions so that molecular weight distributions could be estimated by negative ESI/MS provided there were ionizable alcohol groups in the polycarboxylic acid molecules.

Introduction

Dissolved organic matter (DOM), both natural and anthropogenic in water, is dominated by polycarboxylic acid structures. Aerobic biodegradation of anthropogenic surfactants results in polycarboxylic acid metabolites such as mono-(MCPEGs) and dicarboxylated (DCPEGs) polyethylene glycols that are derived from various polyethylene glycol (PEG) surfactants (*1*), mono-(APECs) and dicarboxylated (CAPECs) alkylphenol polyethoxylates derived from various alkylphenol polyethoxylate (APE) surfactants (*2*), and mono-(SPCs) and dicarboxylated (SPDCs) alkyl benzenesulfonates plus mono-(DATSC) and dicarboxylated (DATSDC) dialkyltetraline sulfonates derived from linear alkyl benzenesulfonates (LAS) surfactants and its coproduct, dialkyltetralin sulfonates (DATS) (*3*). Similarly, aerobic biodegradation of natural organic matter (NOM) results in complex carboxylic acids commonly known as fulvic and humic acids (*4*). Many "emerging contaminants" are polycarboxylic acid metabolites that are being currently discovered as a result of application of electrospray ionization/mass spectrometry (ESI/MS) to water analyses.

When polycarboxylic acids are subjected to either positive or negative ESI/MS, a mixture of molecular aggregates, multiply-charged species, and cationic adducts (H^+, Na^+, K^+, NH_4^+) result (*5,6*). These ions obscure the molecular-weight distributions of DOM. ESI/MS studies (*5*) of poly(acrylic acid) (2,000 dalton average) found that multiple charges formed every 8-10 carbon atoms in this linear polymer. Similarly, PEG's and their carboxylated metabolites formed multiple charges every 12 ethoxy units for the Na^+ adducts when these compounds were analyzed by EIS/MS with positive ion detection (*7*). Small (< 300 daltons) polycarboxylic acid reference standards were found to give predominantly the singly-charged ions for both positive and negative ESI/MS analyses, but glycyrrhizic acid (822 daltons), with three carboxyl groups approximately equally spaced throughout its structure, gave a mixture of doubly and triply charged ions for negative ESI/MS (*5*). Some large polycarboxylic acids such as CAPEC's with widely-spaced carboxyl groups give singly-charged negative ions because of the preferential ionization of the more acidic carboxyl group (the ethoxycarboxyl group) in the molecular structure (*2*).

The objectives of this study are to investigate charge characteristics and fragmentation pathways of certain polycarboxylic acid standards by electrospray ionization/multistage tandem mass spectrometry(ESI/MS/MS). Additionally, methylation of carboxyl groups will be investigated to minimize both multiple charge formation and aggregate formation to obtain better molecular weight distribution information of complex polycarboxylic acid mixtures such as fulvic acid.

Experimental Section

Sample and Standards

The Suwannee River was sampled at its origin at the outlet of the Okefenokee Swamp near Fargo, Georgia, during November 1983. Isolation and characterization of the fulvic acid fraction, which constituted 66% of the dissolved organic carbon, is described in previous reports (8,9). Chemical reference standards discussed in this report were purchased from Aldrich Chemical Co., Milwaukee, Wis. All standards were used as acquired without additional purification except for glycyrrhizic acid (monoammonium salt), which was converted to acid form with a hydrogen-form ion exchange resin. Sample and standard solutions were analyzed unfiltered to eliminate bias from selective sorption onto filter media.

Experiments with ESI/Quadrupole MS

Samples and standards were dissolved in UV grade 50/50 water/methanol and initially analyzed by flow injection analysis on a Hewlett Packard Series 1100 LC/MS with electrospray ionization. The flow rate of the 25/75 water/methanol mobile phase was 0.2 ml/min. The quadrupole mass spectrometer scanned from 100 to 1000 m/z with capillary exit at 70 V for both positive and negative polarity. For electrospray ionization, the drying gas was 350°C at 12 L/min with 35 psi nebulizer pressure, and capillary voltage of 4000 V. The scan range was often increased to 3000 m/z to detect larger mass ions, along with ramped quadrupole voltages to increase high mass ion transmission. Source capillary exit voltage was decreased to 50, based on results with standards, to maximize formation of molecular ions and minimize adducts (positive mode) or fragmentations (negative mode).

Number average (M_n) and weight average (M_w) molecular weight distributions were calculated accorded to accepted methods (10), by summing the product of the intensity as each mass. The spectra used for calculation of molecular weight distribution were background subtracted using blank areas of the same chromatogram, and the peak was summed over the elution time. Chemical and electronic noise in the higher m/z regions of the spectrum were not excluded from the calculation, but there was no dominant chemical and electronic noise in the higher m/z region in the blank areas of the mass spectrometer response between injection of the sample. Reproducibility of these analyses calculated as standard deviations has been reported previously (5).

Experiments with ESI/Ion Trap/MS and Multistage Tandem MS

Standard and fulvic acid solutions were analyzed using a Bruker Esquire ion trap mass spectrometer (Bruker Daltonics, Bellerica MA), equipped with an orthogonal ESI interface (Hewlett Packard Co, Palo Alto, CA). Solutions of standards and Suwannee River fulvic acid were analyzed by infusion directly into the ESI interface at 10 μL/minute using a 25% water/75% methanol mobile phase. The mass range analyzed varied between 100 to 2,000 m/z. After initial experiments, the mass range was shortened to a maximum of 800 to 1,000 m/z to improve dwell time and sensitivity. The source temperature was held at 350°C. The infused solutions were nebulized by coaxial flow of nitrogen through the spray needle at a pressure of 240 kPa. Desolvation was enhanced by a counter current flow of nitrogen gas of 12 L/min. Negative ions were generated and transferred to the ion trap by maintaining a 3 to 3.25 kV potential energy difference between the needle and capillary end cap. The end plate was held at 500V relative to the capillary end cap and the capillary exit varied from –20V to –70V. For each scan, 50,000 ions were accumulated and every 5 scans summed. Ions were accumulated in the trap for up to 0.2 seconds. Helium was introduced into the ion trap at 6 x10^{-6} mbar for the purpose of ion cooling within the trap and to facilitate fragmentation of ions by the application of the RF fragmentor voltages during tandem mass spectrometry (MS/MS) experiments. For the MS/MS experiments, the width of the m/z window for the isolated ion was set between 1 and 3. HPLC/MS spectra of the fulvic acid indicated that there were significant peaks at all masses at unit resolution. After isolation, the supplementary RF voltage applied to the ion trap end caps to induce collision induced dissociation (CID) was varied between 0.2 and 1.4 volts, depending on the ion isolated and the degree of fragmentation achievable or desired. For each ion trap/MS/MS spectrum, the CID conditions were controlled so that after fragmentation, the remaining parent ion was 20-60% of the most intense product ion (base peak) produced.

Methylation

Ethereal diazomethane (3 ml) was generated by reacting N-methyl-N-nitroso-N-nitroguanidine with 5N potassium hydroxide. Five milligrams of the sample to be methylated was dissolved or suspended in 3-mL chloroform. The ethereal diazomethane and sample in chloroform were combined in an ice bath and allowed to react for 1 hr. As the sample methylated, insoluble samples dissolved. Methanol was used to facilitate solubilization of polar samples such as glycyrrhizic acid and Suwannee River fulvic acid. The sample was blown to dryness with filtered N$_2$ at ambient temperature, and dissolved and diluted with the solvent (acetonitrile for Suwannee River fulvic acid and 50% water/50% methanol for glycyrrhizic acid) used for ESI/MS analyses.

Results and Discussion

ESI/ion trap MS spectra of ions in the poly(acrylic acid) spectra are presented in Figure 1. The major ions in spectrum A (Fig. 1) are a homologous series (72 Da difference) of [M-H]⁻ ions with one water loss, but the inferred average MW is considerably less than 2,000 Da because of multiple charging and fragmentation. The low molecular weight bias may also be due to decreasing transmission and sensitivity of the mass spectrometer with increasing m/z ratio. Singly-charged ions can be readily distinguished from doubly charged ions in spectrum B by looking for [M-H$_2$O-H]⁻ as the predominant ion in the MS/MS spectra for singly charged ions and [M-H$_2$O-2H]²⁻ as the dominant ion for doubly charged ions as shown in MS/MS spectrum C. Singly charged ions in Spectrum B are at m/z of 419, and 455 and have greater intensity than the doubly-charged ions. Doubly charged ions in Spectrum B are at m/z of 407, 416, 437, 452 and 461 as determined by the 9 m/z differences from these molecular ions in the MS/MS daughter ion spectra (not shown). The ion at m/z of 443 is likely a mixture of single- and doubly-charged ions because differences of 18 m/z and 9 m/z from the molecular ion were both observed. Spectrum C indicates that water loss by anhydride formation between adjacent carboxyl groups is favored over decarboxylation [M-CO$_2$-2H]²⁻ as from 1 to 2 water molecules (ions at m/z of 443 and 434) are lost before decarboxylation (ions at m/z of 421 and 412) and then an additional water loss occurs (ion at m/z of 403). This suggests that clustered carboxyl groups in the poly(acrylic acid) polymer are successively converted to cyclic anhydrides coupled with water loss.

A fragmentation pathway consistent with the mass spectra of Figure 1 is shown in Figure 2. This fragmentation pathway illustrates how multiple charges are spatially separated in the molecular structure, and how potentially charged carboxyl groups are neutralized by cyclic anhydride formation accompanied by water loss, neutralized by protonation, or are eliminated by decarboxylation. The multiple charge formation as a function of spatial separation in poly(acrylic) acid and PEG compounds (7) is indicative of inductive and electrostatic field effects whereby structurally similar charged groups along a polymer chain exhibit a wide range of pK$_a$ values because the dipole environment of a charged group is affected by adjacent polar and charged groups through both bonds (inductive effect) and space (electrostatic field effect) (11). For poly(acrylic) acids in water, only carboxyl groups with the lowest pK$_a$ values dissociate to form a negatively charged ion. However, less is known about which carboxyl group dissociates in polyelectrolytes such as poly(acrylic) acid.

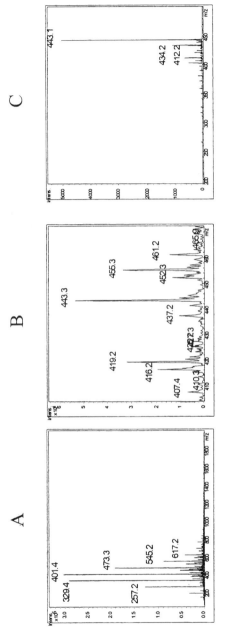

Figure 1. Poly(acrylic acid)(average MW 2000) negative ion electrospray mass spectra A) Full scan MS spectrum. B) MS spectrum of the expanded region between m/z 401-473. C) The MS/MS spectra of 452 m/z ion in spectrum B

318

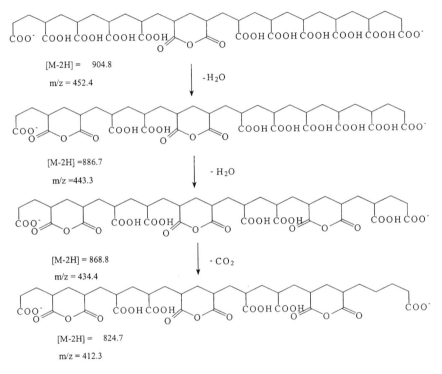

[M-2H] = 904.8
m/z = 452.4

-H₂O

[M-2H] =886.7
m/z =443.3

- H₂O

[M-2H] = 868.8
m/z = 434.4

- CO₂

[M-2H] = 824.7
m/z = 412.3

Figure 2. Hypothetical fragmentation sequence of m/z 452 ion of poly(acrylic) acid by multiple mass spectrometry

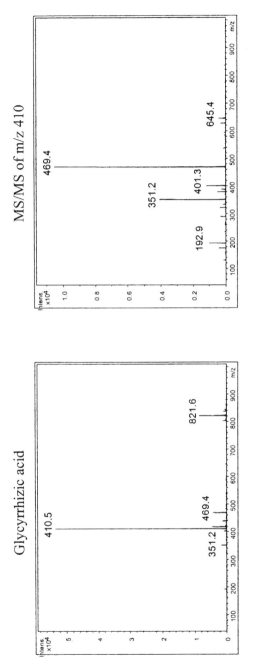

Figure 3. Negative ion electrospray mass spectra of glycyrrhizic acid
(Spectrum A) and MS/MS spectra of 410.5 m/z ion [M-2H]$^{2-}$ (Spectrum B)

To understand charge effects and fragmentation pathways of polycarboxylic acids, where the carboxyl groups are well separated in the molecular structure, glycyrrhizic acid was used as a model compound. The ESI/ ion trap MS spectrum of glycyrrhizic acid is presented in Figure 3A and the MS/MS spectrum of the 410.5 ion [M-2H]$^{2-}$ is presented in Figure 3B. The predominant ion in Spectrum A of Figure 3 is the 410.5 m/z ion [M-2H]$^{2-}$ with a small [M-H] ion at m/z 821.6. The MS/MS spectrum of the doubly-charged 410.5 m/z peak (Spectrum B) gives major fragment ions at 351 and 469 m/z, which were subsequently fragmented (MS/MS/MS, spectrum not shown). A proposed fragmentation pathway for these multiple mass spectra is shown in Figure 4.

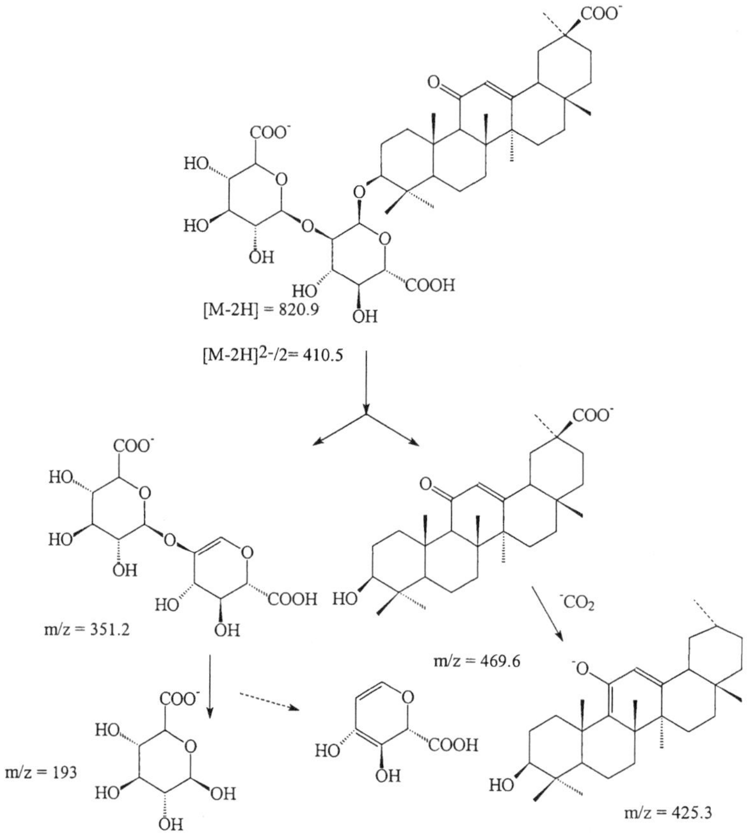

Figure 4 Hypothetical fragmentation sequence of glycyrrhizic acid by multiple mass spectrometry

In contrast to poly(acrylic acid) there is no initial water loss, but glycyrrhizic acid fragments likely at an ether linkage, into two charged fragments. The MS/MS/MS fragmentation involves another ether linkage fragmentation and a decarboxylation in the steroid hydrocarbon fragment. This result implies that when initial water loss is observed with an unknown polycarboxylic acid, the carboxyl groups are likely clustered so as to form cyclic anhydrides. The numerous alcohol groups in glycyrrhizic acid did not dehydrate to cause water loss until minor water loss ions were observed in the MS/MS/MS spectrum of the 351 m/z ion.

The ultimate unknown polycarboxylic acid structures are in fulvic acid mixtures. Most of the MS/MS spectra of various negative ions extracted from the Suwannee River fulvic acid show water and CO_2 losses indicative of singly charged ions, but a few ions also show doubly-charged characteristics. The MS/MS spectrum of the 283 m/z ion mixture of Suwannee River fulvic acid is shown in Figure 5. Initial water loss indicative of cyclic anhydride formation is observed by ions at 274 m/z $[M-H_2O-2H]^{2-}$ and 265 m/z $[M-H_2O-H]^-$. Previous research (12) has shown that carboxyl groups are clustered in Suwannee River fulvic acid, and this clustering facilitates the formation of cyclic anhydrides. Decarboxylation of both doubly- and singly-charged ions is observed by ions at 261 m/z $[M-CO_2 -2H]^{2-}$ and 239 m/z $[M-CO_2-H]^-$. Other significant ions in the MS/MS spectrum of Figure 5 can be explained by successive water losses and decarboxylation losses. If structural characterization of doubly-charged ions found at 283 m/z is desired, multiple MS fragmentations of the 274 m/z $[M-H_2O-2H]^{2-}$ and 261 m/z $[M-CO_2 -2H]^{2-}$ ions can be performed. Similar, studies of singly-charged ions can be performed by multiple MS fragmentations of ions at 265 m/z and 239 m/z. The Suwannee River fulvic acid displays fragmentation characteristic more similar to poly(acrylic acid) than to glycyrrhizic acid suggesting that clustered carboxyl groups are a structural characteristic of this type of DOM.

Selective methylation of carboxyl groups was used to analyze polycarboxylic LAS and DATS metabolites (3) to ensure singly-charged species (charge carried by sulfonate) that could be quantitatively determined. Methylation with diazomethane only methylates carboxyl and the more strongly acidic phenol groups whereas alcohol and weakly acidic phenol groups in complex structures such as fulvic acid are not methylated. Unmethylated alcohol groups are known to ionize in negative ESI/MS as shown by applications to carbohydrate analyses (13). Methylation is also well known to disrupt hydrogen bonding that is responsible for aggregate formation; thus, methylation previous to ESI/MS analyses might be the key to providing molecular weight distributions of primary covalently-bonded structures in complex polycarboxylic acid mixtures such as fulvic acid.

322

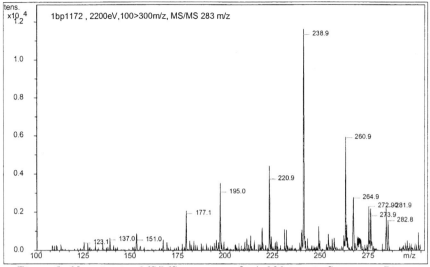

Figure 5. Negative ion MS/MS spectrum of m/z 283 ions in Suwannee River fulvic acid

Negative ion ESI/quadrupole MS analysis of methylated glycyrrhizic acid resulted in a single peak at 863 m/z which is the [M-H]$^-$ ion of glycyrrhizic acid plus three methyl groups added to the carboxyl groups by methylation. However, when methylated poly(acrylic) acid was subjected to same analysis, only a few ions were detected with low signal to noise. This difference in response indicates the necessity of unmethylated hydroxyl groups (present in glycyrrhizic acid and absent in poly(acrylic) acid) to provide ionization sites for negative ion ESI/MS. Positive-ion EIS/MS gave a better response for methylated polyacrylic acid, but it contained both multiply-charged ions and multiple H$^+$, Na$^+$, and K$^+$ adducts.

Methylated Suwannee River fulvic acid was analyzed by negative ion ESI/Quadrupole MS and its calculated molecular weight distribution compared to unmethylated molecular weights reported previously (*5*). Methylation decreased the number average molecular weight (M$_n$) from 591 to 581 Daltons and the weight average molecular weight (M$_w$) from 914 to 695 Daltons. The molecular weight decrease was even greater after correction for added methyl groups because the Suwannee River fulvic acid has an average of four carboxyl groups per molecule and about one phenol group (*4*) that are methylated causing an additional molecular weight decrease of 70 Daltons when subtracted from unmethylated molecular weight values. This decrease in molecular weight, especially the large decrease in M_w, indicates that molecular aggregates rather than multiple charging is responsible for the errors in molecular weight distribution of underivatized fulvic acid as determined by negative ion ESI/MS. This relatively small size of fulvic-acid molecules coupled with inductive and electrostatic field effects favors the formation of singly-charged negative ions. These results should be qualified by the assumptions that each methylated fulvic acid molecule has a hydroxyl group that ionizes and that there are not major relative response differences between nonmethylated and methylated molecules. It is likely that alcohols have lower ionization efficiencies that carboxylic acids, and additional research is required to quantitate the relative response differences between methylated and unmethylated samples.

However, other independent observations [aggregate ions in polycarboxylic acid standards (5); few doubly-charged ions found in the MS/MS spectra; and a similar M_n value of 530 of methylated Suwannee River fulvic acid determined by equilibrium centrifugation in both N,N-dimethylformamide and acetonitrile (9)] strongly support the ESI/MS results. Similar M_n values of Suwannee River fulvic acid were also found by vapor pressure osmometry (M_n = 623-965 Daltons in various solvents) and low-angle X-ray scattering (M_n = 645-616 Daltons) (5).

Conclusions

Characteristic charge characteristics and fragmentation sequences were determined from analyses of polycarboxylic acid standards. Multiple charging was found to be a function of spatial charge separation in molecular structures. Ether and ester cleavage occurs first followed by water loss from cyclic anhydride formation, carbon dioxide loss from decarboxylation, water loss from alcohol dehydration, alkyl group losses, and finally, fragmentation of cyclic anhydrides and lactone esters (5). The presence of clustered carboxyl groups is readily detected by cyclic anhydride formation accompanied by water loss, and these clustered carboxyl groups are responsible for much of the acidity and metal-binding characteristic of NOM in water (4). Lastly, methylation of complex polycarboxylic acid mixtures such as fulvic acid minimizes both multiple charges and aggregate ions that bias molecular weight determinations of these mixtures by ESI/MS.

Disclaimer:
Use of trade names is for identification purposes only and does not constitute endorsement by the U.S. Geological Survey

References

1. Crescenzi, C.; Di Corcia, A.; Marcomini, A.; Samperi, R. *Environ. Sci. Technol.* **1997**, *31*, 2679.

2. Di Corcia, A.; Carvallo, R.; Crescenzi, C.; Nazzari, M. *Environ. Sci. Technol.* **2000**, *34*, 3914.

3. Di Corcia, A.; Capuani, L.; Casassa, F.; Marcomini, A.; Samperi, R. *Environ. Sci. Technol.*, **1999**, *33*, 4112.

4. Leenheer, J.A.; Wershaw, R.L.; Brown, G.K.; Reddy, M.M., *Applied Geochemistry* (in press, 2002).

5. Leenheer, J.A.; Rostad, C.E.; Gates, P.M.; Furlong, E.T.; Ferrer, I. *Anal. Chem.* **2001**, *73*, 1461.

6. Plancque, G.; Amekraz, B.; Moulin, V.; Taulhoat, P.; Moulin, C. *Rapid Commun. Mass Spectrom.* **2001**, *15*, 827.

7. Crescenzi, C.; Di Corcia, A.; Marcomini, A.; Samperi, R. *Environ. Sci. Technol.* **1997**, *31*. 2679.

8. Leenheer, J.A.; Noyes, T.I. *U.S. Geological Water Supply Pap. No 2230*, **1984.**

9. Leenheer, J.A.; Brown, P.A.; Noyes, T.I. In *Aquatic Humic Substances: Influence on Fate and Treatment of Pollutants*, Suffet, I.H., MacCarthy, P., Eds: Advances in Chemistry Series 219; American chemical Society; Washington, DC, 1989, pp. 25-39.

10. Wershaw, R.L., Aiken, G.R. In *Humic Substances in Soil, Sediment, and Water,* Aiken, G. R., MCKnight, D.M., Wershaw, R.L., Eds.; J. wiley and Sons: New York, 1985' pp. 477-492.

11. Perrin. D.; Dempsey, B.; Serjeant. E.P. *pKₐ Prediction for Organic Acids and Bases*, Chapman and Hall, London and New York, 1981.

12. Leenheer, J.A; Wershaw, J.A., Reddy, M.M., *Environ. Sci. Technol.* **1995**, *29*, 399.

13. Cole, R. B. *Electrospray Ionization Mass Spectrometry: Fundamentals, Instrumentation, and Applications*, John Wiley & Sons, New York, 1997, Chapter 13, pp. 459-498.

Chapter 19

The Determination of Anaerobic Biodegradation Products of Aromatic Hydrocarbons in Groundwater Using LC/MS/MS

Michael S. Young[1], Claude R. Mallet[1], David Mauro[2], Sam Fogel[3], Ashok Jain[4], and William Hoynak[5]

[1]Waters Corporation, 34 Maple Street, Milford, MA 01757
[2]META Environmental Inc., 49 Clarendon Street, Watertown, MA 02472
[3]Bioremediation Consulting Inc., 39 Clarendon Street, Watertown, MA 02472
[4]Electric Power Research Institute, 3412 Hillview Avenue, Palo Alto, CA 94304
[5]Northeast Utilities, P.O. Box 270, Hartford, CT 06141

This paper describes chromatographic and spectroscopic analysis in support of an investigation of the natural biodegradation of monocyclic and polycyclic aromatic hydrocarbons (MAH, PAH) in groundwater. Most of the parent aromatic compounds may be readily determined in complex matrices using either GC or LC. Although GC/MS is a preferred technique for organic environmental analysis, it is not amenable to the analysis of highly polar, labile or non-volatile compounds, without cumbersome derivatization. The degradation products of MAHs and PAHs are often acidic or otherwise highly polar in nature and are therefore more easily determined using LC. In this presentation we will discuss the utility of LC/MS and LC/MS/MS for the straightforward analysis of the biodegradation products of MAHs and PAHs.

Introduction

The reduction in mass and/or concentration of a compound in the environment over time or distance as a result of naturally occurring physical, chemical, or biological processes is known as natural attenuation. Utilization of natural attenuation has gained technical and regulatory acceptance as an appropriate remedial approach at some sites. Many natural processes can occur simultaneously to affect the fate of contaminants in soil and groundwater. Among these processes are sorption, precipitation, chemical stabilization, chemical degradation, biological stabilization, and biological degradation. Biological degradation is the most frequent process governing the environmental fate of organic compounds. Recently, anaerobic biodegradative processes of monoaromatic and polycyclic aromatic hydrocarbons have been shown to generate biochemical intermediates. The presence of selected alkylbenzene and PAH metabolites in groundwater and soil provides evidence that biological degradation is reducing the mass of the parent MAHs and PAHs at that site. The objective of this paper is to demonstrate the utility of LC/MS/MS techniques for the straightforward determination of these metabolites.

Manufactured Gas Plant (MGP) Waste Contaminants

The manufactured gas industry produced gas from coal or oil for lighting and heating as well as by-products that served as fuel or feed stocks for the production of chemicals. Over the years, some MGP by-products were released to the environment and exist in soil and sediment or relic plant structures to this day. These by-products primarily included tars, oils, or lampblack, however tars and tar sludge were released in the greatest volumes.

Tar is complex mixture of chemicals formed at high temperature under low oxygen conditions. Those chemicals mostly include aromatic hydrocarbons with one to several rings. However, tars also contain aliphatic hydrocarbons, organic acids, nitrogen-, sulfur-, and oxygen-containing heterocyclic compounds, water, particulate carbon, and other substances. In the environment, tars will disperse as a separate organic phase and the chemicals in tar will partition into the vapor and aqueous phases based on their vapor pressures and aqueous solubilities, respectively. Once in the vapor phase or dissolved in groundwater or surface water, tar chemicals will further disperse and will be degraded by chemical and biological mechanisms. Groundwater was collected from several locations at a former MGP site for use in this study. The groundwater samples contained various amounts of dissolved tar chemicals.

Of the many aromatic hydrocarbons found at MGP site groundwater, naphthalene, methylnaphthalenes, and certain monoaromatic compounds are often among the most abundant. Because they are more water soluble than the higher molecular weight polycyclic aromatic hydrocarbons (PAH), the mono and diaromatic hydrocarbons are usually the predominant contaminants in groundwater at such sites. Consider the GC analysis presented in Figure 1. This chromatogram was obtained from one of the groundwater samples analyzed in this study; naphthalene, methylnaphthalenes, and indan are the most abundant hydrocarbons.

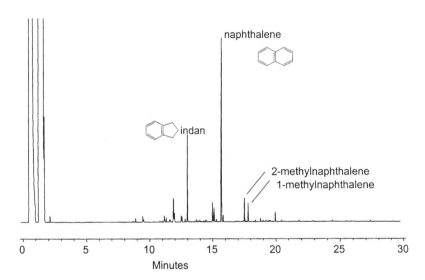

Figure 1. Gas chromatogram of a groundwater sample from an MGP site showing typical aromatic hydrocarbons found in such samples

Anaerobic degradation of naphthalene and other PAH has been demonstrated in microcosm studies (1). The following compounds have been identified as free acid intermediates of the anaerobic degradation of 2-

methylnaphthalene; naphthyl-2-methylsuccinic acid, naphthyl-2-methylenesuccinic acid, 2-naphthoic acid, and 5,6,7,8-tetrahydro-2-naphthoic acid (2). 2-Naphthoic acid and reduced naphthoic acids have also been identified as metabolites of anaerobic naphthalene degradation (3,4). Benzylsuccinic acid, a metabolite of the degradation of toluene, may also be present at PAH degradation sites. Structures for some of these compounds are given in Figure 2. Although the anaerobic biodegradation of indan is not so well documented, this study did identify an unknown compound likely to be a degradate of indan. Note that naphthylmethylsuccinic and benzylsuccinic acids are conjugates to which additional carbon has been added to the metabolite structure according to the biochemical pathways given in reference 2.

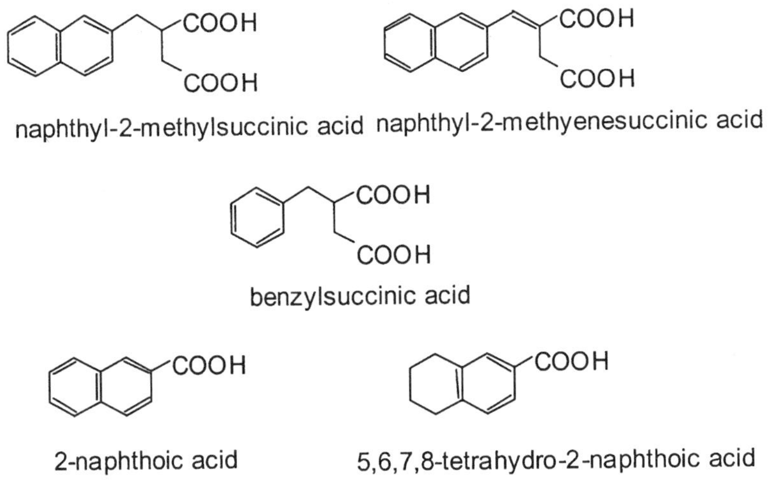

naphthyl-2-methylsuccinic acid   naphthyl-2-methyenesuccinic acid

benzylsuccinic acid

2-naphthoic acid   5,6,7,8-tetrahydro-2-naphthoic acid

Figure 2. Structures of Some Possible Anaerobic Degradates at PAH Sites

Because the metabolites are often acidic and much more polar in nature than the PAH compounds from which they were derived, the analytical methods utilized for determination of PAH in environmental samples are usually not appropriate for determination of the metabolites without modification. For example, derivatization is necessary for the GC/MS analysis of most of these compounds. However, electrospray mass-spectrometry interfaced to liquid chromatograph (LC/MS) is a modern technique that is well suited for the analysis of ionizable compounds such as naphthoic acid. This paper reports the results of the use of LC/MS and LC/MS/MS to demonstrate that degradates of naphthalene or methylnaphthalenes are present in groundwater samples taken at a site contaminated with aromatic hydrocarbons.

Benefits of LC/MS/MS Analysis

There are two main benefits to the use of triple quadrupole MS/MS in multiple reaction monitoring (MRM) mode compared with a single quadrupole MS in single ion recording (SIR) mode. In MRM mode, a specific transition between a precursor and product ion is monitored. The first benefit of this technique is a lower background signal compared with selected ion recording (SIR) using a single quadrupole mass spectrometer; the resulting gain in signal to noise ratio can result in more than a tenfold increase in useful sensitivity. The other main benefit of the MRM analysis is the sure knowledge that the monitored product ion is present only as a result of the fragmentation of the precursor. For confirmation of an analyte identity, multiple MRM transitions are monitored if possible. For benzylsuccinic acid, the transition from m/z 207 (M-1, precursor) to m/z 163 is the primary transition resulting from the loss of the carboxylate functionality. However, using a higher collision energy, the transition from m/z 207 to m/z 91 ($C_7H_7^+$) may be monitored to help confirm the identity of the compound.

Another approach utilized in this study was to monitor a product-ion scan spectrum resulting from a particular precursor ion. Compared with MRM mode, a more complete mass-spectrum is obtained that may be used to obtain qualitative structural information for unknown compounds. However, because the second quadrupole is operated in a full-scan mode, the sensitivity of the product ion mode is less than for MRM mode. For example, a conservative quantitation limit for 2-naphthoic acid is below 0.05 µg/L using MRM mode and approximately 0.5 µg/L using the product-ion scan mode.

Experimental

Solid-Phase Extraction (SPE) of Aqueous Samples

The groundwater sample (400 mL) was adjusted to below pH 2 by addition of 1 mL of concentrated HCl. A reversed-phase SPE cartridge (Oasis HLB, 3cc,

60 mg., Waters Corp.) was conditioned with 1 mL of methanol followed by 1 mL of water. The acidified sample was loaded onto the cartridge at a rate of 5 mL/min. The cartridge was washed with 1 mL of 25:75 methanol/0.1M HCl. The cartridge was then air-dried by vacuum and was eluted with 3 mL of 90:10 methyl-*t*-butyl ether (MTBE)/methanol. A 100 μL aliquot was removed for hydrocarbon analysis by GC/FID. The remaining eluent was evaporated and the residue was reconstituted in 400 μL of 10:90 acetonitrile/water. The recoveries of 2-naphthoic acid and benzylsuccinic acid were better than 80 % for a typical groundwater sample spiked at a level of 5 μg/L.

LC/MS and LC/MS/MS Analysis

The LC/MS system was comprised of a Waters 2690 Separations Module (Waters Corporation, Milford, MA) interfaced to a Micromass Quattro Ultima (Micromass Ltd., Manchester, UK) triple quadrupole spectrometer operated in the negative electrospray mode (ESI-). Analyses were performed using full-scan, MRM and product-scan modes.

The analytical column was an XTerra MSC$_{18}$, 100 x 2.1 mm, 3.5 μM particle size (Waters Corp.). Mobile phase A was 20 mM ammonium formate (pH 4.5) and mobile phase B was acetonitrile. The gradient was 85 % A to 10 % A in 10 minutes. A 20 μL aliquot of the reconstituted eluent was injected for LC/MS/MS analysis; 40μL was injected for full-scan LC/MS analysis.

Full-Scan Mode

Full-scan LC/MS analysis was accomplished using cone voltages of 13 and 22 V.

MRM Mode

Table I summarizes the cone voltages and collision energies utilized for the MRM analysis. These parameters were optimized using actual standard materials, if available. For suspected unknowns, the parameters were estimated from the behavior of similar compounds.

Product-Scan Mode

The cone voltage was maintained at 15 V. Product-ion scans were obtained using collision energies of 13 and 22 eV. The precursor ions chosen for this mode were consistent with those chosen for the MRM analyses (see Table I).

Table I. LC/MS/MS conditions for the MRM analysis of groundwater samples from MGP sites

Compound	Precursor ion (m/z)	Product ion (m/z)	Collision Energy (eV)	Cone Voltage (V)
Naphthoic acid	170.9	126.9	15	20
Tetrahydronaphthoic acid	174.9	130.9	15	20
Benzylsuccinic acid	206.9	162.9	15	15
Methylnaphthoicacid	144.9	140.9	15	20
Naphthylmethylsuccinic acid	256.9	212.9	15	15

GC/FID Analysis

GC analysis was accomplished using an HP 5890 Series II gas chromatograph (Agilent Technologies, Wilmington, DE) equipped with a flame-ionization detector (FID). The analytical column was an Rtx-5, 30 m, 0.32 mm, 0.5 μm *df* (Restek Corp., Bellafonte, PA). The carrier gas was helium at a flow rate of 1.5 mL/min. The oven temperature was held at 35° C for 2 minutes, programmed at 10°/minute to 310°, and then held at 310° for 15 minutes. A 1 μL volume was injected in the split/splitless mode with the purge valve on time set at 0.5 minutes.

Calibration Standards

2-naphthoic acid and 1-naphthoic acid were obtained from Sigma-Aldrich (Aldrich, Milwaukee, WI). Benzylsuccinic acid was obtained from Sigma-Aldrich (Sigma, St. Louis, MO). Aromatic hydrocarbon standards were obtained from Restek (Bellefonte, PA). Standard materials for other suspected metabolites were not commercially available.

Results

The suspected metabolites discussed below were detected in every groundwater sample with measurable aromatic hydrocarbon contamination. The metabolites chosen for quantitation in this study were the naphthoic acids and benzylsuccinic acid. These compounds are known metabolites of the anaerobic degradation of naphthalene, methylnaphthalenes and toluene and standard materials for these metabolites were readily available. Some other suspected metabolites of naphthalene or methylnaphthalene degradation are 5,6,7,8-tetrahydronaphthalene, naphthylmethylsuccinic acid, and methylnaphthoic acid. Although no standard materials were available for these suspected metabolites, efforts were made to identify these compounds or similar compounds using the combination of full-scan LC/MS and product-ion scan LC/MS/MS.

Results are summarized in Table II.

Table II. Results obtained for six groundwater samples taken at an MGP site (μg/L)

Sample	1-Naphthoic acid	2-Naphthoic acid	Unknown MW 205	Total Naphathalenes[1]	Indan[1]
01	0.43	0.37	1.5	2700	1400
02	3.5	6.5	2.1	3200	550
03	0.19	0.59	0.2	650	200
04	n.d.	n.d.	n.d.	n.d.	n.d.
05	1.3	2.5	0.22	3500	1100
06	1.6	0.73	0.40	6500	1100
Spike	0.32 (64%)	0.35 (70%)	---	---	---
Blank	n.d.	n.d.	n.d.	n.d.	n.d.

[1] Result determined using GC/FID

n.d. = not detected

Naphthoic Acids

Both 1-naphthoic acid and 2-naphthoic acids were detected and quantified against pure standard materials. The chromatograms shown in Figure 3 are obtained using LC/MS/MS in the MRM mode. These are the only metabolites found in the groundwater samples that were confirmed against pure standards.

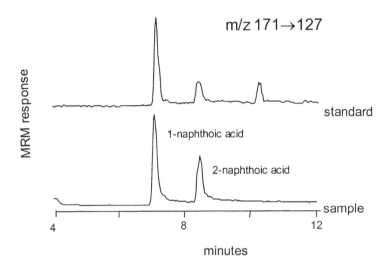

Figure 3. LC/MS/MS (MRM) chromatogram of a groundwater sample from an MGP site showing the presence of naphthoic acids

Indan Metabolite (Similar to Benzylsuccinic Acid)

Benzylsuccinic acid (BSA) was not identified in these samples. However, a compound was identified that is apparently very similar in structure to BSA. This suspected metabolite is present at concentrations similar to that of the naphthoic acids. Figure 4 shows the product spectrum obtained at 13 eV for the unknown compared with the spectrum obtained for BSA under the same conditions.

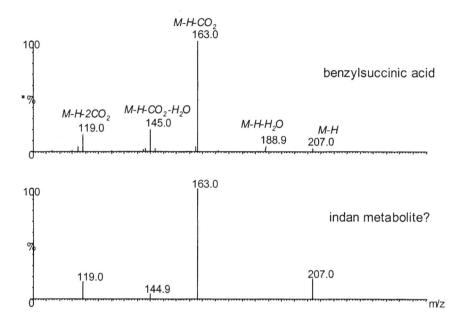

Figure 4. LC/MS/MS product-ion mass-spectrum obtained for possible indan metabolite showing the similarity with spectrum obtained for benzylsuccinic acid

5,6,7,8-Tetrahydronaphthalene

5,6,7,8-Tetrahydronaphthalene was not available at the time of this study. One substance was identified in the extracts with a mass-spectrum possibly consistent with this compound.

Naphthlymethylsuccinic acid

A number of compounds were identified at very trace levels with mass-spectra that may be consistent with this compound. The target ions are m/z 213 ($M-H^+-CO_2$), m/z 169 ($M-H^+-2CO_2$) and m/z 141 ($C_{11}H_9^+$). The apparent concentrations of these compounds are well below 100 μg/L.

Methylnaphthoic acid (or Naphthylacetic acid)

A compound was identified with a product-ion mass spectrum consistent with methylnaphthoic acid or naphthylacetic acid. The mass-spectrum presented in Figure 5 was obtained with collision cell energy of 13 eV.

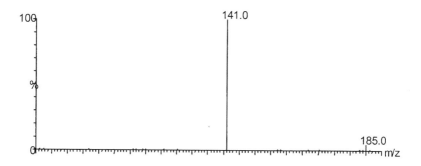

Figure 5. LC/MS/MS product-ion mass-spectrum (products of m/z 185) obtained for suspected metabolite of methylnaphthoic acid or naphthylacetic acid

Confirmation of Metabolites at a Second Site

Groundwater samples obtained from a second MGP site have been analyzed using the methods described in this paper. These samples contained the two naphthoic acids and most of the suspected metabolites seen in samples from the first site.

Discussion

LC/MS/MS Compared With GC/MS

The determination of acidic compounds in groundwater is certainly facilitated by the use of liquid chromatography compared with gas chromatography for several reasons. First, because the metabolites are present at concentrations 2 or 3 orders of magnitude lower than the parent hydrocarbons, it may be necessary to perform a class fractionation to remove hydrocarbon interferences from the acidic metabolites prior to GC/MS analysis. This step is often labor-intensive and adds opportunities for analyte losses and other quality control problems. Second, methylation or other derivatization often is required for the GC/MS determination of these compounds, further complicating the sample preparation.

The use of LC for the analytical separation eliminates the requirement for derivatization. The use of the atmospheric pressure interface (electrospray mode in this study) reduces or eliminates the need for any class fractionation prior to the LC analysis. The non-polar aromatic hydrocarbons present in the sample have little or no response in electrospray mass-spectrometry. Consequently, a much more rapid and straightforward analysis is available for the determination of acidic compounds in aqueous samples. Moreover, the use of LC/MS methodology is highly complimentary to GC/MS for biodegradation monitoring. While all of the parent compounds and some of the metabolites are reliably measured with GC/MS, metabolites that are too polar or labile for GC methods are usually well suited for LC/MS methods.

Using GC/MS in the full-scan mode, much structural information is available to the analyst. Usually full-scan quadrupole GC/MS is significantly more sensitive than full-scan quadrupole LC/MS. More importantly, a high degree of fragmentation is usually seen using electron-ionization GC/MS in the full-scan mode. Even the absence of such fragmentation provides strong structural evidence of particularly stable molecular ions. The generation of fragment (product) ions is essential for the use of mass-spectrometry in the identification of unknowns. For electrospray-ionization at atmospheric pressure, cone voltages and collision cell energies must be optimized for each individual compound in order to produce product ions. Moreover, the expected precursor ion must be known. In this study the identity of target compounds was known. Even for the cases where no standard materials were available, reasonable estimates for ionization conditions could be obtained from the behavior of similar compounds. Consequently, a great deal of information regarding the presence of suspected metabolite in groundwater samples was obtained using LC/MS/MS much more rapidly than could reasonably be expected using derivatization GC/MS.

Future Study

Efforts are underway to obtain more standard materials for the suspected metabolites. Efforts will be made to identify the exact structure of the suspected indan metabolite. These efforts will include further LC/MS and LC/MS/MS analysis and also GC/MS analysis after derivatization with diazomethane. The analytical methodologies discussed in this paper will soon be applied to groundwater samples at a number of other MGP sites.

Conclusions

- LC/MS/MS is a powerful analytical tool useful for the straightforward determination of polar acidic compounds in complex matrices.
- SPE on a hydrophilic/lipophilic balanced polymeric sorbent allows for the efficient and simultaneous extraction of non-polar contaminants and polar metabolites from groundwater samples.
- Naphthoic acids and other metabolites have been found in both MGP sites investigated to date. There is good evidence that these may be considered 'signature metabolites'; the presence of these signature metabolites is indicative of anaerobic degradation at the site.

References

1. Rockne, K.J., and Strand, E. *Environ. Sci. Technol.* **1998,** *32,* 3962-3967.
2. Annweiler, E., Materna, A., Safinowski, M., Kappler, A., Richnow, H., Michaelis, W., and Meckenstock, R. *Appl. Environ. Microbiol.* **2000,** *66,* 5329-5333.
3. Meckenstock, R., Annweiler, E., Michaelis, W., Richnow, H., and Schink, B. *Appl. Environ. Microbiol.* **2000,** *66,* 2743-2747.
4. Zhang. X., and Young, L.Y. *Appl. Environ. Microbiol.* **1997,** *63,* 4759-4764.

Chapter 20

Analysis of Halogenated Alkylphenolic Compounds in Environmental Samples by LC/MS and LC/MS/MS

Mira Petrovic and Damià Barceló

Department of Environmental Chemistry, IIQAB-CSIC, c/Jordi Girona 18–26, 08034 Barcelona, Spain

The state-of-the-art in the LC-MS and LC-MS-MS analyses of halogenated alkylphenolic compounds formed during wastewater and drinking water chlorination is reviewed. Single stage LC-MS with ESI ionization showed limitations in the analysis of real-world samples, resulting from the presence of isobaric interferences when only deprotonated molecules are monitored. With collision-induced dissociation halogenated NPs and NPECs gave specific fragments formed by the cleavage of the alkyl moiety or by the loss of (ethoxy)carboxylic group, respectively. Characteristic pattern of isotopic doublet signals of these specific transitions coupled with formation of [Br]$^-$ and [Cl]$^-$ permitted unequivocal identification of halogenated alkylphenolic compounds in environmental samples. Limits of detection by LC-MS and LC-MS-MS were in the range of 10 and 1 ng/l, respectively.

Alkylphenol ethoxylates (AP$_n$EOs, n=number of ethoxy units) are widely used nonionic surfactants. The findings on toxicity and weak estrogenicity of some of APEOs persistent metabolites have raised concern over their environmental and health effects *(1-4)*. As shown in a simplified way in Figure 1 APEOs may be transformed in wastewater treatment plants (WWTP) or in the environment by: (i) shortening of the polyethoxy chain producing mainly AP_1EO, AP_2EO and fully deethoxylated alkylphenols (AP); and (ii) carboxylation of the terminal EO units or/and the alkyl side chain resulting in the formation of alkylphenol carboxylates (APEC) and carboxyalkyl phenol carboxylates (CAPEC), respectively. These metabolites are susceptible to *ortho*-halogen substitution and they can be further transformed into halogenated by-products during chlorination in water treatment works.

While the occurrence of non-halogenated metabolites in the environment and their ability to mimic the endogenous hormone, 17β-estradiol, is well documented *(1,2)*, only little data exist on the occurrence and ecotoxicology of brominated and chlorinated alkylphenolic compounds. One of the reasons for this is the low relative abundance of these compounds (generally less than 10% of the total pool of alkylphenolic compounds) and lack of appropriate analytical methods for their unequivocal identification and quantification.

Early methods included fast atom bombardment-mass spectrometry (FAB-MS) *(5-9)*, which proved to be a reliable tool for the identification of halogenated metabolites in raw and drinking water, but not for their quantification. Recently, several authors reported the analysis of halogenated alkylphenolic compounds by gas chromatography-mass spectrometry (GC-MS), reversed-phase liquid chromatography with electrospray mass spectrometry (LC-ESI-MS) or tandem mass spectrometry (LC-ESI-MS-MS) (Table I).

Table I. Survey of analytical methods for quantitative determination of halogenated alkylphenolic compounds

| Compound | Matrix | Sample preparation | | Analytical method | Ref |
		Extraction	Clean-up		
XAP, XAPEC	Water Sludge	SPE-C_{18} PLE (MeOH-Acetone, 1:1)	- SPE-C_{18}	LC-MS-MS	*(10)*
XAP, XAPEC, XAPEO	Water Sludge Sediment	SPE-C_{18} Sonication (MeOH-DCM, 7:3)	- SPE-C_{18}	LC-MS	*(11)*
XNP	Sediment	Continuous-flow sonication (MeOH)	SPE-NH_2 RP-HPLC	LC-MS	*(12)*
XNPEC XNPEO	Water	SPE-C_{18}	derivatization to methyl-esters	GC-MS	*(13)*
BrNP, BrNPEO	Water	SPME	-	GC-MS	*(14)*

340

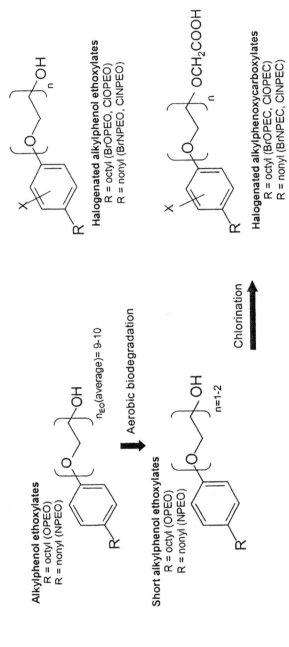

Alkylphenol ethoxylates
R = octyl (OPEO)
R = nonyl (NPEO)

$n_{EO}(average) = 9\text{-}10$

Aerobic biodegradation

Short alkylphenol ethoxylates
R = octyl (OPEO)
R = nonyl (NPEO)

n=1-2

Chlorination

Halogenated alkylphenol ethoxylates
R = octyl (BrOPEO, ClOPEO)
R = nonyl (BrNPEO, ClNPEO)

Halogenated alkylphenoxycarboxylates
R = octyl (BrOPEC, ClOPEC)
R = nonyl (BrNPEC, ClNPEC)

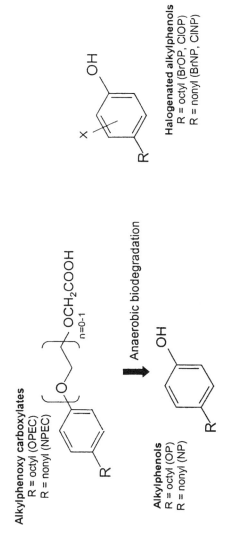

Figure 1. Biodegradation pathway and by-products formation of alkylphenol ethoxylates

This review will focus on the current state-of-the-art in the LC-MS and LC-MS-MS analyses of halogenated alkylphenolic compounds. It also summarizes knowledge on the occurrence and ecotoxicity of this group of contaminants and addresses future research needs in this area.

Analysis of halogenated alkylphenolic compounds

LC-MS

Owing to the presence of chlorine and bromine atoms, respectively in the molecules, halogenated alkylphenolic derivatives yield a characteristic pattern of isotopic doublet signals. This provides a highly diagnostic fingerprint for this group of compounds.

Halogenated APEOs were detected using an ESI-MS under positive ionization (PI) conditions *(11)*. Like their non-halogenated analogs, halogenated APEOs show a great affinity for alkali metal ions, and they produce exclusively evenly-spaced ($\Delta 44$ Da) sodium adduct peaks $[M+Na]^+$ with no further structurally significant fragmentation. Problem arise from the fact that the chlorinated derivatives ($ClAP_nEO$) have the same molecular mass and they gave the same ions as brominated compounds with one ethoxy group less ($BrAP_{n-1}EO$) and chromatographic separation of these two groups of compounds, which is quite difficult to obtain, is a prerequisite for their quantitative determination. However, the characteristic doublet signal of brominated and chlorinated compounds, respectively due to different contribution of their isotopes ($^{79}Br:^{81}Br=100:98$ and $^{35}Cl:^{37}Cl=100:33$, respectively) provides additional confirmation of their presence. An example of the identification of halogenated NPEOs by LC-ESI-MS is shown in Fig. 2. A similar ion series at m/z from 409/411 to 851/853 revealed the presence of ClNPEOs (n_{EO}=3-13) and BrNPEOs (n_{EO}=2-12), respectively. In addition halogenated OPEOs were also identified.

Several authors reported determination of halogenated NPs *(11,12)* and NPECs *(11)* by LC-MS under negative ionization (NI) conditions. Using an ESI interface, halogenated NPs gave a doublet signal of the $[M-H]^-$ ions at m/z 297/299 for BrNP and at m/z 253/255 for ClNP (Fig. 3A and B). Halogenated NPECs gave two signals, one corresponding to deprotonated molecule and another to $[M-H-CH_2COOH]^-$ in the case of XNP_1ECs or $[M-H-CH_2CH_2OCH_2COOH]^-$ for XNP_2Ecs (Fig. 3C-F).

However, real application showed that with "soft ionization" LC-MS, giving solely deprotonated molecule or very limited number of fragments, the identification of halogenated compounds is quite difficult. Using a single stage of mass selectivity in the analysis of real-world samples, $ClNP_1EC$ was obstructed by a severe isobaric interference of linear alkyl benzene sulfonate ($C_{11}LAS$), which is often found in environmental and wastewater samples in concentrations several orders of magnitude higher than those of halogenated alkylphenolic compounds *(11)*. All attempts to separate $ClNP_1EC$ and $C_{11}LAS$ using gradient elution with standard mobile phases for reversed-phase separation (methanol/water or acetonitrile/water) failed. Determination of $ClNP_1EC$ was achieved by monitoring fragment ion at *m/z* 253/255, which also suffered the isobaric interferences in some real samples.

LC-MS-MS

Driven by the estrogenic potency of some compounds and their low environmental concentrations, the detection limits required for the monitoring of endocrine disruptors are being pushed from the microgram to the nanogram or even to below nanogram per liter range. Currently, a breakthrough is occurring for analytical methods based on LC-MS-MS. However, although considered as one of the most powerful techniques for structure interpretation and quantification, LC-MS-MS has been seldom used in the analysis of acidic and neutral NPEO metabolites and currently only a single paper describes the application of ESI-MS-MS in the analysis of their halogenated derivatives *(10)*.

ESI-MS-MS permitted unambiguous identification and structure elucidation under negative ionization conditions (halogenated NPECs and NPs produced signal), while NPEOs under positive ionization conditions produced no fragmentation. As a result these compounds were analyzed using a single stage MS monitoring $[M+Na]^+$ adduct ion in SIM mode.

With collision-induced dissociation (CID) under NI conditions, halogenated NPECs and NPs undergo fragmentation with few major pathways depicted in Fig. 4. For halogenated NP_1ECs and NP_2ECs the predominant reaction was loss of CH_2COOH and $CH_2CH_2OCH_2COOH$, respectively that resulted in intense signals at *m/z* 253/255 for ClNPECs and *m/z* 297/299 for BrNPECs. Further fragmentation of $[ClNP]^-$ occurred primarily on the alkyl moiety leading to a sequential loss of *m/z* 14 (CH_2 group), with the most abundant fragments at *m/z* 167 for ^{35}Cl and *m/z* 169 for ^{37}Cl with the relative ratio of intensities of 3.03. Fragment corresponding to the $[Cl]^-$ ion was observed only when sufficient collision energy was applied. The intensity of this ion was not very pronounced, but nevertheless remained useful for the identification of chlorinated NP. The

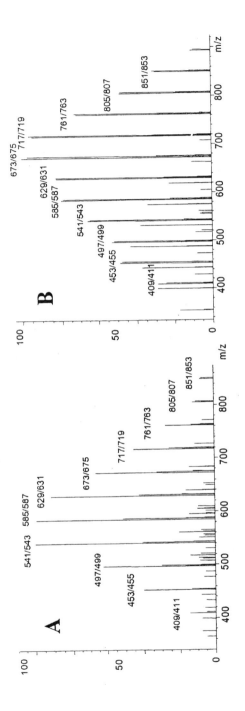

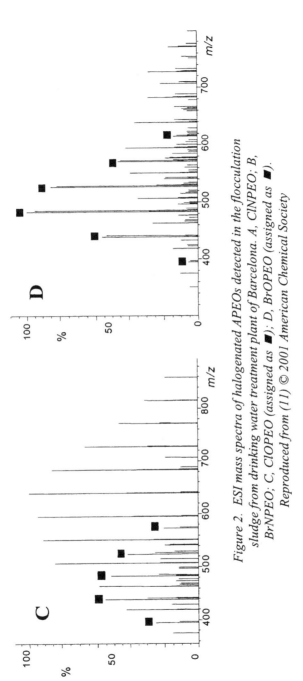

Figure 2. ESI mass spectra of halogenated APEOs detected in the flocculation sludge from drinking water treatment plant of Barcelona. A, ClNPEO; B, BrNPEO; C, ClOPEO (assigned as ■); D, BrOPEO (assigned as ■). Reproduced from (11) © 2001 American Chemical Society

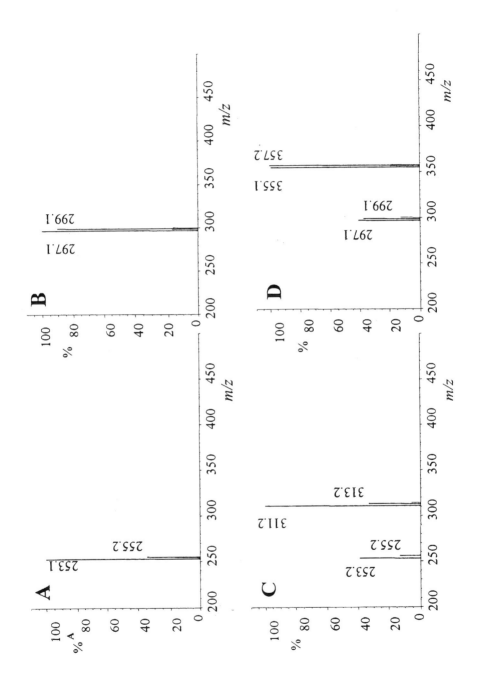

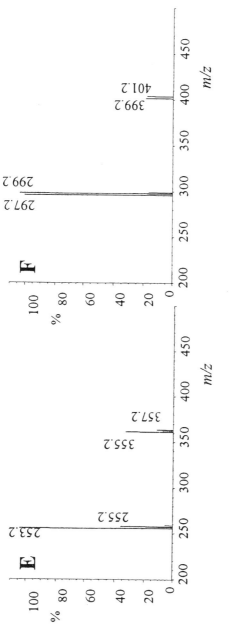

Figure 3. ESI mass spectra of standards of halogenated alkylphenolic compounds detected under NI conditions. A, ClNP; B, BrNP; C, ClNP₁EC; D, BrNP₁EC; E, ClNP₂EC; F, BrNP₂EC. Reproduced from (11) © 2001 American Chemical Society

brominated compounds showed a markedly different fragmentation pathway. Even at low collision energy [BrNP]⁻ (*m/z* 297/299) yielded intense signals at *m/z* 79 and *m/z* 81 corresponding to the [Br]⁻ (ratio of isotopes 1.02), while the fragmentation of the side chain was suppressed and resulted just in a low-intensity fragment at *m/z* 211/213 resulting from loss of C_6H_{14}. Such difference in the mechanism of fragmentation of chlorinated and brominated compounds is presumably the consequence of the lower energy of a $Br-C_6H_5$ (benzene) bond compared to a $Cl-C_6H_5$ bond.

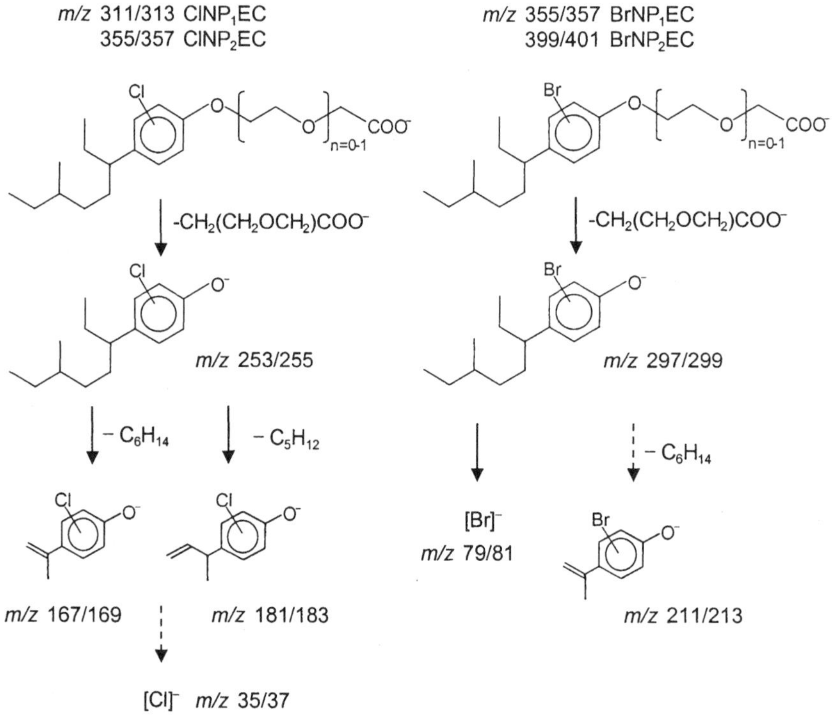

Figure 4. Tentative fragmentation mechanisms for halogenated alkylphenolic compounds in the ESI-MS-MS

A summary of all the ions formed under CID-ESI conditions is shown in Table II.

From the observed ion fragmentation pathways a reliable and sensitive quantification method, that overcomes the main drawbacks on existing LC-MS methods, was developed. For each compound, two specific multiple reaction monitoring (MRM) channels were chosen for quantitation purposes. Generally, [M-H]⁻ ions where chosen as precursor ions, with the exception of halogenated NP₂ECs. Even at low cone voltages (5-10 V), in-source CID was obtained for these compounds and their spectra displayed low abundance of the deprotonated molecules. Thus, for these compounds the most intense MRM channels were those monitoring the fragmentation of "the first generation" products (e.g. 297/299 → 79/81 for BrNP₂EC and 253/255 → 167/169 for ClNP₂EC).

Table II. Fragments formed with CID in ESI-MS-MS

Compound	Precursor ion (m/z)	Product ions (m/z)
BrNP	297/299 [M–H]⁻	79/81 (100)ᵃ; 211/213 (10)
BrNP₁EC	355/357 [M–H]⁻	297/299 (100); 79/81 (90)
BrNP₂EC	399/401 [M–H]⁻	297/299 (100); 79/81 (85); 211/213 (7)
ClNP	253/255 [M–H]⁻	167/169 (100); 181/183 (20); 195/197 (10); 35/37 (10)
ClNP₁EC	311/313 [M–H]⁻	253/255 (100); 167/169 (45); 181/183 (15);
ClNP₂EC	355/357 [M–H]⁻	253/255 (100); 167/169 (45); 181/183 (20); 195/197 (15)

ᵃ relative abundance at collision energy of 30 eV using argon as collision gas

The high selectivity of the MS-MS detection of halogenated nonylphenolic compounds is shown in Figures 5 and 6, which display extracted MRM chromatograms for halogenated NPECs and NPs in chlorinated river water and in the DWTP flocculation sludge, respectively. The specificity of MRM mode permitted unequivocal identification and quantification of isobaric target compounds (e.g. BrNP₁EC and ClNP₂EC) and eliminated interference from co-eluting isobaric no-target compounds (e.g. C₁₁LAS) observed using a LC-MS in SIM mode.

Positive identification detection criteria were based on: a) LC retention time of the analyte compared to that of a standard (± 2%), and b) the ratio of abundances of two specific precursor ion → product ion transitions (within 10% of the ratios obtained for the standard). For halogenated compounds a highly diagnostic criteria was the abundance ratio of characteristic isotope ratios (⁷⁹Br :

^{81}Br=1.02; ^{35}Cl : ^{37}Cl=3.03), and for each transition reaction two specific channels, corresponding to the two isotopes of interest, were monitored and the ratio of their abundances calculated. The detection limits of the LC-MS-MS method *(10)* fell down to 1-2 ng/L for the analysis of halogenated NPECs and NPs in water samples (after SPE preconcentration) and to 0.5-1.5 ng/g in sludge samples (target compounds were extracted using pressurized liquid extraction), which is significant improvement in comparison to LODs reported previously for GC-MS *(14)* and LC-MS *(11)* methods.

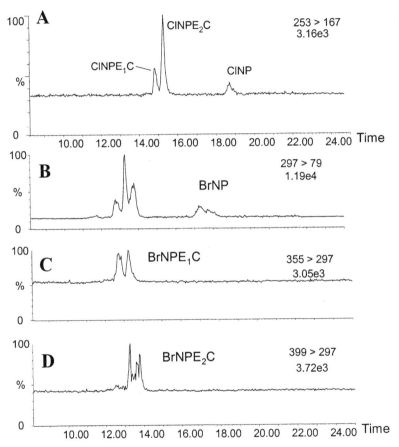

Figure 5. LC-MS-MS analysis of halogenated alkylphenolic compounds in chlorinated river water (the Llobregat river, Spain). MRM mode: A) 253 → 167 for detection of ^{35}Cl nonylphenolic compounds; B) 297 → 79 for ^{79}BrNP; B) 355 → 297 for ^{79}BrNP$_1$EC; D) 399 → 297 for ^{79}BrNP$_2$EC

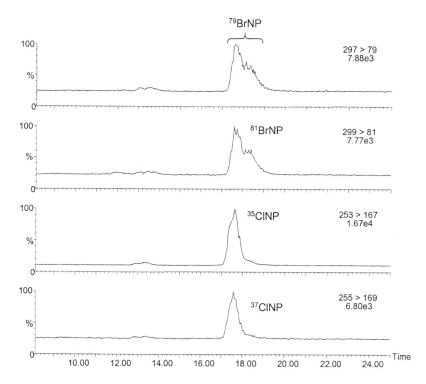

Figure 6. LC-MS-MS analysis of halogenated nonylphenols in the flocculation sludge from the DWTP Sant Joan Despí in Barcelona (Spain). MRM mode: 297 → 79 and 299→ 81 for BrNP; 253 → 167 and 255 → 169 for ClNP

Occurrence in the Environment

Studies to date have largely focused on short chain NPEOs, NPECs and NP, and only few reports have included halogenated metabolites. Table III lists the reported data on the levels of halogenated alkylphenolic compounds in environmental and wastewater samples.

Fujita et al. *(16)* detected halogenated NPEOs and NPECs in the secondary effluents and in final effluents of 25 out of 40 WWTP studied across Japan. They were produced via chlorination at concentrations up to 52.4 µg/l, for the most abundant BrNPECs, accounting for up to about 10% of nonylphenolic compounds.

In another study *(12,15)*, halogenated NPs were found in a sewage-impacted urban estuary sediments (Jamaica Bay, USA). The authors concluded that ClNP and BrNP are likely not of major toxicological concerns due to their

Table III. Reported levels of halogenated alkylphenolic compounds in environmental and wastewater samples

Matrix	Compound	Concentration	Ref.
Estuary sediment (Jamaica bay, NY, USA)	ClNP BrNP	<0.2 – 18.7 ng/g <0.2 – 26.9 ng/g	(12,15)
River water (the Llobregat river, Spain)[a]	BrNP BrNP$_1$EO BrNP$_2$EO	n.d.	(14)
Prechlorinated water (after sand filtration)	BrNP BrNP$_1$EO BrNP$_2$EO	0.41 μg/l 0.78 0.41	
Tap water	BrNP BrNP$_1$EO BrNP$_2$EO	<0.18 – 0.32 μg/l <0.15 – 1.2 <0.3 – 0.64	
River water (the Llobregat river, Spain)	XNP XNP$_n$EO (n=1-15) XNP$_n$EC (n=1-2)	n.d.	(10,11)
Prechlorinated water	XNP XNP$_n$EO (n=1-15) XNP$_n$EC (n=1-2)	0.025-0.21 μg/l 4.0 0.42-0.50	
Tap water	XNP XNP$_n$EO (n=1-15) XNP$_n$EC (n=1-2)	n.d n.d. <0.005-0.01	
Flocculation sludge	XNP XNP$_n$EO (n=1-15) XOP$_n$EO (n=1-15) XNP$_n$EC (n=1-2)	220-250 ng/g 2690 55 20	
Secondary effluent (40 WWTP all over Japan)	BrNP$_n$EO (n=1-2) BrNP$_n$EC (n=1-2) ClNP$_n$EO (n=1-2) ClNP$_n$EC (n=1-2)	n.d. – 0.5 μg/l n.d. – 0.9 (0.1)[b] n.d. – 6.5 (0.6) n.d. – 1.4 (0.2)	(16)
Final effluent (40 WWTP all over Japan)	BrNP$_n$EO (n=1-2) BrNP$_n$EC (n=1-2) ClNP$_n$EO (n=1-2) ClNP$_n$EC (n=1-2)	n.d. – 0.1 μg/l n.d. – 52.4 (1.8)[b] n.d. – 4.0 (0.5) n.d. – 3.3 (0.3)	

[a] influent to DWTPs Abrera and Sant Joan Despí
[b] average value
n.d. – not detected

low relative abundance (less than 1% of the total NPEO metabolite pool), however they may prove useful as tracers of NPEO sources that have undergone chlorination (e.g. pulp and paper mill bleaching processes, chlorine disinfection of wastewaters).

Several works *(5-11,14,17)* have focused on the identifying and monitoring APEOs and their halogenated derivatives in the Llobregat River (Catalonia, NE Spain). This river is one of the most important sources of drinking water for the Barcelona area with a population of 3.2 million people. The drinking water treatment plant (DWTP) Sant Joan Despí is situated downstream of densely industrialized area near the mouth of the river. Since 1960, the effluents from tannery, textile, pulp and paper industries, situated along the basin, have discharged into the river. It was previously reported that the tap water of Barcelona contains high concentration of brominated disinfection by-products *(18)* due to high concentration of bromide ions (average 1.3 mg/L, maximum 3.5 mg/L) *(7)* in raw water that arises from daily dumps of salt mining effluents into the upper course of the river.

Halogenated alkylphenolic compounds were found to be formed during the potabilization process that consisted of prechlorination, flocculation with $Al_2(SO_4)_3$, sand filtration, ozonation, granular activated carbon filtration and a final chlorination. In chlorinated river water, halogenated NPECs, NPEOs and NP were found at concentrations ranging from sub µg/l to low µg/l levels, accounting for 10-25% of total alkylphenolic compounds *(11,14)*. In the treated water leaving the DWTP, halogenated derivatives were identified, but their concentration rarely exceeded 10 ng/L. Normally, chlorinated derivatives occurred in minute amounts compared to brominated compounds.

Environmental Fate and Ecotoxicology

Biotransformation of halogenated alkylphenolic compounds has been studied under aerobic and anaerobic conditions *(19,20)*. It seems likely that halogenation render alkylphenolic molecules even more refractory and more lipophilic, which could potentially cause greater persistence in the environment and possibly accumulation in the food chain.

Furthermore, there is more recent concern about their toxicity, mutagenicity and possible estrogenicity. The occurrence of mutagens has been documented in secondary and tertiary treated effluents as well as in drinking water (e.g. brominated haloforms, haloacetic acids). Reinhard and co-workers *(21)* suspected that occurrence of mutagenicity in wastewater is also correlated with the formation of brominated alkylphenolic by-products, however their preliminary experiments using the Ames test, failed to confirm this hypothesis.

Maki et al. *(13)* reported that neither NPECs nor the brominated derivatives showed mutagenicity by the *umu* assay, but they determined that, both BrNPEOs and BrNPECs, show higher acute toxicity to *Daphnia magna* than their non-brominated precursors NPEOs and NPECs. The 48-hours acute toxicity (EC50, 50% effective concentration) was reported to be 67 μg/l for BrNPEOs and 141 μg/l for BrNPECs, which was remarkably higher than for their non-halogenated precursors (148 and 990 μg/l, respectively).

On the other hand, no data is available on the estrogenic activity of halogenated alkylphenolic compounds. A recent, still unpublished, study *(22)* employing recombinant yeast assay (RYA) and enzyme linked receptor assay (ELRA) showed that halogenated compounds retained a significant affinity for the estrogen receptors, although their activity was slightly weaker than of their non-halogenated precursors. These findings suggest that they may be still able to disturb the hormone imbalance of exposed organisms. This was especially clear for halogenated NPECs, which acted as true anti-estrogens in the RYA.

Conclusions

LC-MS and LC-MS-MS methods have proven to be reliable and robust tools that can be used for routine analysis of potentially estrogenic halogenated NPECs and NPs formed during wastewater and drinking water chlorination. However, further research is needed to develop an appropriate method for the analysis of halogenated APEOs (e.g. using APCI-MS-MS or TOF-MS), since ESI-MS-MS resulted in no fragmentation.

The occurrence of halogenated alkylphenolic compounds in the aquatic environment (surface waters, sediment) has been reported, however they were found to constitute a relatively minor component of the persistent pool of APEO metabolites. However, due to their presence in drinking water it seems necessary to investigate possible reproductive and developmental effects in more detailed studies.

Acknowledgements

This work has been supported by the EU Program Copernicus (EXPRESS-IMMUNOTECH) contract Number ICA2-CT-2001-10007 and by the Spanish Ministerio de Ciencia y Tecnologia (PPQ2001-4954-E)

References

1. Jobling, S.; Sheahan, D.; Osborne, J. A.; Matthiessen, P.; Sumpter, J. P. *Environ. Toxicol. Chem.* **1996**, 15, 194-202.
2. Jobling, S.; Sumpter, J.P. *Aquatic Toxicol* **1993**, 27,361-372.
3. Sonnenschein, C.; Soto, A.M. *J Steroid Biochem Molec Biol* **1998**, *65*, 143-150.
4. Servos, M.R. *Water Qual. Res. J. Canada*, **1999**, 34, 123-177.
5. Ventura, F.; Figueras, A.; Caixach, J.; Espadaler, I.; Romero, J.; Guardiola, J.;Rivera, J. *Wat. Res.* **1988**, 22, 1211-1217.
6. Ventura, F.; Figueras, A.; Caixach, J.; Espadaler, I.; Romero, J.; Guardiola, J.; Rivera, J. *Wat. Res.* **1988**, *10*, 1211-1217.
7. Ventura, F.; Caixach, J.; Figueras, A.; Espadaler, I.; Fraisse, D.;Rivera, J.; *Wat. Res.* **1989**, 23, 1191-1203.
8. Ventura, F.; Fraisse, D.; Caixach, J.; Rivera, J. *Anal. Chem.* **1991**, *63*, 2095-2099.
9. Ventura, F.; Caixach, J.; Romero, J; Espadaler, I.; Rivera, J.; *Wat. Sci. Tech.* **1992**, 25, 257-264.
10. Petrovic, M.; Diaz, A.; Ventura, F.; Barceló, D. *Anal. Chem.* (submitted)
11. Petrovic, M.; Diaz, A.; Ventura, F.; Barceló, D. *Anal. Chem.* **2001**, *73*, 5886.
12. Ferguson, P.L.; Iden, C. R.; Brownawell, B. J. *Anal. Chem.* **2000**, 72, 4322-4330.
13. Maki, H.; Okamura, H.; Aoyama, I.; Fujita, M.; *Environ. Toxicol. Chem.* **1998**, 17, 650-654.
14. Diaz, A.; Ventura, F.; Galceran, M. T. *J. Chromatogr. A* **2002**, 963, 159-167.
15. Ferguson, P.L.; Iden, C. R.; Brownawell, B. J. *Environ. Sci. Technol.* **2001**, *35*, 2428-2435.
16. Fujita, M.; Ike, M.; Mori, K.; Kaku.H.; Sakaguchi, Y.; Asano, M.; Maki, H.; Nishihara T. *Wat. Sci. Technol.* **2000**, *42*, 23-30
17. Paune, F.; Caixach, J.; Espadaler, I.; Om, J.; Rivera, J. *Wat. Sci.* **1998**, *32*, 3313-3324.
18. Cancho, B.; Ventura, F.; Galceran, M. T. *Bull. Environ. Contam. Toxic.* **1999**, 63, 610-617
19. Fujita, Y.; Reinhard, M. *Environ. Sci. Technol.* **1997**, *31*, 1518-1524.
20. Ball, H.A.; Reinhard, M.; McCarty, P.L. *Environ. Sci. Technol.* **1989**, *23*, 951-961.
21. Reinhard, M.; Goodman, N.; Mortelmans, K. E. *Environ. Sci. Technol.* **1982**, 16, 351-362.
22. Garcia-Reyero, N.; Requena, V.; Petrovic, M.; Fisher, B.; Hansen, P.D.; Barceló, D.; Piña, B. (submitted).

Chapter 21

Liquid Chromatography/Electrospray Ionization Tandem Mass Spectrometry and Derivatization for the Identification of Polar Carbonyl Disinfection By-Products

C. Zwiener, T. Glauner, and F. H. Frimmel

Engler-Bunte-Institut, Wasserchemie, Universität Karlsruhe (TH), Karlsruhe, Germany

Derivatization with 2,4-dinitrophenylhydrazine (DNPH) in combination with LC-ESI/MS/MS was successfully applied to investigate polar disinfection by-products with carbonyl groups. The collision-induced dissociation mass spectra (CID-MS) show distinct fragments of the DNPH moiety which facilitates selective measurements of different carbonyl groups by tandem mass spectrometric experiments like precursor ion scans (e.g. m/z 163 for aldehydes; m/z 152 for ketones, m/z 182 for dicarbonyls and hydroxycarbonyls). Several multifunctional carbonyls could be tentatively identified in treated water samples by putting together all information of HPLC separation, derivatization, precursor ion scans, and CID-MS. However, DNPH derivatization is not a suitable reagent for halogenated carbonyl compounds due to side reactions. The chlorine is substituted by DNPH. O-(carboxymethyl) hydroxylamine (CMHA) is a suitable alternative. CMHA derivatization is specifically aimed at the carbonyl group and does not affect the chlorine. The formed oxime derivatives reveal abundant signal intensity in ESI and meaningful CID-MS with common fragments of the CMHA moiety and diagnostic fragments of the carbonyl moiety.

Introduction

Disinfection is required for drinking water production to minimize microbial activity and to exclude health effects caused by microorganisms. Chlorine and chlorine containing chemicals are the most common disinfectants used today. In addition, ozonation plays an increasing role in water treatment. Apart from the intended effect the disinfectants can also react with organic and inorganic water constituents to form disinfection by-products (DBP). In the following only organic DBPs will be considered. In the case of chlorine disinfectants trihalomethanes (THM) are in general the most abundant class of DBPs. THMs were first found by Rook (1, 2) and are the only DBPs regulated in the Drinking Water Directive in Europe to a level of 50 µg/L (3). Therefore only THMs serve as indicators for organic DBP formation in drinking water regulations. The U.S. Environmental Protection Agency (EPA) regulates THMs and five haloacetic acids (HAA). The maximum contaminant levels were set to 80 µg/L for THMs and to 60 µg/L for five HAAs (4).

Further efforts to identify additional DBPs after application of chlorine, and ozone –based chemicals have been made and led to long lists of compounds (5, 6). Moreover, there are still about 50% of the halogenated DBPs unknown as a mass balance of the identified DBPs and the total organic halogen (TOX) reveals. This applies also to the assimilable organic carbon (AOC) with respect to ozone DBPs (7). A toxicological estimation cannot be performed concerning this missing gap. In particular, the occurrence of the strong mutagen MX ((3-chloro-4-dichloromethyl-5-hydroxy-2($5H$)furanone) (8) reveals the importance of identifying even traces of DBPs and of minimizing those in drinking water treatment. Epidemiological studies give indication on effects of DBPs on human reproduction and development, e.g. low birth weight or spontaneous abortion (9). Further studies concerning DBPs and health effects have been undertaken (10).

New research is in particular required for highly polar and high-molecular weight compounds. Since the DBP formation primarily occurs by reaction of the disinfectant with natural organic matter (NOM) the properties of the products and intermediates may be expected to range from those of NOM itself (polar, high molecular weight) to those of THMs (volatile, non-polar, low molecular weight). NOM is a polydisperse and heterogeneous mixture of natural aliphatic and aromatic compounds (11). The reaction with disinfectants yields unknown breakdown products and for example the well-known THMs (Figure 1). The breakdown of NOM during oxidation and chlorination could be shown by size exclusion chromatography (SEC) with organic carbon detection (12). The halogenated fraction can be measured as TOX. The focus of our work is on carbonyl compounds as intermediates and end-products of reactions between disinfectants and NOM. Reaction pathways of THM and MX formation show the importance of carbonyl compounds as intermediates (13, 14). Furthermore,

358

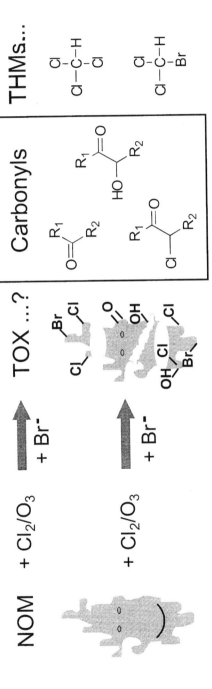

Figure 1. DBP formation by the reaction between natural organic matter (NOM) and chlorine or ozone in the presence of bromide ions

carbonyl compounds occur as end-products (DBPs) of chlorination and ozonation in water treatment.

For carbonyl compounds derivatization with 2,4-dinitrophenylhydrazine (DNPH) and HPLC/UV is a well-known method in aerosol science (*15*). Recently detection with atmospheric pressure chemical ionization (APCI) mass spectrometry was used for DNPH derivatives (*16, 17*). DNPH was also successfully applied to detect DBPs of ozonated aqueous samples with electrospray ionization (ESI) mass spectrometry (*18*). The application and optimization of ESI tandem mass spectrometry for the analysis of DNPH derivatives was shown in literature (*19, 20*). The capabilities and limits of tandem mass spectrometry to identify the derivatives of carbonyl DBPs will be discussed in this work.

Methods

Derivatization

DNPH derivatization was carried out by a modified method according to literature (*19*) by adding 500 µL of reagent solution (6.7 mM DNPH) to 50 mL aqueous sample. The reagent solution was prepared by dissolving 20 mg DNPH (Fluka) in 15 mL HCl/water/acetonitrile 2:5:1 (v/v) according to literature (*21*). The reaction time was at least 12 h at room temperature.

CMHA derivatization was carried out by adding 1 mL of a 60 mM aqueous solution of O-(carboxymethyl) hydroxylamine hemihydrochloride (Fluka) to 50 mL aqueous sample. The reaction time was at least 10 h at room temperature.

For both reagents standard solutions of pure carbonyl compounds were prepared in water-acetonitrile (Aldrich), 50:50 (v/v).

Extraction and preconcentration of aqueous samples

The acidified samples were preconcentrated by SPE on Oasis HLB cartridges (200 mg sorbent, Waters). Elution was done with acetonitrile resulting in preconcentration factors of 25 to 250. More details are described elsewhere (*19*).

Analysis

Analysis was performed using an Agilent 1100 HPLC system coupled with an electrospray ionization source (TurboIon Spray, Applied Biosystems Sciex) to a triple-quadrupole mass spectrometer (API 3000, Applied Biosystems Sciex). Samples (50 µL) were injected onto a reversed-phase column (Xterra MS, 150 mm x 2.1 mm, 5 µm particles; Waters) at a flow rate of 300 µL/min. A gradient elution with acetonitrile/water both with 1 mM ammonium acetate was applied from 30 % to 100 % in 25 min. ESI was used with a nebulizer gas flow of 1.3 L/min, a spray voltage of 5000 V and a dry gas temperature of 450 °C. The mass spectrometer was run in different modes at unit resolution. CID-MS were produced by collision with nitrogen gas molecules at a collision gas thickness of 2.19×10^{17} molecules/cm^2. The values of the collision energies were chosen to decrease the abundance of the precursor ion at least to one third of its original value. Further details are described elsewhere (19).

Samples

Samples were collected from tap waters (20), swimming pools (19) and Lake Hohloh, a bog lake in the Black Forest near Karlsruhe (11). Ozonation was done at a concentration of 1.2 mg/L for 2 h at room temperature. Subsequent chlorination was carried out with a solution of hypochloric acid (1.8 mg/L chlorine) for 72 h at room temperature in the dark. The residual chlorine was quenched after reaction with an excess of ammonium chloride (Merck).

Results and Discussion

Derivatization with DNPH

The well-known derivatization of carbonyl compounds with 2,4-dinitro-phenylhydrazine (DNPH) yields the corresponding hydrazones, which have several advantages for the subsequent analysis compared to the parent carbonyl compounds (Figure 2). The hydrazone derivatives are not charged but easy to ionize by negative electrospray ionization (ESI) yielding the negative deprotonated molecules [M-H]$^-$. Due to a kind of labeling of analytes with carbonyl functional groups with the DNPH moiety they show common fragmentation patterns in tandem mass spectrometry. This can be used to measure selectively DNPH derivatives by tandem mass spectrometric experiments like precursor ion scans. Furthermore the hydrazones are easy to preconcentrate by solid-phase extraction and to separate on reversed-phase HPLC columns.

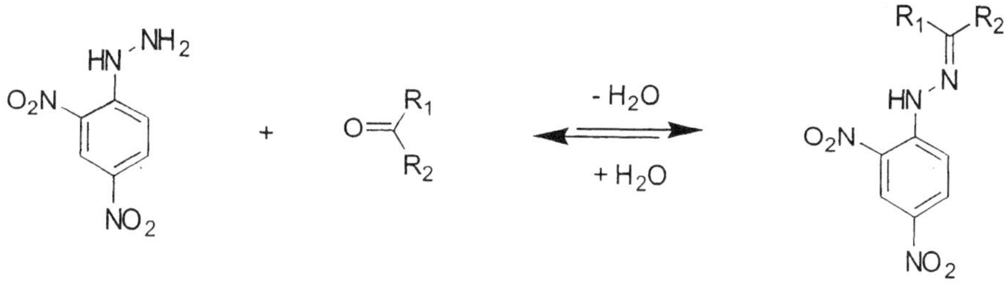

Figure 2. Derivatization reaction between 2,4-dinitrophenylhydrazine (DNPH) and a carbonyl compound (R₁ and R₂ represent hydrocarbon moieties)

Tandem mass spectrometric properties of DNPH derivatives

The collision-induced dissociation mass spectra (CID-MS) of the DNPH derivatives of an aldehyde, a ketone, a hydroxyketone, and an oxoacid are shown in Figure 3. In general most of the abundant mass fragments below m/z 200 can be assigned to mass fragments of the DNPH moiety only. In this mass range common fragments for different compound classes can be found, like m/z 152 for carbonyl compounds in general, m/z 163 for aldehydes, m/z 182 for α-hydroxyaldehydes and dicarbonyls (Figure 4). In particular dicarbonyl compounds show much less fragments than other carbonyl compounds. Aldehydes and ketones show also abundant fragments of the neutral loss of NO (e.g. m/z 263 in Figure 3a) and sometimes of NO_2. Oxoacids show the neutral loss of CO_2 (e.g. m/z 281 in Figure 3d). CID mass fragments of further aldehydes and ketones are compiled in Table I. The data reveal that aldehydes in general show abundant ions at m/z 163 and at m/z 152, whereas ketones show only ions at m/z 152. The ions at m/z 205 and m/z 220 represent aldehydes and ketones with 4 or more carbon atoms. The CID mass fragments of the more polar carbonyls are shown in Table II. Hydroxyketones show abundant fragments at m/z 152 and at m/z 182, the dicarbonyls with both carbonyl groups derivatized show almost only one fragment at m/z 182, the ketoacids ions at m/z 182 and a neutral loss of CO_2. This general scheme gives the possibility to use precursor ion scans for ions at m/z 152, m/z 163 or m/z 182 to selectively measure the DNPH derivatives of different classes of carbonyl compounds in complex samples.

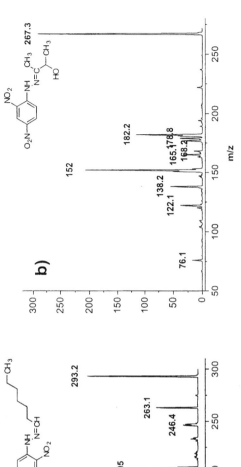

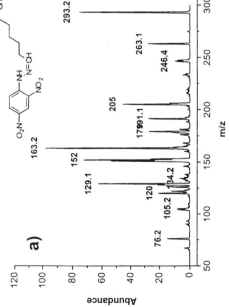

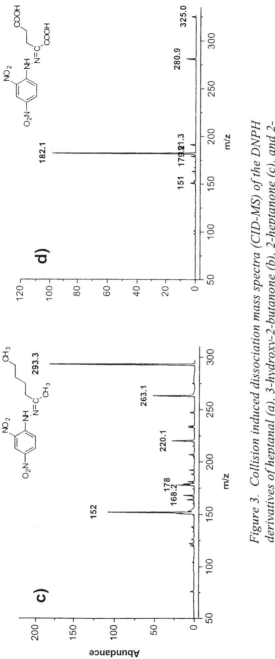

Figure 3. Collision induced dissociation mass spectra (CID-MS) of the DNPH derivatives of heptanal (a), 3-hydroxy-2-butanone (b), 2-heptanone (c), and 2-oxoglutaric acid (d)

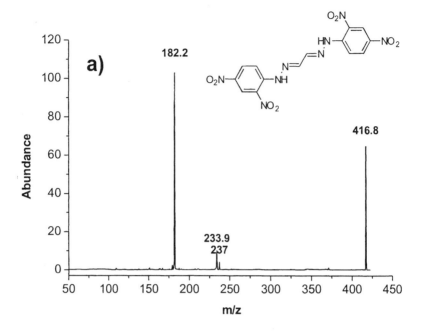

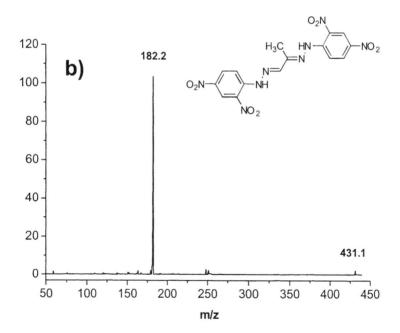

Figure 4. CID-MS of the DNPH derivatives of glyoxal (a) and methyl glyoxal (b)

Table I. Selected collision-induced dissociation mass spectrometric (CID-MS) fragments of DNPH derivatives of aldehydes and ketones.

Compound	Molecular mass Amu	[M-H]⁻ ion of DNPH derivative m/z	Selected CID-MS fragments of the derivatives m/z (abundance)
Butanal	72.1	251.1	163 (100), 152 (90), 151 (60), 191 (40), 221 (35), 205 (14)
Pentanal	86.2	265.1	163 (100), 152 (94), 151 (51), 191 (40), 205 (52), 235 (38)
Hexanal	100.1	279.1	163 (100), 152 (95), 151 (50), 205 (49), 191 (33), 249 (24)
Heptanal	114.2	293.2	163 (100), 152 (70), 151 (53), 205 (45), 191 (26), 263 (26)
2-Butanone	72.1	251.1	178 (31), 221 (44), 151 (20), 181 (20)
2-Pentanone	86.1	265.1	152 (80), 205 (56), 235 (42), 178 (23)
3-Pentanone	86.1	265.1	152 (35), 205 (100), 235 (16), 178 (15)
2-Hexanone	100.2	279.1	152 (75), 249 (42), 220 (25), 178 (20)
2-Heptanone	114.2	293.1	152 (60), 263 (28), 220 (15), 178 (13)
4-Heptanone	114.2	293.1	152 (95), 263 (32), 206 (30), 179 (14)

Table II. Selected collision-induced dissociation mass spectrometric (CID-MS) fragments of DNPH derivatives of hydroxycarbonyls, dicarbonyls, and ketoacids.

Compound	Molecular mass amu	[M-H]- ion of DNPH derivative m/z	Selected CID-MS fragments of the derivatives m/z (abundance)
Hydroxyacetone	74.1	253.0	152 (100), 179 (14), 182 (13)
1,3-Dihydroxyacetone	90.2	269.0	182 (50)
3-Hydroxy-2-butanone	88.1	267.3	152 (100), 182 (60)
5-Hydroxy-2-pentanone	102.2	281.0	152 (100), 252 (25), 179 (11), 220 (10)
Glyoxal	58.0	417.1	182 (100), 234 (10)
Methylglyoxal	72.1	431.0	182 (100), 248 (5)
Dimethylglyoxal	86.1	445.2	182 (100), 265 (15)
Glyoxylic acid	74.0	253.0	182 (100), 209 (42)
2-Oxopropionic acid	88.0	267.0	182 (100), 223 (40)
Ketomalonic acid	118.0	297.0	182 (100), 152 (8)
2-Oxoglutaric acid	146.1	325.0	182 (100), 281 (8), 151 (4), 163 (4)

Table III. Retention times of the DNPH derivatives of polar carbonyl compounds in the HPLC chromatogram of a standard solution.

Compound	Retention time(min)
Glyoxylic acid	2.1
2-Oxoglutaric acid	2.1
2-Oxopropionic acid	2.2
Ketomalonic acid	2.3
Hydroxyacetone	7.8
3-Hydroxy-2-butanone	9.9
5-Hydroxy-2-pentanone	10.1
Acetaldehyde	11.9
Glyoxal	17.5
Pentanal	18.1
Methylglyoxal	18.9
Dimethylglyoxal	20.4
4-Heptanone	20.9
3-Octanone	22.4

Selective measurement and identification

The chromatograms of precursor ion scans of m/z 182 are shown in Figure 5 for aqueous samples of a bog lake after chlorination and after ozonation plus chlorination. The mass to charge ratio of several abundant peaks could be determined and the peaks at 17.5 min (m/z 417) and at 18.9 min (m/z 431) could be identified as glyoxal and methylglyoxal by comparison of the retention times (Table III) and the CID-MS with a standard. The following shows as an example how DBPs may be identified based on tandem mass spectrometry of DNPH derivatives. Comparison of the retention times for the peak at 7.9 min would lead to the suggestion of hydroxyacetone, but the mass to charge ratio does not fit. Further information was obtained from the CID-MS (product ion scan of m/z 461) of the peak at 7.9 min (Figure 6). The CID-MS shows as expected an abundant ion at m/z 182 indicating a dicarbonyl compound or a hydroxycarbonyl and a neutral loss of 44 indicating a carboxyl group. Putting together a carboxyl group and two carbonyl groups 2,3-dioxopropionic acid and

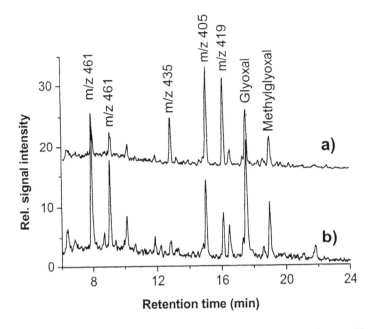

Figure 5. Chromatograms of precursor ion scans at m/z 182 of treated brown water from a bog lake. – a) Chlorinated; b) Ozonated and chlorinated

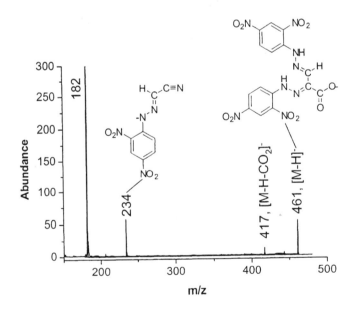

Figure 6. Production scan of m/z 461 and tentatively identified DNPH derivative of dioxopropionic acid

the DNPH derivative given in Figure 6 would result. The fragment occurring at m/z 234 could be assigned to a structure, which may confirm the suggested compound.

Information collection and doing the jigsaw was in general done due to the following scheme for compound identification based on tandem mass spectrometry of DNPH derivatives of carbonyl compounds. Information on peak identity can be obtained from:

- Retention time of HPLC separation (elution order of different compound classes)
- Mass to charge ratio, mainly representing [M-H]⁻ ions (e.g. nitrogen rule, isotope cluster, mass range)
- Selection of the compound class by derivatization and precursor ion scan (e.g. at m/z 182 for dicarbonyls)
- CID-MS may show fragment ions which give structural hints (e.g. neutral loss of CO_2 for acids

After putting together all pieces of information normally one or more structures can be suggested. Final identification and verification of the suggested structures is in general based on the retention time and CID-MS of an authentic standard compound. In case the compound is not commercially available it has to be synthesized. Of course this slows down the identification process dramatically. Further possibilities of compound identification may result from methods like LC/NMR or accurate mass measurements (HRMS). However, these methods were not applied in this work.

Prospects and limits of DNPH derivatization

The limitations of compound identification based on DNPH derivatization are quite obvious due to CID-MS with only a few fragments giving only limited structural hints about the analyte moiety of the derivative rather than the DNPH moiety. In particular there are no satisfying suggestions for the peaks at m/z 405, m/z 419, and m/z 435 in the chromatogram of Figure 5. Nevertheless the application of the method revealed to be very useful to get more information on the occurrence of polar DBPs in water treated with ozone and chlorine. In Table IV examples for oxoacids, hydroxydicarbonyls, and dicarbonyls in treated water samples are given.

A major drawback of DNPH derivatization however are several side reactions of DNPH. Our investigations showed that above all chlorine atoms in halogenated carbonyls can be substituted by DNPH as well. So derivatization of

3-chloro-2-butanone resulted in the formation of two derivatives due to the reaction of DNPH with the site of the chlorine and of the carbonyl carbon (Figure 7). The chlorine atom was not kept in any derivative since the chlorine substitution seems to be the preferential pathway. Several other derivatization reagents with hydrazine as functional group seem to show the same effect. Therefore the search for an alternative derivatization reagent with another functional group was necessary.

Table IV. Tentatively assigned DNPH derivatives of polar carbonyl compounds found in treated water by precursor ion scans.

[M-H]⁻ ion of DNPH derivative	Carbon atoms	Retention time (min)	Precursor at m/z 182	Precursor at m/z 163	Precursor at m/z 152
Oxoacids					
253.4	2	1.4	X		X
267.3	3	1.4	X		X
281.4	4	2.4	X		
295.4	5	3.1			X
309.3	6	4.0			X
Hydroxy dicarbonyls					
461.5	4	7.9 / 9.1	X		
489.6	6	10.1 / 10.6	X		
503.4	7	11.1	X		
Dicarbonyls					
445.5	4	15.6 / 17.7	X		
459.5	5	18.5 / 21.1	X		

Derivatization with carboxymethyl-hydroxylamine

Our search brought us to another well-known functional group for the derivatization of carbonyls. The hydroxylamine group is known from pentafluorobenzyl hydroxylamine (PFBHA) being used for analysis of carbonyls by GC/ECD. To meet the requirements of ESI O-(carboxymethyl) hydroxylamine (CMHA) turned out to be the best choice. The acidic functional group is easy to ionize in ESI and therefore guarantees highly abundant signals for the CMHA derivatives. The derivatization of a carbonyl compound with CMHA yields the according oxime (Figure 8). SPE and reversed-phase HPLC

371

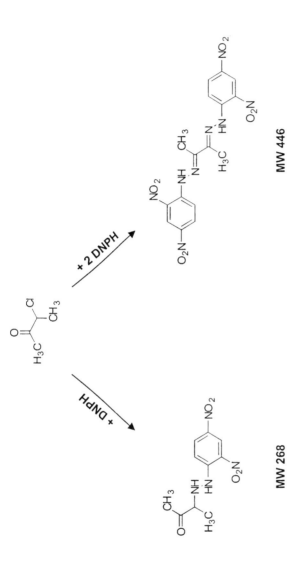

Figure 7. Formation of two derivatives for the reaction of 3-chloro-2-butanone with DNPH

Figure 8. Derivatization reaction between O-carboxymethyl hydroxylamine (CMHA) and a carbonyl compound (R₁ and R₂ represent hydrocarbon moieties)

separation have to be done in acidic solution. Most important however is the property of the derivatization reaction with CMHA to leave the C-Cl bond unchanged. Furthermore it should have comparable capabilities like DNPH to selectively form products, and by this to lead to meaningful CID-MS results.

The CID-MS of the CMHA derivatives of 3-chloro-2-butanone and 1,1-dichloroacetone are shown in Figures 9 and 10. The respective fragmentation is shown for each molecular ion of the isotope cluster of chlorine. Similar to the DNPH derivatives we observe for the CMHA derivative of 3-chloro-2-butanone fragments of the CMHA moiety, which can be used for selective measurement. Furthermore there are fragments of the carbonyl moiety, which can give structural information (Figure 9). Common fragments are found at m/z 73 (glyoxylic acid anion) and m/z 75 (hydroxyacetic acid anion). Diagnostic ions are at m/z 68 due to cleavage of the N-O bond of the oxime and at m/z 142 due to the neutral loss of HCl. Also the CID-MS of 1,1-dichloroacetone shows the common fragments at m/z 73 and m/z 75 (Figure 10). Diagnostic ions are at m/z 85 (dichloromethyl anion) and at m/z 126 due to cleavage of the N-O bond of the oxime. Comparison of the fragmentation patterns of different isotopes of the molecular ion gives additional information about the whereabouts of the chlorine in the fragment ions. The occurrence of the common fragments at m/z 73 and m/z 75 and of the diagnostic ions due to cleavage of the N-O bond of the oxime could be confirmed for ketones as well as for aldehydes. Therefore precursor ion scans at m/z 73 and at m/z 75 could be used for selective measurement of the CMHA derivatives similar to the procedure shown for the DNPH derivatives. In general the CID-MS of CMHA derivatives give more structural information compared to DNPH derivatives. In conclusion, CMHA has revealed a very suitable alternative derivatization reagent and the only suitable reagent for halogenated carbonyl compounds. Further work will focus on method optimization and application of CMHA derivatization for further identification of polar carbonyl disinfection by-products.

Conclusions

In conclusion, derivatization with DNPH in combination with HPLC-tandem mass spectrometry revealed as a useful method for the identification and measurement of non-halogenated polar carbonyl compounds in treated water. The derivatization reaction renders derivatives which enable better separation on HPLC and better ionization by ESI compared to the parent compound. Common mass fragments facilitate the selective measurement of different compound classes like aldehydes, ketones, hydroxyketones, oxoacids, and dicarbonyls by tandem mass spectrometry. Further functional groups especially the polar ones

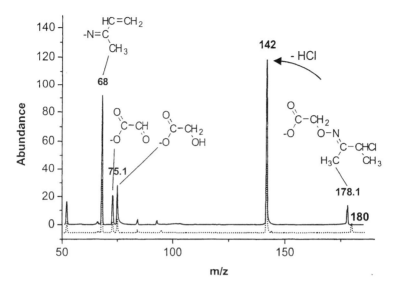

Figure 9. CID-MS of the CMHA derivative of 3-chloro-2-butanone. – Fragmentation patterns of the molecular ions at m/z 178 (solid line), and at m/z 180 (dotted line)

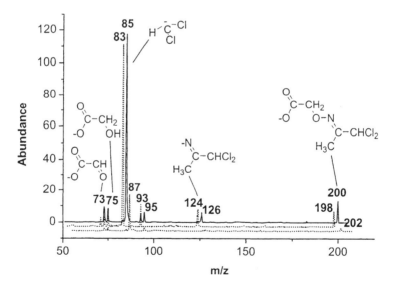

Figure 10. CID-MS of the CMHA derivative of 1,1-dichloroacetone. – Fragmentation patterns of the molecular ions at m/z 200 (solid line), at m/z 198, and at m/z 202 (dotted lines)

do not need to be derivatized as for gas chromatography since polar analytes are amenable to liquid chromatography. The method is therefore suitable for highly polar analytes with multiple functional groups. The occurrence of oxoacids, hydroxycarbonyls, hydroxydicarbonyls, and dicarbonyls in treated water samples could be shown applying this method.

DNPH derivatization however is not suitable for halogenated carbonyls due to side reactions in which the chlorine is substituted. For halogenated carbonyls as well as for other carbonyls the derivatization with CMHA proved to be very useful. The chlorine atoms remain untouched during derivatization with CMHA and the mass spectrometric fragmentation patterns of the CMHA derivatives show common fragments for selective measurement and diagnostic ions for structure identification.

The derivatization methods and tandem mass spectrometric measurements can be applied to get more knowledge on DBP formation in water treatment. They therefore play a key role for the understanding of the function of the different water treatment steps. With this knowledge specific process technologies concerning minimization of DBP formation can be developed.

Acknowledgments

We are grateful to the Bundesministerium für Bildung und Forschung (BMBF; Federal Ministry of Education and Research) and to the Deutsche Forschungsgemeinschaft (DFG; German Research Foundation) for financial support. Finally, we would like to thank D. Rousova for her contribution to the CMHA derivatization, S. Stahl, S. Seeger, and B. Dietrich for their technical assistance.

References

1. Rook, J.J. *Water Treatmt. Examin.* **1974,** *23,* 234-243.
2. Rook, J. J. *Environ. Sci. Technol.* **1977,** *11,* 478-482.
3. Council Directive 98/83/EC on the Quality of Water Intended for Human Consumption; *Off. J. Europ. Commun.* **1998,** *No L330.*
4. National Primary Drinking Water Regulations: Disinfectants and Disinfection by-Products; *Fed. Regist.* **1998,** *63,* 69389.
5. Richardson, S.D. In *Encyclopedia of Environmental Ananylsis and Remediation,* Meyers, R.A. Ed.; Vol. 3, Wiley, New York, 1998, pp 1398-1412.

6. Richardson, S. D.; Thruston, A. D.; Caughran, T.V.; Chen, P. H.; Collette, T. W.; Floyd, T. L.; Schenck, K. M.; Lykins, B. W.; Sun, G.-R.; Majetich, G. *Envrion. Sci. Technol.* **1999**, *33*, 3368-3377.

7. Weinberg, H. *Anal. Chem.* **1999**, *4*, 801A-808A.

8. Kronberg, L.; Vartiainen, T. *Mutation Res.* **1988**, *206*, 177-182.

9. Waller, K.; Swan, S. K.; DeLorenze, D.; Hopkins, B. Epidemiology **1998**, *9*, 134-140

10. Richardson, S. D.; Simmons, J. E.; Rice, G. *Environ. Sci. Technol.* **2002**, *36*, 188A-205A.

11. *Refractory organic substances in the environment;* Frimmel, F. H.; Abbt-Braun, G.; Heumann, K. G.; Hock, B.; Lüdemann, H.-D.; Spiteller, M., Eds.; Wiley-VCH, Weinheim, 2002.

12. Frimmel, F. H.; Hesse, S.; Kleiser, G. In *Natural organic matter and disinfection by-products: characterization and control in drinking water;* Barrett, S. E.; Krasner, S. W.; Amy, G. L., Eds.; American Chemical Society, Washington, DC, 2000; pp 84-95.

13. Boyce, S.D.; Hornig, J. F. *Environ. Sci. Technol.* **1983**, *17*, 202-211.

14. Långvik, V.-A.; Hormi, O.; Tikkanen, L.; Holmborn, B. *Chemosphere* **1991**, *22*, 547-555.

15. Fung, K.; Grosjean, D. *Anal. Chem.* **1981**, *53*, 168-171.

16. Kölliker, S.; Oehme, M.; Dye, C. *Anal. Chem.* **1998**, *70*, 1979-1985.

17. Grosjean, E.; Green, P. G.; Grosjean, D. *Anal. Chem.* **1999**, *71*, 1851-1861.

18. Richardson, S. D.; Caughran, T. V. ; Poiger, T. ; Guo, Y.; Crumley, F. G. *Ozone Sci. Engng.* **2000**, *22*, 653-675.

19. Zwiener, C. ; Glauner, T. ; Frimmel, F. H. *Anal. Bioanal. Chem.,* **2002**, *372*, 615-621.

20. Zwiener, C. ; Glauner, T. ; Frimmel, F. H. *Proc. IWA World Water Congress,* Melbourne, Australia, April 8-11, 2002.

21. Kieber, R. J. ; Mopper, K. *Environ. Sci. Technol.* **1990**, *24*, 1477-1481.

Chapter 22

Identification of Homologue Unknowns in Wastewater by Ion Trap MS^n: The Diagnostic-Ion Approach

Imma Ferrer[1,3], Edward T. Furlong[1], and E. M. Thurman[2]

[1]U.S. Geological Survey, P.O. Box 25046, MS 407, Denver, CO 80225
[2]U.S. Geological Survey, 4821 Quail Crest Place, Lawrence, KS 66049
[3]Current address: immaferrer@menta.net

A novel approach to identify unknown surfactants in wastewater samples was developed using LC/MS/MS and the concept of diagnostic and double diagnostic ions. The diagnostic ion is defined as a fragment ion found in all members of a family of surfactants that is characteristic of the family. The double diagnostic ions are fragment ions found only using MS^3 and are unique to ion trap mass spectrometry. In this work, different homologues of three families of surfactants (polyethylene glycol, nonylphenol ethoxylates and alcohol polyethoxylates) were identified in environmental wastewater by using this approach. Several diagnostic ions were obtained for polyethylene glycol homologues using MS^2. They included m/z 89, 133, and 177, which are ions formed by the fragmentation and cyclization of the ethylene-glycol chain. Double diagnostic ions for nonylphenol ethoxylates and alcohol polyethoxylates were obtained by MS^3. This approach is uniquely suited to ion trap MS and illustrates the power of LC/MS^n for identifying unknowns in complex environmental wastewater samples.

Introduction

A high percentage (around 90%) of the organic matter of an environmental sample is not known. To date only a fraction of the individual components in wastewater effluents have been identified and quantified. Usually the identification of unknowns has been the task of gas chromatography/mass spectrometry (GC/MS). A typical approach to GC/MS identification is first to compare mass spectra in a sample extract with a reference mass spectrum library, to analyze the suspected authentic standard and, finally, to compare the retention time plus the mass spectrum. This protocol has been used in many environmental applications over the years (1). Even higher confidence in identification can be obtained by GC/MS with accurate mass measurement high resolution mass spectrometry to calculate the exact mass and propose an molecular formula for ions in the mass spectrum. In contrast to this well-known approach, for LC/MS analyses there is still no protocol for unknown identification since there are no accepted standard spectral libraries and there are no recognized approaches for how to identify unknowns in a sample. Typically, the analyst matches the ions and retention time of specific standards to the unknown compound, but this approach only works when the analyst has *a priori* knowledge of what to look for in the sample or if there is a substantial peak in the chromatogram, and the ion in that case corresponds to the molecular weight of the suspected compound.

One of the shortcomings in HPLC/MS using a single quadrupole MS is the lack of sensitivity when running in full-scan mode (typically the analyte must be 10 µg/L or higher in the sample for a good spectrum). This lack of sensitivity results from scanning a broad mass range. This disadvantage usually has been overcome by the use of selected ion monitoring (SIM) conditions, in which the mass spectrometer dwells on a few ions of interest. But again, these ions must be known before analyzing the sample. This permits selectivity and sensitivity for specific compounds, but does not allow identification of new compounds. In the past few years, the use of ion trap technology has improved the detection limits of LC/MS by two orders of magnitude (2). Several applications using ion trap systems with MS/MS features have been reported for the determination of polar organic compounds in environmental matrices, including pesticides, surfactants, pharmaceutical compounds, and other contaminants of emerging environmental concern (3-6). The advantage of these techniques relies on the low limits of detection achieved under selected conditions (4) and the unequivocal identification of many compounds by MS/MS fragmentation (6).

However, the introduction of LC/ion trap/MS/MS is fairly recent and its implementation in routine and in research laboratories is only now beginning.

In general, no unknown identifications have been made with LC/MS techniques due in part to the complexity of the environmental samples and the limited sensitivity of this technique. In this work we took advantage of one of the most important features of ion traps which is the high sensitivity obtained still when working under scan conditions. This allows detecting the full-scan spectrum of any peak in a chromatogram, and represents a high value for the identification of unknowns in the samples because we get information of all the ions (molecular + fragments) generated by a specific compound. The sensitivity offered by this type of instrument is typically one to two orders of magnitude higher than a single quadrupole system operated in full-scan mode. This advantage is important because the initial analysis is always carried out under full-scan conditions. A second advantage of our approach is the higher fragmentation possible using an ion trap and, a third advantage is the capability of performing multiple MS/MS experiments with a single isolated ion to determine structural information. We have combined these unique advantages of the ion trap and applied them to identify unknowns in complex mixtures, such as wastewater.

This paper provides a practical methodology for identifying unknown compounds in LC/MS chromatograms in the absence of spectral libraries and substitutes in its place the concept of the diagnostic ion approach. Examples are shown for identification of three families of surfactant compounds in wastewater samples.

Experimental Section

Chemicals

Polyethylene glycol (PEG), nonylphenol polyethoxylates (NPEO), and alcohol polyethoxylates (AEO) were purchased from Union Carbide (Danbury, CT). High-performance liquid chromatography (HPLC) grade acetonitrile, methanol, and water were purchased from Burdick and Jackson (Muskegon, MI). Ammonium formate and formic acid were obtained from Aldrich (Milwaukee, WI).

Sampling

Water samples were collected from different wastewater treatment plants and river sites near the plants. All samples were taken downstream from wastewater-effluent sites. The samples were collected with a DH81 Teflon sampler by using equal-width-increment methods. Collected samples were composited in a glass churn prior to filtering through a 0.45-μm filter (Whatman GF/F, Whatman Inc., Clifton, NJ). The sampler and churn were cleaned prior to sampling with three rinses of tap water, three rinses of water prepared by reverse osmosis, three rinses of deionized water, one rinse of analytical grade methanol, one rinse of organic-free water, and three rinses of the field sample. Prior to filtering the sample, the filter was pre-wetted by passing 1 L of organic-free water through the filter. Filtered samples were stored at 4°C until shipment to the laboratory.

Sample preparation

Water samples were preconcentrated using an off-line approach onto Oasis cartridges. First, the cartridges were conditioned sequentially with 6 mL methanol and 6 mL HPLC-grade water. Then, a 1-L of water sample was preconcentrated through the cartridge at a flow rate of 15 mL/min. Finally, the compounds trapped on the sorbent were eluted with 6 mL of methanol and 6 mL of methanol 0.09% TFA. The extract was reduced to 100 μL under nitrogen and brought up to a volume of 1 mL with ammonium formate buffer solution.

LC/MS^n

Liquid chromatography/electrospray ionization/ion trap tandem mass spectrometry (LC/ESI/MS/MS), in positive ion mode of operation, was used to separate and identify the surfactant homologues. The analytes were separated by using a series HP 1100 HPLC (Hewlett-Packard, Palo Alto, CA) equipped with a reversed-phase C18 analytical column (Phenomenex RP18, Torrance, CA) of 250 by 3 mm and 5-μm particle diameter. Column temperature was maintained at 25 °C. The mobile phase used for eluting the analytes from the SPE and HPLC columns consisted of acetonitrile and 10 mM ammonium formate buffer, at a flow rate of 0.6 mL/min. A gradient elution was performed as follows: from 15% A (acetonitrile) and 85% B (10 mM ammonium formate) to 100% A and 0% B in 40 minutes. This HPLC system was connected to an ion trap mass spectrometer, an Esquire LC/MS/MS (Bruker Daltonics, Bellerica,

MA) system equipped with an electrospray ionization (ESI) interface. The electrospray nebulizer was orthogonal to the inlet. Selected operating conditions of the MS system were optimized in full-scan mode (m/z scan range: 50–400) by flow injection analysis of each selected compound at 10-μg/mL concentration. The maximum ion accumulation time was set at 200 ms.

Results and Discussion

The diagnostic ion approach by LC/MS/MS

The first step in the identification of unknown compounds was to analyze the sample with a slow liquid chromatographic gradient capable of reasonable separation of all the major peaks that contained ionizable compounds. Second, we set the fragmentor voltage, also referred to as capillary exit voltage, at an intermediate value (70 V) so the production of a major molecular ion and some fragment ions in the electrospray source is favored. Third, the spectrum of each chromatographic peak in a series of samples was scrutinized.

Figure 1 shows the full-scan chromatogram of a wastewater sample analyzed by ion trap LC/ESI/MS in positive ion mode. Four observations were made. First, we observed that several sequentially eluted chromatographic peaks close in retention time had apparent protonated molecules $[M+H]^+$ and that masses increased by 44 units from peak to peak, suggesting either a CO_2 group or an ethoxy unit (m/z = 239, 283, 327, 371...). The second observation was the presence of the same three ions at m/z 89, 133, and 177 in each one of the chromatographic peaks, suggesting a homologue relation between them. The third observation was that the suspected protonated molecules $[M+H]^+$ were odd numbered, thus suggesting either the absence or an even number (2, 4, 6...) of nitrogens in the chemical structure, because even electron atoms with an even number of nitrogens have odd protonated molecules and an odd number of nitrogen atoms have even protonated molecules (7). The nitrogen rule is useful because protonated molecules and fragment ions in LC/MS produce even electron ions. Related to this last observation, we could also assess the presence of peaks at 17 amu higher than the suspected protonated molecule in each one of the spectra (m/z = 256, 300, 344, 388...). Because these masses corresponded to even masses, we could conclude that an ammonium adduct (adding one nitrogen to the ion) was formed in the electrospray source. This result was not unexpected because ammonium formate was used in the mobile phase. The next step was to isolate the molecular ion and determine if the same diagnostic ions appear. Finally, fragmentation (increasing the fragmentor voltage) was used to verify that the diagnostic ions observed were likely derived from the appropriate molecular ion via collision induced dissociation (CID) fragmentation. Once this result was verified, a literature search for similar reported masses narrowed the candidate compounds to ethoxylated surfactant compounds.

Several references (8, 9) report the presence of a m/z 133 fragment ion for polyethylene glycol (PEG) surfactants. The other two diagnostic ions (89 and 177), however, were not reported by these authors. Only one paper (10) reported the same fragment ions at m/z 89, 133, and 177 and attributed them to cyclic structures of PEG surfactants. On the basis of this information, a standard of PEG was analyzed by ion trap and the diagnostic ions were verified as arising from the PEG homologues. Figure 2 shows the spectrum of a standard mixture of PEG's clearly showing the presence of the three diagnostic ions under CID conditions. We also observed the homologous molecular ions and their respective ammonium adducts, as in the spectra from the environmental samples (Figure 1). Once the molecular ions were identified for these compounds, we proceeded to the last step for final identification using the MS/MS capability of the ion trap. An isolation and fragmentation MS/MS experiment was carried out on both the standard and the sample to confirm this compound. The PEG standard matched the elution time and diagnostic ions in wastewater samples. Figure 3 shows the MS/MS spectrum of the molecular peak at m/z 283 corresponding to PEG_6. The three diagnostic ions were unique and characteristic of this compound, as shown in the spectrum.

In the next section we further define the concept of diagnostic ions and we apply the same approach to other families of homologue surfactants that have been identified in other studies of environmental water samples (11, 12). In this work we focused on the identification of surfactants in wastewater samples. The purpose of this study was to determine if other families of surfactants produced diagnostic ions to apply the approach developed in the previous section for the identification of such compounds.

Diagnostic ions and fingerprint chromatogram

A diagnostic ion thus is defined as a fragment ion found in all members of a family of compounds, which is characteristic only of that family. For example, if the main fragmentation ions formed by the same family of compounds are identical, then they may be considered "diagnostic ions" of the family. Previous

382

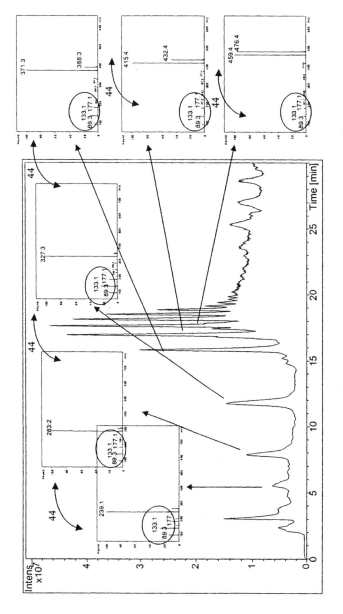

Figure 1. Full-scan chromatogram of a wastewater sample analyzed by ion trap LC/ESI/MS in positive ion mode. Spectra for each peak are also shown.

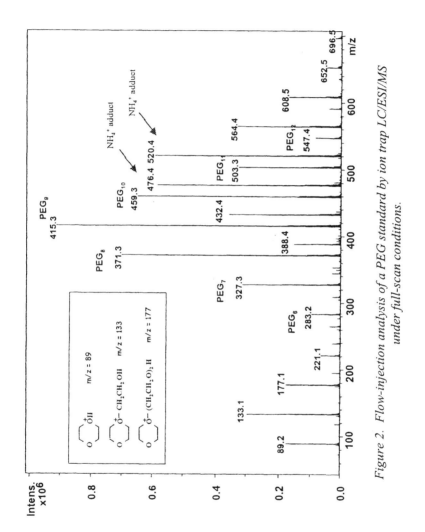

Figure 2. Flow-injection analysis of a PEG standard by ion trap LC/ESI/MS under full-scan conditions.

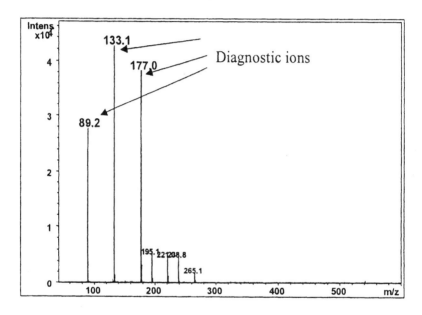

Figure 3. MS/MS spectrum of molecular peak of PEG6 at m/z = 283.

authors (1) have indicated that characteristic product or neutral loss ions could be used in class-specific analyses. Our approach differs in how diagnostic ions are first observed and identification developed from characterization of the diagnostic ion. Molecular ions may be combined with the diagnostic ions to determine unknowns within the family. Furthermore, the diagnostic ions may be displayed as a "fingerprint chromatogram" using the extracted ion chromatogram, which is a powerful tool for identifying unknowns in water samples. A more specific chromatogram showing the peaks of interest results from using diagnostic ions to produce an extracted ion profile or chromatogram. This "fingerprint chromatogram" approach is useful in that it will be different for each family of surfactants because the ions extracted and the retention times for each one of the homologues will be different. This idea can be used for a rapid identification of homologues between the same compound family. In this way, once the specific ion chromatograms are extracted, molecular ions or adducts and fragment ions can be identified. Diagnostic ions may be different for every family of compounds, and these could include surfactants, pharmaceuticals, and pesticides (6).

Application of the diagnostic ion approach by LC/MS^n

A discovery of the double diagnostic ion was made when the next family of surfactants, the nonylphenol polyethoxylates (NPEO), was studied with the same approach. Figure 4 shows the flow-injection analysis of an NPEO standard by ion trap LC/ESI/MS under full-scan conditions. The first set of diagnostic ions previously observed for PEG homologues at m/z 89, 133, and 177 was also observed for the NPEO's, but, in this particular case, a second set of three new diagnostic ions at m/z 121, 165, and 209 was also observed. MS/MS experiments were carried out for each molecular ion to determine that the particular fragmentation was characteristic of each NPEO homologue. Surprisingly, when using MS/MS, only one peak was obtained in each case corresponding to the fragmentation of the alkyl chain (C_9H_{19}) in the aromatic ring (see Figure 5a). Thus, a neutral loss of 126 was obtained for each homologue. Interestingly, when MS^3 experiments were carried out for the main fragment ion, the diagnostic ions at 121, 165, and 209, previously observed under CID conditions, were obtained (see Figure 5b). Moreover, not only was this second set of diagnostic ions observed but the first set of diagnostic ions identified for PEG homologues was also obtained. This result can easily be explained by considering the similar chemical structure of the ethoxylated chain on the phenol side of the molecule.

Based on these results, we introduce here the idea of the "double diagnostic ions" by using the MS^n (in this case MS^3) capability of the ion trap. Because of the MS^n potential of the ion trap, we can get double confirmation on diagnostic ions, which is helpful when analyzing complex matrices, since only one specific

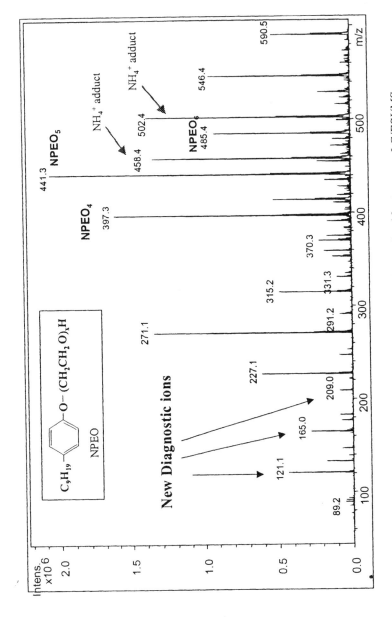

Figure 4. Flow-injection analysis of a NPEO standard by ion trap LC/ESI/MS under full-scan conditions.

ion is isolated and only unique fragment ions for that specific ion are obtained. This result is a special feature of the ion trap, which makes it unique compared to all other MS/MS instruments. Figure 6 shows a 3-dimensional plot summarizing the MS^3 experiment carried out for the NPEOs (Figures 4 and 5) and shows the complete "ion package information" that is obtained by using the MS^3 feature of the ion trap.

This approach then was applied to another family of surfactants: the alcohol polyethoxylate homologues. Figure 7 shows the flow-injection analysis of an AEO standard by ion trap LC/ESI/MS under full-scan conditions. It is important to note the presence of only ammonium adducts instead of protonated molecules. In this case, under CID conditions, we obtained the same first set of diagnostic ions as those observed in PEG homologues at m/z 89, 133, and 177. When MS/MS experiments were performed (see Figure 8a), the ammonium adduct ion fragmented yielding the protonated molecule. But the main base peak fragment corresponded to the loss of the alkyl chain, thus yielding ions with the same molecular weight as PEG (m/z = 371) because, after fragmentation, the fragment presents the same chemical structure as a PEG. Not surprisingly, the diagnostic ions obtained in the MS^3 (Figure 8b) are the same as in the PEG homologues (m/z 89, 133 and 177). In this case only one set of diagnostic ions is obtained but it occurs specifically in MS^3. This example indicates how the unique capabilities of the ion trap (full-scan and MS^n) are required to confirm unambiguously the identification of homologue unknowns, especially in complex samples.

The diagnostic ions obtained for each family of surfactants in both CID and MS^n modes of operation are listed in Table I. An extensive amount of information is obtained using CID and MS^n modes in an ion trap and is highly useful for the identification of some complex homologue surfactants in environmental samples. Compounds between the same family will have common diagnostic ions that can be rapidly extracted and identified in a complex chromatogram. Matrix interferences are avoided when using MS/MS and MS^3 modes of operation, thus giving a characteristic fingerprint spectrum for every compound. Looking for these diagnostic ions in a real sample will increase the chance of identifying more unknowns as well. Within a class, this approach can be easily applied to other families of organic compounds, such as pesticides and pharmaceuticals. Future work will include the exploration of diagnostic ions for other families of compounds.

Acknowledgments

Imma Ferrer acknowledges the financial support from the U.S. Geological Survey (USGS) Toxic Substances Hydrology Program, the USGS National

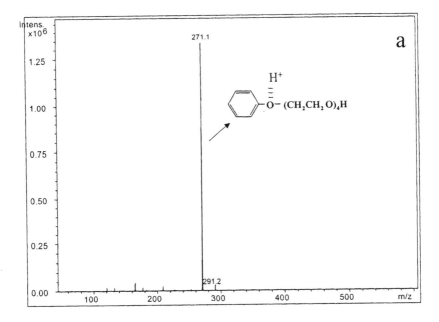

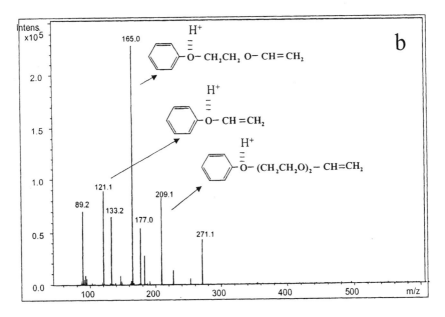

Figure 5. Double diagnostic ions for NPEO's: (a) MS/MS spectrum of molecular peak at m/z = 397 and (b) MS³ of fragment at m/z = 271.

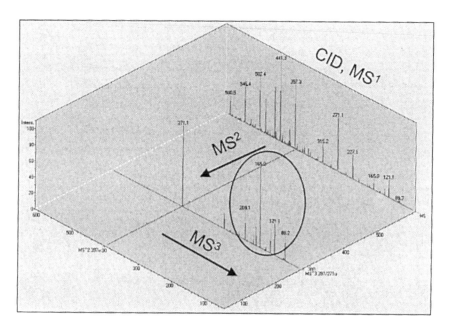

Figure 6. 3-D plot showing the CID spectrum, as well as the MS² and MS³ spectra.

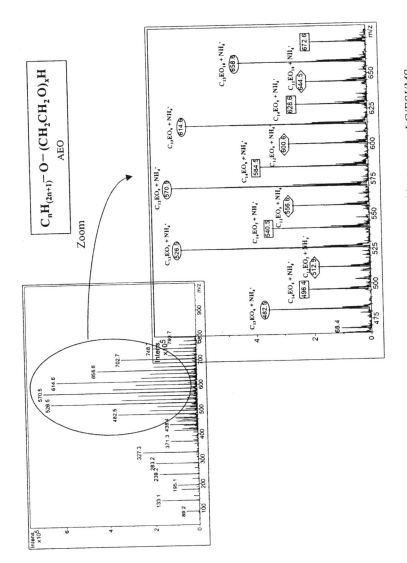

Figure 7. Flow-injection analysis of an AEO standard by ion trap LC/ESI/MS under full-scan conditions.

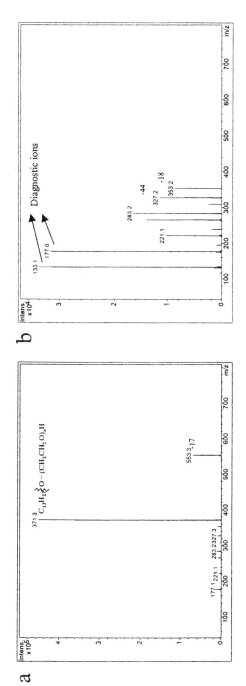

Figure 8. Double diagnostic ions for AEO's: (a) MS/MS spectrum of the ammonium adduct peak at m/z = 570 (corresponding to protonated molecule m/z = 553) and (b) MS³ of fragment at m/z = 371.

Table I. Main ions and fragments obtained by ion trap LC/MS/MS

	Ions obtained by ion trap LC/MS		
	CID conditions	*MS/MS*	*MS³*
Polyethylene glycols (PEGs)	$[M+H]^+$ $[M+NH_4]^+$ 89, 133, 177	-18, -44 89, 133, 177	-18, -44 89, 133, 177
Nonylphenol polyethoxylates (NPEOs)	$[M+H]^+$ $[M+NH_4]^+$ 121, 165, 209	-126 ($-C_9H_{19}$)	89, 133, 177 121, 165, 209
Alcohol polyethoxylates (AEOs)	$[M+NH_4]^+$ 89, 133, 177	-17, $-C_nH_{n+1}$	-18, -44 89, 133, 177

Water-Quality Assessment Program, and the USGS National Water Quality Laboratory's Methods Research and Development Program. The use of trade, product, or firm names in this article is for descriptive purposes only and does not imply endorsement by the U.S. Government.

References

1. Busch, K. L.; Glish, G. L.; McLuckey, S. A. Mass Spectrometry/Mass Spectrometry: Techniques and Applications of Tandem Mass Spectrometry; VCH: New York, 1988.
2. March, R. E. Quadrupole Ion Trap Mass Spectrometer, Encyclopedia of Analytical Chemistry, Ed Robert A. Meyers, John Wiley & Sons Ltd., Chichester, 2001.
3. Castro, R.; Moyano, E.; Galceran, M. T. J. Chromatogr. A, **2001**, 914, 111-121.
4. Ferrer, I.; Furlong, E. T. Environ. Sci. Technol., **2001**, 35, 2583-2588.
5. Furlong, E. T.; Ferrer, I.; Cahill, J. D. **2002**, J. Chromatogr. A, in press.
6. Thurman, E. M.; Ferrer, I.; Furlong, E. T. this volume, chapter 8, **2003**.

7. McLafferty, F. W.; Turacek, F. Interpretation of Mass Spectra, 4th ed.; University Science Books: Sausalito, CA, **1993**.
8. Castillo, M.; Barcelo, D. Anal. Chem., **1999,** 71, 3769-3776.
9. Petrovic, M.; Barcelo, D. Anal. Chem., **2000,** 72, 4560-4567.
10. Seraglia, R.; Traldi, P.; Mendichi, R.; Sartore, L.; Schiavon, O.; Veronese, F. M. Analytica Chimica Acta, **1992,** 262, 277-283.
11. Crescenzi, C.; Corcia, A. D.; Marcomini, A.; Samperi, R. Environ. Sci. Technol., **1997,** 31, 2679-2685.
12. Castillo, M.; Martinez, E.; Ginebreda, A.; Tirapu, L.; Barcelo, D. Analyst, **2000,** 125, 1733-1739.

Indexes

Author Index

Subject Index

Highlights from ACS Books

Desk Reference of Functional Polymers: Syntheses and Applications
Reza Arshady, Editor
832 pages, clothbound, ISBN 0–8412–3469–8

Chemical Engineering for Chemists
Richard G. Griskey
352 pages, clothbound, ISBN 0–8412–2215–0

Controlled Drug Delivery: Challenges and Strategies
Kinam Park, Editor
720 pages, clothbound, ISBN 0–8412–3470–1

A Practical Guide to Combinatorial Chemistry
Anthony W. Czarnik and Sheila H. DeWitt
462 pages, clothbound, ISBN 0–8412–3485–X

Chiral Separations: Applications and Technology
Satinder Ahuja, Editor
368 pages, clothbound, ISBN 0–8412–3407–8

Molecular Diversity and Combinatorial Chemistry: Libraries and Drug Discovery
Irwin M. Chaiken and Kim D. Janda, Editors
336 pages, clothbound, ISBN 0–8412–3450–7

A Lifetime of Synergy with Theory and Experiment
Andrew Streitwieser, Jr.
320 pages, clothbound, ISBN 0–8412–1836–6

For further information contact:
Order Department
Oxford University Press
2001 Evans Road
Cary, NC 27513
Phone: 1-800-445-9714 or 919-677-0977
Fax: 919-677-1303

Bestsellers from ACS Books

The ACS Style Guide: A Manual for Authors and Editors (2nd Edition)
Edited by Janet S. Dodd
470 pp; clothbound ISBN 0–8412–3461–2; paperback ISBN 0–8412–3462–0

Writing the Laboratory Notebook
By Howard M. Kanare
145 pp; clothbound ISBN 0–8412–0906–5; paperback ISBN 0–8412–0933–2

Career Transitions for Chemists
By Dorothy P. Rodmann, Donald D. Bly, Frederick H. Owens, and Anne-Claire Anderson
240 pp; clothbound ISBN 0–8412–3052–8; paperback ISBN 0–8412–3038–2

Chemical Activities (student and teacher editions)
By Christie L. Borgford and Lee R. Summerlin
330 pp; spiralbound ISBN 0–8412–1417–4; teacher edition, ISBN 0–8412–1416–6

Chemical Demonstrations: A Sourcebook for Teachers, Volumes 1 and 2, Second Edition
Volume 1 by Lee R. Summerlin and James L. Ealy, Jr.
198 pp; spiralbound ISBN 0–8412–1481–6
Volume 2 by Lee R. Summerlin, Christie L. Borgford, and Julie B. Ealy
234 pp; spiralbound ISBN 0–8412–1535–9

The Internet: A Guide for Chemists
Edited by Steven M. Bachrach
360 pp; clothbound ISBN 0–8412–3223–7; paperback ISBN 0–8412–3224–5

Laboratory Waste Management: A Guidebook
ACS Task Force on Laboratory Waste Management
250 pp; clothbound ISBN 0–8412–2735–7; paperback ISBN 0–8412–2849–3

Good Laboratory Practice Standards: Applications for Field and Laboratory Studies
Edited by Willa Y. Garner, Maureen S. Barge, and James P. Ussary
571 pp; clothbound ISBN 0–8412–2192–8

For further information contact:
Order Department
Oxford University Press
2001 Evans Road
Cary, NC 27513
Phone: 1-800-445-9714 or 919-677-0977